MAN IN SYSTEMS

MAN IN SYSTEMS

Edited by

MILTON D. RUBIN

President,
The Society for General Systems Research

GORDON AND BREACH SCIENCE PUBLISHERS

NEW YORK LONDON PARIS

Gordon and Breach, Science Publishers, Inc.
440 Park Avenue South
New York, N.Y. 10016

Editorial Office for the United Kindom

Gordon and Breach, Science Publishers Ltd
12 Bloomsbury Way
London W.C.1

Editorial Office for France

Gordon & Breach
7-9 rue Emile Dubois
Paris 14e

Library of Congress Catlog number 70-110816. ISBN 0 677 140606.

PREFACE

I

This volume is based on the papers delivered at the annual meeting of the Society for General Systems Research which took place December 26-30, 1968 in Dallas, Texas. While the Society is concerned with all types of systems, the papers presented are concerned largely with social systems, hence the title Man in Systems.

The Society was founded in 1954 in response to a need for an organization to present a forum to those sensitive to the idea that many concepts have applicability over a broad class of disciplines. A general system is thus one which is applicable to more than one field of inquiry. While not all the papers in this book develop this notion, nevertheless, all the authors are sympathetic to it. They are involved with the development of theory in one field or more, and are potentially if not actually practitioners of general systems research. The general systems notion of applying an idea from one field to another is of course, a major component of creativity.

The meeting and the volume present a forum for a wide range of opinions as to what constitutes a system, and as to what constitutes general systems research, and needless to say, the editor does not necessarily agree with all the views expressed in the papers.

This book will be the first proceedings of the Society to be published in a unit. The other publications of the Society are the Yearbook and the quarterly Bulletin. Papers from previous meetings have appeared in the Yearbooks, and in one case the papers from a symposium at an annual meeting were published as a book (Positive Feedback, edited by John H. Milsum, Pergamon Press, 1968). It is hoped that a Proceedings can be published every year and with a shorter time interval between presentation and publication. In the case of the present volume, the decision to publish was not made until after the meeting commenced, so that there was considerable delay in the acquisition, review and revision of the papers. In fact, not all the papers presented orally at the meeting are in this volume.

The Program Chairman of the meeting was Milton D. Rubin, who was at the time of the meeting President of the Society, and whose Presidential Address is the first section (I) in this volume. The success of the meeting is owed greatly to Richard F. Ericson, Associate Chairman of the meeting and Secretary-Treasurer of the Society,* and to Lawrence L. Schkade, Arrangements Chairman, and his staff from North Texas State University.

The remaining sections of the book were based on the respective sessions of the meeting. Edgar Taschdjian, Chairman of the Biology Department, St. Francis College, Brooklyn, N.Y. chaired the session (II) on Meta-Language Dialogues: General Systems and Education, organized jointly with Jere W. Clark, Director, Center for Interdisciplinary Creativity, Southern Connecticut State College. Bertram M. Gross, Director, Center for Urban Studies, Wayne State University, Detroit,** and now Vice-President of the Society, organized and chaired the session on Social Systems in the General Systems Spectrum (III). Other sessions than this one of course also covered aspects of the social system, and in fact, while most papers dealing with education were included in the preceding session (II), there were several papers on education in this particular session (III). Lawrence L. Schkade, now Associate Dean of the School of Business Administration, University of Texas, organized and chaired the session on Systems Research in Organization and Management (IV). Stuart Wright of the General Systems Research Section, National Institutes of Health, organized and chaired the session on the Analysis and Evaluation of a Scientific Field (V). A session on Conflict Resolution and Arms Control was organized by Lawrence B. Slobodkin, now President of the Society. However, since only one written paper from this session was received, that by Chase, it was associated with the paper written by Kaplan for the Social Systems session, in a final section entitled International Systems (VI).

It is hoped that this book will be of great interest to readers in many disciplines, and provide a useful documentation of some current thinking with respect to general systems.

---

*Professor of Management, School of Government and Business Administration, George Washington University.

**Presently Distinguished Professor of Urban Affairs and Planning at Hunter College of the City University of New York.

II

There is displayed herein a wide variety of views towards systems. No stringent doctrine governed the submisson or acceptance of these papers, nor was there any attempt at agreement on the definition of system. In fact, most definitions of system seem to miss several important points of the systems concept. Perhaps the term should be considered a primitive and be undefined. However, there are several threads that are important and should be mentioned. (A general reference set of source books of course is the Yearbooks of the Society). Simply to say that a system is a set of elements and the relationships among them leaves too much unsaid. Adding that this group of elements has a goal then in general says too much.

The first subject that is left unsaid is that of rationality. The systems approach must to a great degree be a rational approach, although particular systems one deals with may have a large degree of irrationality from a scientific point of view of the apparent nature of the content of the system. However, there may be a symbolic structure to the material which may have a large rational element, even though the content of the statements may or may not be in agreement with reality or true in any scientific sense. If nothing rational whatsoever can be found in an entity (or statements about an entity), then it can hardly be dignified by the title of system. It would be desirable to establish a quantitative scale of systemicity, or systemness, with a number of dimensions, one of which would be rationality. We thus come to the view that an entity is not either a system or not a system, but may be a system of a very low degree or on some scale, of higher degree.

The second quality that inheres in the concept of system, or the second dimension of systemness, as we are now looking at it, might be invariance. We can hardly call an entity a system unless there is some degree of invariance or constancy about it. This of course does not mean that a system is static; it only means that something about it must be static. The planets of the solar system are moving with respect to each other, yet they constitute a dynamic system which can be abstractly represented very closely over a long period of time by a set of equations based on Newton's laws of acceleration and gravitation.

Abstract systems can even be unstable. An actual physical system that is unstable cannot exist for long although it could change into another system. If it alternates among several unstable states, then this performance should be incorporated in the system description rather than being called unstable. It has been found in psychiatry that the symptom neuroses so prevalent in the early days of psychoanalysis have given way to character disorders*. Perhaps in some way then the "system" has changed which indicates that a not broad enough framework has been chosen for a high enough degree of invariance in the systemness of the system.

Thirdly is the dimension of analyzability as discussed by the editor in the January 1965 Newsletter of the Society, or of decomposability as discussed by Toda and Shuford in the 1965 Yearbook of the Society. One of the qualities of systemness then is the degree to which an entity can be analyzed or decomposed. The paper by Kuhn in this volume presents further useful categorization of systems, primarily from the point of view of social systems.

One of the elusive missions of social systems authors is to try to give us a picture of the human social system, not just that of our modern technological state, nor of a primitive tribe, but what if any there is common to all. However, studies in ethology, or animal behavior, have begun to have an impact, and tentative generalizations are beginning to be made that should have a profound effect on social systems thinking. It is a matter of surprise to this editor that ethology with its wealth of experimental evidence has had so little impact on psychiatry as well as on social systems. The only reference to the subject in this volume is that in the paper by the editor. It would seem that an integration could be made

---

*See Allen Wheelis "The Quest for Identity" 1958 (Norton) page 40.

of results from studies of animal behavior as well as human behavior to construct a new theory of social systems that would not necessarily change fundamentally while we talked about it, and it is hoped that we will see such formulated particularly in the areas of management and psychiatry. This application we consider to be general systems research even if it is not published under that title. Such a theory is important, in that it should reveal the constraints in possible social system construction, because we cannot freely construct any conceivable idealized social system, because of fundamental systemic considerations that if neglected will not allow the system to operate. It is likely then that a viable framework for an inclusive human social systems theory would also have to include elements of animal social systems.

One final comment we may make with regard to systems, and in particular with regard to the Society and its membership and their interests. Classification, theory construction and thinking in terms of systems in order to be able to reduce large amounts of information to manageable form has perhaps been an important element in man's development. Different people exhibit in widely differing degrees this systems thinking activity, because of varying ability and need to do system thinking. We hope in the meeting and publication of the Society to give an adequate forum for discussion of the substantive uses of systems thinking as well as discussion of the very systems thinking propensity itself or absence of it, and that we may continue to furnish a meeting place for people with strong interests, needs and abilities in this area.

The Editor

MAN IN SYSTEMS

TABLE OF CONTENTS

SECTION I

THE GENERAL SYSTEMS PROGRAM: WHERE ARE WE GOING?

THE GENERAL SYSTEMS PROGRAM: WHERE ARE WE GOING?

Milton D. Rubin*

It has been said recently that the reason the new theatre is so fragmented is that life is so fragmented, and that the purpose of the theatre is to put a mirror to the world and show it to us as it really is. Is this a new cry, is it a matter that has some special significance at the present time? After all, one of the pieces of classical systems poetry was Heine's, written over one hundred years ago:

> This world and this life are so scattered,
> they try me,
> And so to a German professor I'll hie me.
> He can well put all the fragments together
> Into a system convenient and terse;
> While with his night-cap and dressing robe
> tatters,
> He'll stop up the chink of the wide
> Universe.

Perhaps life seems so fragmented now to many because they do not accept any particular system, whether of re-

---

*Milton D. Rubin is a Consulting Scientist at the Raytheon Company, Wayland, Mass. 01778. He is Past President of the Society for General Systems Research, and this article is based on his Presidential Address to the Society.

ligion or science or philosophy. Of course it is not solely our job to reconstruct the new world picture, but isn't it part of our program to formulate possible patterns for world pictures, to study the nature and limitations of models of the world, to construct abstract systems, to develop and present methodology for system building? To me, this is the global aspect of general systems. We are the system builders, whether there is system in the world, or whether we put it there. We are the structuralists rather than the existentialists, and the ecological thinkers rather than the millenial thinkers. However, we accept that there are anti-system people as well as system people, and, as an institution we are not just partisans for the systems approach, but provide a forum for pro- and anti-systems. In fact, we can provide a meeting-ground for conservative and liberal, right and left, in systems language reducing some old antinomies, perhaps absorbing them in the existing one between system and anti-system.

Our Society is now in its second decade. Our aims and objectives were defined originally and primarily as the investigation of isomorphy of concepts, laws and models in various fields, and the facilitation of useful transfers from one field to another. A secondary aim was the encouragement of the development of adequate theoretical models in the fields which lack them. We have done more in the latter area of encouraging development of theoretical models in fields which lack them, than in the former area of isomorphy between fields. We should now either reiterate and reinforce our primary aim of the investigation of isomorphies in various fields, or else reconsider our aims and reformulate them to include formally other aspects of the systems approach. We should review the past and plan a program for the future, or, as a minimum, outline a direction more clearly, as well as some early steps in that direction, hopefully under the guidance of some longer-range planning. We might also consider how we can have more impact both on the world of intellect and the world of action.

First of all then, what could we conceive as a long range program (or at least a forecast) for general systems, before we try to plan a program for the Society? We certainly cannot make a rigid plan for the intellectual future, but a general systems program could be a guideline for our near-future efforts towards a program for the Society, which would logically be the next thing to discuss. A Society program would then be a basis for planning any Society actions and organization. Therefore, I would like to talk especially about the general systems program, and comment on our present organization so a dialogue can be commenced with respect to a general systems program, a Society program, and future Society organization. Can we actually follow this tightly-ordered, logical schema in the organization of our Society? In practice, I think we shall see difficulties, but as systems people we have some kind of commitment at least to try to lay out a plan constructed on a theoretical framework.

Opening his 1968 article in the International Encyclopedia of the Social Sciences, Anatol Rapoport has described general systems theory as a program in contemporary philosophy of science rather than as a theory, and he states that all the interpretations have the common aim: the integration of diverse content areas by means of a unified methodology of conceptualization or research. This does not mean unification of content of all areas, but a unified or at least an integrated view of many areas, and structural isomorphy is one way of accomplishing this objective, as brought out toward the end of his article, where he emphasizes what he believes is the key task of general systems theory, to show how the organismic aspect of a system emerges from its mathematical structure.

As indicated by the opening paragraphs of this paper, I think that a large part of the attraction of general systems is its relation to a philosophy of life as well as to philosophy of science. Perhaps it is a way of relating philosophy of science to a philosophy of life, so I would like to take a somewhat broader view of general systems than that mentioned in Anatol Rapoport's article, and follow up on the implications of a program, to define it a little better for action purposes.

The major effort of our program then would involve building theories giving views of large cross-disciplinary areas. Now this may turn out to be possible only to a limited degree, or to a great degree. Nevertheless, it is probably the important binding force to our membership. Furthermore, in this search we are not tied to anyone's particular theory of how this should be done, and there may be various ways in which it could be done, although our prime motif is structural isomorphy.

At this point, I would list some issues about our broader aims. I have alluded to some and will return again. They should be considered with respect to the breadth of interests of the Society.

1. How far are we interested in, and how far should we encourage general theories of systems? Of course, we are interested in theories of systems, but how general can the theories be? Can there be such a thing as a general theory of which all systems are special cases? What theoretical limitations do we run into in generality of theories?

2. How far is it our business to be interested in generating a general unified theory of the world? Some of our members have this interest. The physicists are interested in unified theories of the *physical* world, but unless we are complete reductionists, these theories even if accomplished, and they perhaps look even further off than ever, will not lead to a general unified theory of the world as a whole. Who is working on theoretical and empirical limitations of such general theory? If one is generated, can it be proven empirically true or false, or more in the spirit of modern science, can it be proven useful?

3. How far are we interested in aiding in the generation of a consolidated world outlook? In an integrated view of the world? How far is such a view related to the previous question of a general unified theory of the world? How far can structural isomorphy carry us in generating such an integrated view? A general systems statement was made by Ralph Waldo Emerson, unbeknownst to him one hundred years ago, that he could be so credited: He who has mastered any law in his private thoughts is master to that extent of all men whose language he speaks, and of all into whose language his own can be translated.

The philosophers are interested in unity. Cassirer says that the philosopher desires to apprehend the world as an absolute unity, and hopes ultimately to break down all diversity. But what is special about the general systems philosopher? I think that we are primarily interested in concepts with a scientific foundation and that are structural in nature, although I think that there are many of us who can look at a written article and say that it is a good general systems article or that it is not (the article may be a good article, but not necessarily a good general systems article, whether it was intended to or not, perhaps without our quite being able to explain why it is or isn't a good general systems article).

4. How far are we interested in the integrated man? An integrated view of the world may not make its possessor an integrated person. However, the purposes of this paper cannot include an examination in any profundity of the integrated man and the integrated life. Nietzche said that Socrates aimed at the integrated life; the previous philosophers aimed at a high degree of exact knowledge. McLuhan says "The clown ventures to perform the specialized routines of the society, acting as integral man. But integral man is quite inept in a specialist situation".

5. We can end this series of questions by asking, if we are able to answer some of the previous questions in the affirmative, how far we want to educate for an integrated view of the world, and how far we want to educate for the integrated person? One of the functions, or attributes of science is the integration of experience into a coherent whole. Primitive man organized his experience through magic, myth and primitive systematization. However, we must not force all science together; in fact, it may not fit, although we have a faith that it does: nature must fit even though science may not. Many of us are happy to find areas in which things fit together, and ways in which to relate various diverse areas; others of us must see ways in which everything fits together. Many people do not seem to care if anything fits together. One might also ask about the relationship of a coherent world picture to sanity. But for a mentally healthy world picture are we not concerned to a great extent with the social system? Isn't it the social system then of which we want a coherent picture more than anything else? Many people talk about changing the social system rather than understanding it. But are there not invariants we must understand in order to make most meaningful changes?

Whatever we can do toward working on these issues, I believe that this is one of the very few groups in the world before whom I could stand and ask such questions and seriously expect that these questions might be thought about with the idea that constructive plans could eventuate, although I must say that in reading the descriptions of other sessions at this overall AAAS meeting, more sound like possible sessions for our Society meeting than ever before.

There are people who think that our major business is to supply models to the world. Of course, we are not the only model makers; but if we are specialists, we are specialists in model-making. But why do we often tend to consider ourselves as generalists? Well, it is that theoretical model-making is so generally applicable to all aspects of our view of the world that we can consider ourselves as generalists. However, we must not forget that there are different kinds of generalists, rather broad-gauge people in particular areas, such as the business generalist, liberal arts humanist generalist, or engineering generalist. We are talking about the general systems generalist, although one might make a point that any generalist should see things as systems. However, all generalists probably suffer to some extent from the problem of not having a defined and accepted ecological niche, and I believe that it is one of our duties to aid in the development of such a niche, at least for the general systems generalist.

We pride ourselves on being interdisciplinary, but we must observe some working discipline. We don't have a monopoly on interdisciplinary work, but our main claims to interdisciplinarianism are the transfer of concepts from one field to another, and the exposition of the isomorphy of concepts over many fields. To apply a concept from one field to another, it must first be correctly understood in its original field. We must be firm about this. If there is one way in which we can earn clear demerits, it is by misusing a concept through misunderstanding it in its original field, and this is one of the occupational hazards of the generalist, because in developing into a generalist of any kind, he must of necessity sacrifice some specialized proficiency. An obvious error is thus presented to the practitioners of a field which they can exploit with glee. Only after we understand a concept in its original field, can we address ourselves to the transfer of the concept to some other field.

Since our central interests are the isomorphy of systems and the transfers of concepts, laws and models from one field to another, we should study the matter of isomorphy and the rigorous transfer of concepts from one field to another. Cannot rules be established to help in judging the suitability of transfer from one field to another? A metaphor is less rigorous than an analogy and an analogy is perhaps less rigorous than an isomorphy, so the same degree of rigor is not demanded for all situations. But this matter of degree of rigor with relation to the type of transfer also needs to be studied. The very loosest types of metaphors are bandied about as analogies on the political platform. We should attempt to establish criteria for such transfers from one field to another, and hopefully when we have rationalized and formalized this matter to a reasonable degree, it should become important material for use in the educational system. Samuel Butler said years ago "though analogy is often misleading, it is the least misleading thing we have". I think it is a major part of the general systems program to study the logical rigor of the transfer of concepts from one field to another so that bounds of error can be established and the technique be made more useful and reliable. Please don't ask me how far we can go, or should go at this stage. More rigor is necessary, and it should be studied. We are not looking for a rigor that will eliminate vigor, but we need more than we have.

We must also review our relations with the practical systems people, with systems analysis, systems engineering, planning-programming-budgetting, and operations research. Do we really help supply these people with models? Or can we be more helpful to people in the social systems areas? We hear more and more of the systems approach. A new book on marriage (Lederer and Jackson) stresses the systems concept saying it is the most useful way of viewing marital interaction that has yet been developed, with statements giving due credit to von Bertalanffy and Wiener. Industry is selling systems instead of equipment. Yet systems analysis has not been as pervasive as one would think. To quote a recent article (Schlesinger), "Vietnam reflects more the *absence* than the *failure* of analysis. From the standpoint of systems analysis, it represents almost an anthology of classic errors". Many government projects surely have proceeded with negligible systems analysis. There still has been too little rather than too much, even in defense and aerospace areas. Systems analysis and systems engineering are being applied as routinized procedures where they may not be applicable, and are not being used creatively where they might be.

Perhaps just the idea of systems is powerful when it gets across. A large part of the battle is getting the concept accepted. It is important to educate people to be systems-oriented, even if a great deal of theory does not get across, just so that they appreciate the interactions of members of a system and the environment, and to be on the watch for such interactions. In some people this education can come in a flash; other people may not absorb it in a lifetime. If there is a meaningful systems approach in many fields, there have to be concepts in common across these many fields, and the interdisciplinary structural content is part of general systems theory.

With this background it becomes appropriate to discuss the organization of the Society. While the Society is generalist in orientation, we must communicate with the disciplines, and further, most of our members are also members of such disciplines with a higher career priority than their interests in general systems. We therefore proposed several years ago a series of substantive committees with the notion of relating general systems to the various disciplines. In addition, we also proposed committees in administrative areas such as membership, meetings, finance, etc. It turned out that there was much more interest in the substantive committees, and in fact it was difficult later to even get more than the slightest response from those members who had already expressed interest in the administrative committees. The active committees are the Task Force for General Systems Education, the Organization and Management Studies Group, and the Committee on Social Systems. These are all actually as much Colleges or Study Groups as committees, and they have all had meetings, and as a matter of fact it is very possible that active and interested people may actually emerge through this channel to take part in administrative committee work. The committees or groups evolved as a result of a questionnaire which indicated these areas as most popular and interesting, and were thus initiated experimentally. Now in a Society such as ours, one would like to see some systemic basis for organization in the long run, so what should be the systemic basis or perhaps taxonomy? Should the basis arise from a systemic theory of knowledge, and if so, whose theory of knowledge? Should the basis arise from some planned purposive action in the real world? I did suggest on an earlier page that if we could lay out an intellectual program, we could plan an action program and then organize for this action program. This procedure seems very logical, but we will be unlikely to definitively accomplish these steps in succession although it should be a long run framework for our planning. Even if we proceed with a very low degree of activity, if the Society is to exist at all, we should do some planning.

One scheme which has some support is the division into physical systems, biological systems, and social systems. In point of fact, the major committee areas of interest, education, management, and social systems, all fall into the social systems area; and we might search for a better taxonomy for the social systems area than our present committee structure, bearing in mind always that we also have the aim of addressing existent coherent areas of interest. One suggested taxonomy is with respect to the size of the social units being studied. Actually we are not far from this, in fact I believe that our present Social Systems Committee is probably more concerned with social units approaching the size of a city, a state, a megalopolis or larger, whereas our present Organization and Management Studies Group is probably more concerned with social units of the size of a corporation or smaller.

Our education committee is interested in education in general systems, as well as in introducing general systems into the educational process. It has taken a broad view, and certainly has found a niche in the social systems education area, as for example the recent general systems education conference at Southern

Connecticut State College where the theme was social science curriculum. One of the major problems though is that there is not an appealing modern social theory yet, and I will come back to this later.

Among the next groups in popularity in the questionnaire answers, were engineering and medicine, so we see that there was interest in physical systems and biological systems. Another interest area was that of urban problems. I believe that this belongs primarily under social systems, but in any case I think that many people, whatever their committee area desire was, were primarily interested in the societal aspects, so we perhaps might do well to review our membership interests as to the concentration in the social systems area as compared to the others. Also with regard to the classification scheme, there can be various theories as to how these systems fit together, and certainly interface problems, so I think that we want to be thoughtful with regard to choice of a systemic basis of organization.

We certainly want to continue our interests in the physical and biological systems not only for their inherent interest, but, even if our interests should be more heavily weighted toward social systems, because of the relations between concepts in all these fields. In fact, perhaps it is timely to have a better theory of social systems to give us a framework for appreciating biological and physical systems from a human orientation. As we consider the social system, we shall want to consider how far we want to go as an organization in dealing with real-world tough and simultaneously sensitive problems which might benefit by treatment from a general systems point of view.

But where does the social system exist? Is it not only a network of relationships that exist only in our minds? Is it then an abstraction if it exists only in our minds? Curiously then we are in the social system yet the social system is in us. I am not talking about a Theory of The Social System which of course would be an abstraction and in us, but about The Social System, if there is such a thing.

Again, considering the physical system as a supersystem, the biological system is a system of this supersystem and the social system is a subsystem of the biological system. On the other hand, considering a social system as an entity, biological systems are part of this supersystem and physical systems are subsystems of these biological systems. Which way do we go? These problems are partly solvable through semantic considerations of course. In fact, one author in the area of social theory (Engelmann) avoids part of these problems by not going beyond behavioral systems, and by not introducing the term social system.

However, I have gone through this matter because it brings out the problems in organizing our Society activities in a systemic way. Whatever we do or don't do, we don't want to get bogged down organizationally for a period of years while we try to settle theoretical questions. These questions should be discussed by all means, but meanwhile the Society will have to proceed pragmatically where we do not have reasonably generally accepted theory to guide us in generally approved directions.

We should consider several more substantive committees besides the ones already mentioned. Cybernetics is an important interdisciplinary force, and some people may attach to it approximately the meaning we attach to general systems. In any case, there are people who identify their interest with cybernetics rather than general systems; and I think that we should have a committee to interface with organizations in cybernetics. We should also consider a committee on computer simulation. Operating or planned systems are conceptually abstracted and modelled in digital computer simulation programs. Computer simulation is being performed on social systems, in particular an urban simulation game and an international system simulation are discussed respectively in the House and the Kaplan papers in this volume. Computer programs are in a gray area conceptually: are they abstract or concrete? Such a question is much more than semantic; our discussions on the subject might be of great interest to the patent offices. General purpose simulation programs have been devised which can be adapted to a variety of operating systems, and we should examine our interests in these programs. Furthermore, should not general systems concepts be important in the large computer-oriented area of information retrieval?

We have a Publications Committee reviewing the publication policy of the Society. We should consider the possibility of supporting the publication of books in general systems. There are books which are too interdisciplinary for commercial scientific publishers, but apparently of not broad enough interest for general commercial publishers. We could consider setting up a separate institute to publish such books. The publishers themselves may not know whom to call on for editorial aid in reviewing material of this nature. They rely on the existing disciplines or on the star system. We do know that papers are being received by disciplinary journals, which are too broad-gauge for them to publish, but which might be satisfactory for us to publish if the authors knew of our existence. We also have to think of an expanded journal publication policy.

Since I have mentioned the theory of the social systems so much, I will say a little more. One of the major problems in such a theory is to account for the apparent lack of wisdom in the conduct of affairs of such an almost rational creature as man.

In Proverbs Chapter 8, the scribe asked thousands of years ago: "Doth not wisdom cry? and understanding put forth her voice? She standeth in the top of high places, by the way in the places of the paths. She crieth at the gates, at the entry of the city, at the coming in at the doors." Today we have enormously greater resources of knowledge, yet is there any more wisdom, is it heard any more clearly? If not, why not? This problem exasperated Walter Pitkin so much about 25 years ago that he wrote a book entitled "A Short Introduction to the History of Human Stupidity". His thesis was that stupidity, the opposite of wisdom, was so egregious in the affairs of the world that it must be considered as an essential separable factor in the human makeup, that people had a specific urge toward unwisdom, stupidity. Of course, one could also point out a parallel to Freud's death instinct.

A book review by Tiger recently described the year 1968 as International Aggression Year because of all the publications on the subject. With all the work on animal ethology providing sources for isomorphic theories about human behavior, the resulting writings have tended to cluster about the concept of aggression, which I believe is a misplaced emphasis. The big point is that aggression and other matters seem at last to be able to fall into some kind of a system; not a complete system, we will never know enough for such a system to be complete. But the time is at hand for a new grand synthesis, which will of course in time display inconsistencies and be replaced with another. However, I do not think that anyone has put together as complete a system, a theory, as is now possible, which can not only contribute to a new wisdom, but must also explain unwisdom.

Whatever the theory of social systems, we should consider that anthropology, social psychology, sociology and economics are looking at what might be considered different aspects of the same social system, and a more comprehensive and meaningful view can be had if all these aspects can be blended, and we should support such interdisciplinary efforts in every way we can. This is needed in particular in education. What are we educating for? The objectives of our educational system hinge on the requirements of our social system. Unfortunately, a vocal element of our youth in college disapproves of our social system as well as our educational system. I think that one of the problems is that they have been promised too much. They have been taught the fuzzy idea that the function of education is to cause us all to grow to reach our full potentialities, and of course this is part of its function, but to state it this way without any qualification is either unenlightened or dishonest. We must take a general systems view on growth. All the trees in the forest cannot grow to reach the sun; many must perish on the forest floor. As there are more people, there will be less opportunity for uninhibited growth unless more ecological niches are opened up. Systems considerations will enter more and more as life gets more complex.

To my mind a new social theory would have to be based on isomorphisms from biology rather than from physics, and perhaps its successor would be based on neither.

Drawing basically on concepts of Calhoun, Engelmann, Wynne-Edwards, Mayr, Welch, McBride and others, in rough outline such a theory would be based on the idea that interactions between creatures within a species population, with successive dominance-submission encounters, build up a status hierarchy. Too much interaction produces a too rigid hierarchy or social breakdown. Too little leads toward sensory deprivation. This in*tra*-species competition must be considered of course embedded in a historical and evolutionary background. Speciation results in milder in*ter*specific competition, and perhaps the whole theoretical system can be synthesized with minimum emphasis on the concept of aggression. There are useful organizational analogies to be made - specialization can be considered as a form of speciation, and takes place not only for improved expertise and efficiency, but for reduction of competition by breaking up a group into many non-competitive and hopefully complementary ecological niches.

A number of papers have been presented at the 1968 AAAS meeting, on animal ethology, taking up the question of successive encounters in space and time, but primarily oriented toward in*tra*specific aggression. Yet it is interesting that this area has received so much attention from zoologists (although still controversial among them) but has had so little influence in psychiatry and sociology. I should like to emphasize it because of our interests in transferring concepts across disciplinary lines.

If evolution is primarily oriented to the survival of the species rather than to the survival of the individual, then we may see some light on the operations of society which seem antithetic to the individual. Individuals may be psychosomatically sacrificed for the stability of the society, and the wise man may benefit his society but be forgotten. Ernst Mayr asked why man's brain size expanded for hundreds of thousands of years and then stopped. Was there some evolutionary disadvantage to larger brain size? Lemurs (in the opinion of Jolly) evolved monkey-type societies without evolving monkey-type intelligence. Is there a limit on the value of intelligence to a single individual in a given society? Once he reaches a given certain level, is there any advantage in more intelligence for intraspecific competition, other things being equal? Is more intelligence primarily of value in interspecific competition (and with nature in general) and therefore to species survival rather than individual survival? I will leave you there with regard to social systems and unwisdom. Also with regard to the Society, I do not make a specific recommendation as to which social system theory we should use in our design of this Society.

To summarize a program as I see it, we should encourage the development of interdisciplinary isomorphic social system models, with application to varieties of individual social systems, covering the scale from whole cultures to working organizations, and encourage a broad view of social systems in educational curricula. However, we should not neglect the biological and physical systems, particularly their interfaces and isomorphies with social systems. We must encourage more rigor in isomorphy, and emphasize precise meanings of basic concepts used in isomorphy.

Organizationally, we should encourage our present substantive committees on social systems, management and education, and others where necessary. We also should have a planning committee concerned with the future program and organization of the Society. We have had a temporary publications committee, and we should also have a meetings committee, and we should form committees to consider our interests and relations with cybernetics and simulation.

It is hoped that a dialogue can continue on these matters, leading to productive research in general systems.

BIBLIOGRAPHY

1. Rapoport, A., *Systems Analysis: General Systems Theory*, International Encyclopedia of the Social Sciences, MacMillan Company & Free Press (1968) p. 452.

2. Cassirer, E., *The Philosophy of Symbolic Forms*, Vol. 3 (The Phenomenology of Knowledge), New Haven, Yale University Press (1957, 1965) Introduction, p. 1.

3. McLuhan, M., *Understanding Media*, New York, McGraw Hill Publishing Co. (1964) p. 167.

4. Lederer, W.J., and Jackson, D., *The Mirages of Marriage*, New York, W.W. Norton and Company (1968).

5. Schlesinger, J.R., The "Soft" Factors in Systems Studies, *Bulletin of The Atomic Scientists*, November 1968, p. 12.

6. Engelmann, H.O., *Sociology, a Guided Study Text*, Dubuque, Iowa, Wm. C. Brown Book Co. (1969).

7. Pitkin, W.B., *A Short Introduction to The History of Human Stupidity*, New York, Simon and Schuster (1932).

8. Tiger, L., Two Cheers for Aggression, a review of a book by Anthony Storr entitled Human Aggression, *New Republic*, Sept. 14, 1968.

9. Calhoun, J.B., *Mammalian Populations*, Part I, Mayer, William V., and VanGelder, Richard G., Editors, Vol. 1 of Physiological Mammalogy, New York, Academic Press (1963).

10. Wynne-Edwards, V.C., *Animal Dispersion in Relation to Social Behavior*, New York, Hafner Publishing Co. (1962).

11. Mayr, E., *Animal Species and Evolution,* Cambridge, Mass., The Belknap Press of Harvard University Press (1963).

12. Welch, Bruce L., *Psychophysiological Response to the Mean Level of Environmental Stimulation,* A Theory of Environmental Integration, Laboratory of Population Ecology, College of William and Mary, Williamsburg, Virginia.

13. Jolly, Alison, Lemur social behavior and primate intelligence, *Science, Vol. 153,* 501-506, 29 July 1966.

14. Brun, Jean, *Socrates,* New York, Walker and Company (1962), p. 98.

15. McBride, G., Social Organization and Stress in Animal Management, *Proc. Ecol. Soc. Aust. 3:* 133-138, 1968.

SECTION II

META-LANGUAGE DIALOGUES: GENERAL SYSTEMS AND EDUCATION

Although much has been done during the last three decades to improve the hardware, software, and methods of education within the age-old structure of knowledge, the real need of the day is to restructure knowledge. The information explosion, the growing degree of specialization among scientists, and the growing complexity and interdependence of specialties in the scientific world all demand that knowledge generally be simplified, unified, and fortified with operational content if our civilization is to continue to advance.

The first and perhaps most essential step toward achieving such an ambitious goal as that of remodeling knowledge is to devise a meta-language - i.e., a common language which would apply to all fields of thought - whereby the various scientific specialists could communicate with each other.

The need for a meta-language in science might be likened to the need for a commonly acceptable medium of exchange or monetary unit in the economic world. For example, in a simple tribal economy consisting exclusively of self-sufficient tribes with few if any product interchanges, the need for a monetary unit as we know it today was slight. As the tribes began to specialize and to exchange surpluses of products, however, the need for a common monetary unit became crucial to the success of the entire economic system.

Likewise in science, when there were only general scientists, each of whom gathered his own facts and developed his own principles and models, the need for a commonly used medium of communications interchange was small. Today, this common medium is especially important because the number of specialties now numbers in the hundreds (thousands?). The success of each scientist depends more and more on the interchange of ideas with other specialists. Without some common communications medium for exchanging scientific products or findings, the system is subject to breakdown.

Once a common language has been established, the next great scientific advance will come from identifying the isomorphic universality of many more or less universal principles. Establishing the identity of the cybernetic process in physiology, economics, engineering, and physics would greatly increase the value of the label, cybernetic process, for example.

Still another step forward would be to develop a common paradigm to approach and view the analogous or isomorphic processes and principles in the different sciences. Even if the paradigm were applicable only to the most elementary phases of each and every scientific discipline, this would provide a tremendous advantage. Each specialist would then be able to "get into" and "out of" other specialties more readily when the need arose. Still another contribution would be made by devising ways to incorporate into the meta-language paradigm operationally defined values and value systems.

Given the basic elements of a meta-language paradigm incorporating an operationally defined value system, and a common approach to (and exit from) many scientific fields, the stage is set for meta-language dialogues to take place. This may involve as few as two persons or as many as several dozen. Regardless of the number of persons, however, the meta-language dialogue not only makes it possible for each specialist to make a contribution to solving the problem of the day, but also makes it possible for him to receive numerous contributions to his own thinking from the other participating specialists. Even if a specialist should not learn a single new fact or principle in the interchange, he would, provided his mind was open, receive two types of contributions: (1) A figurative mirror to reflect the degree of his progress in learning to reduce his relevant insights to language any intelligent person can understand; and (2) An idea of what this problem looks like from the view-point of several other specialists.

The general idea of the meta-language dialogue is symbolized by the accompanying schematic diagram.

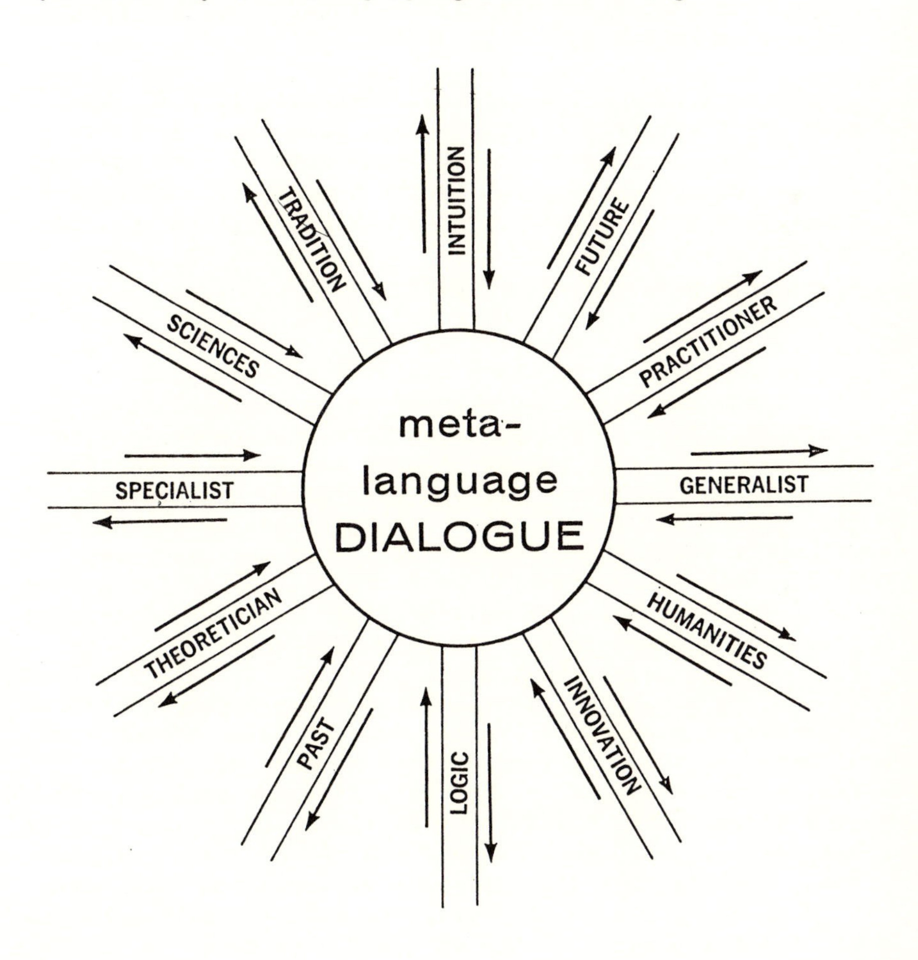

SECTION II

This section comprises five papers.

The Clark paper expands on the nature of meta-language dialogues and general systems in education. It emphasizes the need for further active promotional effort in this area, as well as deepening intellectual content.

Moss discusses specialization in all fields, but particularly in American medical practice. He emphasizes the growing need for the general practitioner or team practice, and the need for the generalist in all fields, as compared to the trends of our recent past to more and more minute specialization and fragmentation.

The paper by Kingsley et al. contains an introduction to graph theory, and proposes graph theory as one possible meta-language. An example is given of an application and there is discussion of other work in the same area, particularly with regard to the structuring of knowledge.

Firestone discusses further the problem of super-specialization, with more specific regard to the social sciences and their teaching in the university. He shows how this specialization is related to the university structure and describes new interdisciplinary approaches with special emphasis on the general systems approach.

The final paper in this section, by Helvey, contains a description of a curriculum for a graduate course in cybernetics, with an interdisciplinary view of the systems encountered primarily in the natural sciences.

Jere W. Clark

DEVELOPING AND UTILIZING META-LANGUAGE DIALOGUES

The Role of the SGSR Task Force on General Systems Education

Jere W. Clark*

ABSTRACT

This paper discusses General Systems Education, its objectives and its implementation. Pursuant to this, the Task Force for General Systems Education of the Society for General Systems Research is described, with emphasis on its role in promoting this educational philosophy.

Implied in this entire paper is the view that the Task Force is strategically located in the space-time panorama of history to accelerate the process of developing and utilizing meta-language dialogues. Several questions which are raised in the paper might be summarized with another question dealing first and foremost with attitudes toward the future. That is, will the Task Force view itself collectively (along with the SGSR) as an observer, analyst and commentator on the course of events, or will it in addition to that of the general systems theorist, adopt the attitude of the general systems practitioner and consider itself to be one of the guiding sub-systems toward the future?

---

* Jere W. Clark is Professor and Chairman, Department of Economics, and Director, Center for Interdisciplinary Creativity, Southern Connecticut State College, New Haven, Conn.

In the latter case, the Task Force might have the potential of becoming one of the major controlling subsystems in an international society. If we may be permitted the use of one more analogy, we might suggest that the distinction between these two attitudes is essentially the distinction between the thermometer and the thermostat. Even if we assume the role of the thermostat we can concentrate primarily on predicting the future as many other groups of interdisciplinary thinkers have done, or we can advance still one step further into the control process.

There seems to be considerable sentiment leaning toward the last and broader view of our role. If that is a correct reading of attitudes then we should plan for the task of designing the kind of future we want to live in and the related task of *designing promotional strategies* for realizing that dream.

1. GENERAL NATURE AND PURPOSE OF THE TASK FORCE

A brief statement of the nature and purpose of the Task Force on General Systems Education of the Society for General Systems Research (SGSR) will provide a useful background against which to sketch the role of the Task Force in promoting the development and use of metalanguages.

The Task Force consists of approximately sixty members of the Society for General Systems Research and represents a wide variety of educational, governmental, industrial and military institutions from coast to coast, and most of the traditional academic subject areas (broadly conceived) as well as the various systems groups, including those in cybernetics, operations research, instructional systems technology, game theory, simulation modeling, and optimization theory.

The general purpose of the SGSR Task Force is to cooperate with other systems groups in an effort to promote the general systems approach wherever appropriate in education generally, from kindergarten through graduate school in formal education, and in varied aspects of industrial, military, governmental and adult education. The top priority of the Task Force for the next three years is generating promotional strategies, developmental models, innovative materials, and other prerequisites to success in meeting this nationwide need for systems education. A major effort will be made in the hope of achieving some major "breakthroughs" in restructuring knowledge and basic curricular patterns and in applying systems techniques practically to educational, social, and urban problems.

By calling attention to the systems education needs of the nation and by outlining suggestive types of projects to meet them, the Task Force may be able to encourage the development during the next few years of a network of centers and institutions for General Systems Education, each with its own emphasis or focal point. In these and other ways the Task Force will seek to facilitate the coming transition of our educational system from a base of mechanical specialization to the slowly emerging base of originality, self-direction, personal involvement, and synthesis, utilizing systems concepts and perspectives.

2. THE PROMOTION OF META-LANGUAGES

Perhaps the most obvious role of the Task Force in promoting the development and utilization of meta-languages is that of encouraging and vitalizing training and educational programs designed to develop general systems theorists and practitioners. The requirement that one person acquire even a general working orientation to many different branches of the social, biological, and physical sciences could hardly be met meaningfully without some common body of principles, processes, and ideas couched in one functional language which could transcend the discipline boundaries. A generalized cybernetic nomenclature, for example, would help both a student and a practitioner to know within any given relevant situation what information to look for, how to recognize it when he sees it, how to use it once he has acquired it, and how to communicate his findings meaningfully to other specialists.

The general systems practitioner is thought to be an intelligent, creative person with some degree of intellectual sophistication. Therefore, it may be reasonable to expect him to invest the time and energy necessary to acquire such a set of universal tools. Within this context, we believe it is only natural that our Task Force should strive to promote the development and utilization of meta-language dialogues.

A less obvious but equally important role of the Task Force is to work toward using general systems concepts and processes to democratize interdisciplinary science. By democratizing science we mean the process of teaching its fundamentals effectively to all citizens. We are approaching a period of technological development when the main fruits of the technological revolution of the last few decades cannot be harvested by society unless the so-called common man is involved in the planning of many of our most important scientific and social ventures.

We are learning the hard way that a bolt or a nut in a space vehicle is fundamentally different from a man who lives in a slum and is to be rehabilitated. The best of plans can be vetoed by slum inhabitants if they believe the plan will not work properly for them, and/or if they distrust the planners. However wrong their beliefs may be, they, being human, base their actions on their beliefs rather than on what the planners think they ought to believe. Unless people are brought into the planning act early so as to acquire the necessary information from them and to get them meaningfully involved in the planning process, the plans may be wasted. Without some simple, common language to be used by the planners and the potential benefactors of the plans, such involvement may be virtually impossible.

In planning educational programs in the past, our education system has generally operated on the assumption that professional persons are by and large the only ones who have a real need for abstract, systematic thoughts, or theories, or models. It has been assumed also that when it comes to interdisciplinary thinking, it is mainly the mental elite who have the necessary mental equipment to do the required creative thinking. Both of these assumptions are now being challenged by research and by events. The weakness of the first assumption becomes obvious when we raise a pair of analogous questions regarding money and intellectual wealth. With regard to money, we might ask, who needs an extra dollar more, a millionaire or an unemployed pauper? Likewise we might ask, who needs the mental leverage of systematic and/or general systematic thought, the intellectual millionaire or the intellectual pauper?

With regard to the second assumption - that dealing with the ability to do original thinking - a growing body of research and experience is suggesting that a creative challenge often makes the difference between success and failure for culturally handicapped people just as is true of other people. This is especially true if the handicapped persons are taught how to develop whatever creative potential they may have.[1]

---

[1] More and more psychological research is indicating that creativity respects no artificial restraints such as race, economic and social status, amount of formal education, or even the I.Q. scores (for people with average or better I.Q. scores). There is a tremendous amount and variety of practical material for helping individuals and groups nurture whatever creative talents they may have. Alex F. Osborn's textbook, *Applied Imagination* (Scribner's 1965) has been for fifteen years and still is the main textbook used. For a catalog of books, pamphlets, workbooks, teacher's guides and audiovisual teaching materials which supplement Osborn's text, write to: Dr. Sidney J. Parnes, President, Creative Education Foundation, Inc., State University College (Buffalo), 1300 Elmwood Avenue, Buffalo, New York 14222.

For an article surveying the nature and rationale of

3. AN ANALOGY

One of the simplest ways to begin a discussion of the nature and importance of meta-language dialogues in our society is to consider the pressure-temperature-volume interrelationships of gases in the physical world. We are quite aware that the greater the number of molecules within a given volume (at a given temperature) the greater is the expected number of collisions among molecules. Likewise, as our human population increases within a relatively fixed geographic area, the chances of collision of interests among people are increased. Also, within a given number of molecules, other things remaining constant, the higher the temperature the greater the speed of molecules, and hence the greater the chance of collision. Advances in transportation and communications generally in society tend to produce somewhat analogous results. Now if we were to increase the number of molecules (or people) and increase temperature (advance communications technology) at the same time, we further increase the chances of physical and social collisions or conflicts of interests.

Is this not essentially what is happening today in the social world? More and more people are acquiring more and better means of communicating with other people. As this process continues, the pressures are surely building up toward some one or a combination of the following two alternatives:

1. Relying on a high degree of regimentation to guide or "program" the actions of each person;

2. Relying on programs designed to increase the

---

this type of creativity training, see Jere W. Clark, "Creativeness - Can It Be Cultivated?", *The Business Quarterly*, University of Western Ontario, Spring, 1965, pp. 29-39. (A digest of this article appeared in *Management Review*, American Management Association, June, 1965, pp. 51-54).

degree of flexibility and ability of individual persons to anticipate collisions of interests and to adapt their actions accordingly, thereby making possible the survival of democracy.

Under the regimented alternative just mentioned, a major effort would be made to program peoples' minds until political revolution intervened, or until our civilization had been otherwise destroyed or seriously paralyzed. If our nation would choose the second alternative - that of emphasizing democracy - we should need to restructure our educational system so as to make change a virtue rather than a villain. We should have to put a premium on some of the easily taught, critically important, creative mental skills which have thus far been neglected. In doing so, might we be able to come closer to George Leonard's ideal of education and ecstasy, using creativity as a source of both motivation and insight which mutually interact in a positive feedback fashion. We might come to recognize that the general systems approach is as much a *set of attitudes* as it is a set of intellectual concepts and processes. We might indeed be pleased to find that what the layman needs first - if not mainly - is to acquire the general systems attitudes. These attitudes might then be used as keys to unlock opportunities for further conceptual development.

4. ORGANIZATIONAL, PROMOTIONAL IMPLICATIONS

What does all this mean in terms of the Task Force and meta-language dialogues? I want to suggest that there may be a major promotional effort that must be made in the USA and around the world if the next major round of benefits from space age technology is to be realized. It may mean that today every city and town in our nation needs a competently staffed center or institute for general systems education. These centers could operate to give guidance to school and college administrators and teachers and students striving to acquire general systems attitudes, concepts, and perspectives. If we had only one such center for each 2,000,000 people, that would come to two hundred centers and similar agencies.

To get an idea of what is involved in establishing 100 centers, we might note the record of the Joint Council on Economic Education. It was founded in 1949 through the joint efforts of the American Economic Association, the National Education Association, labor and management agencies, and other interested groups to promote economic education in elementary and secondary schools. After nineteen years of rather progressive pioneering work, the Joint Council on Economic Education has succeeded in establishing a total of fifty-four collegiate centers for promoting economic education.

Suppose it were to take nineteen years to get the first fifty-four agencies for promoting general systems education into operation. At that rate, might we be likely to realize George Orwell's 1984 before the chronological year, 1984?

The establishment of fifty-four (or even 100) centers would indicate that the cadre who are to teach the field cadre had already been educated. Then how many more years would we need to orient the field cadre? The students themselves?

The big question in my mind is this: Are we, the members and friends of the Society for General Systems Research, and more specifically, the members of the SGSR Task Force, going to let nature take its course for another three to six years until things become so chaotic that non-rational pressures will be put on the Federal Government to enact legislation establishing a crash program of general systems education? In that case, what would happen if we should not already have had in operation an orderly and systemic program programmed for educating at least the cadre of the field cadre? Might a pre-crisis, planned program provide some of the means whereby a wiser and earlier bit of legislation might be passed? After all, Senator Gaylord Nelson gave as a chief - if not the chief - reason for Congress' failure to approve the recent bi-partisan $125 million proposal for experimenting with the application of systems techniques to social problems, the simple fact that "the systems boys" failed to make their case convincing.

Specifically, the Task Force should consider these and related questions. In particular, has the SGSR not the responsibility to society to do what it can to see that this function is performed? Is this type of promotional program, with its missionary connotations, consistent with the SGSR policies and bylaws? If not, can the Society be adapted through constitutional processes and otherwise in time? If not, should the Task Force (in cooperation with the SGSR generally and other interested groups) take the initiative to try to identify and bring together representatives of interested agencies for an organizational meeting to explore mutually advantageous forms of cooperative action?[2]

---

[2] Additional information regarding the need for using general systems attitudes and concepts to democratize science is presented in Jere W. Clark, "Facing the Crisis of Intellectual Poverty," *Speech Journal*, Southern Connecticut State College, Spring, 1968.

SUPER-SPECIALIZATION: THE TREND AND ITS IMPACT IN AMERICAN MEDICINE

N. Henry Moss*

ABSTRACT

Specialization and superspecialization have increased in recent years in medicine as well as in other fields. It has been reported that only two percent of doctors enter general practice, and this is related to the fact that the cost of medical care in the United States will jump two and a half times in the next eight years unless steps are taken. We must develop in medicine, as well as in other fields, generalists or polymaths. These people should have a background in humanities and history, as well as technical know-how, and there must be an upgrading and glamorizing of this type of person. He should be a leader and should be amply rewarded. A possible alternative mechanism is the team approach as practiced in various fields and is known as group practice in medicine. In any case, it is time for the pendulum to swing back from the depth of the super-specialist to the breadth of the generalist.

---

* N. Henry Moss, M.D., F.A.C.S. is Associate Clinical Professor of Surgery and Attending Surgeon, Temple University Health Sciences Center and Albert Einstein Medical Center, Philadelphia, Pa.

Alfred North Whitehead outlined three stages of progression in the educational process: the stage of romance, when knowledge is novel and exciting; the stage of specialization, when information and techniques toughen and test the intellect; and finally, the stage of generalization, when freedom again dominates, and the mind fuses culture with expertise. Few physicians attain this third plateau; nine to fifteen years of wearing academic blinders in our traditional undergraduate college and medical school education and subsequent graduate training programs narrow the accepted definition of "getting an education" down to passing examinations, increasing and briefer exposure to a greater number of departments and subdepartments, writing papers on increasingly fragmented subject matter, and becoming more vulnerable to specialization at earlier and earlier phases of the educational process. Concomitant with these developments has been the unabated progression of departmentalization and subdepartmentalization into which the young trainees neatly fit and from which they rarely emerge to attempt an assault on the third plateau of education.

Today, the crisis in American Medicine and in the delivery of health care to the public is in large measure due to the dramatic trend toward superspecialization of the past two decades and the obvious failure of our system to fill the void created.

Accelerating specialization, a mere embryo in the 1920's and early thirties, blossomed after World War II in the late forties. It has become further subdivided in the fifties and dramatically apparent in the sixties. Today almost everyone is a specialist, many are superspecialists and some are specialists of even greater magnitude.

Whereas the impact of increasing specialization has its rewards in individual achievement and expanding knowledge, one must also look at the toll. In our pursuit of knowledge one may logically ask is it synthesis or fragmentation that reaps the greatest reward? Or is one without the other as inadequate as a ship without a sail?

Specialization and fragmentation are not new developments in medical history. However, the increasing frequency with which it is occuring creates inherent dangers that must be set directly. The cardiologist is now more commonly seen as a coronary artery occlusionist or a congenital valvular expert or one devoting his time and effort to cardiac catheterization. The day may not be far off when the specialist of the right coronary artery will meet the specialist of the left coronary artery and in greeting each other will encounter the specialist on the collateral circulation at the interventricular septum.

Medicine has helped lead in this development that is present in a wide array of other scientific disciplines, from the specialization of the last thirty years to the numerous subspecialties of the current decade. One out of every six physician specialists has narrowed the scope of his practice in some important way since 1960. How frequently we hear the benefits of the latest diagnostic or therapeutic triumph, which has improved the scientific level in the practice of medicine, yet has left the poor patient fragmented by the multiplicity of specialists or groping for appropriate guidance when the clinical problem is subtle and difficult. More and more medical schools today ask their students to select their choice of future career after only two years of primarily basic science training and prior to an appropriate exposure to the broad fields of medicine and surgery. These students must make this selection when they are ill prepared for the choice. All too often the subspecialist in internal medicine or surgery no longer feels competent to properly manage a patient presenting a problem a notch or two beyond the confines of that subspecialty.

How often do we find the individual really capable of accurately and effectively encompassing a broad perspective beyond his own special area, a person with extensive and variegated knowledge who can see the forest for the trees, the strengths and weaknesses of each narrow subcategory within his own major discipline, and even beyond. Instead of intensive preoccupation with one small body of information, such a person ranges wide and consequently sees the relationships and cross-fertilization potential of one realm of knowledge to another, as did some of the giants of yesteryear. He is competent to engage in interdisciplinary discussion and investigation, or to work at the interface of his field. He identifies neglected areas of study and encourages other investigators to become interested in them.

Where are our great synthesizers of knowledge? Unlike the emerging nations of Africa that so desperately need specialists in all fields of science and medicine, we in this country have successfully developed copious numbers of such specialists, a multitude of analyzers in one field after another, penetrating into smaller and smaller crevices and seeking finer and finer fragments of learning. I am reasonably sure that industry, government, universities, and medical schools have need of and eagerly seek capable people who can effectively give the broad sweep, recognize the important from the unimportant, and maintain a reasonable understanding of the literature in multiple medical and scientific areas. As our nation grows, not only in its gross national product, but also in its population, its technological products, and its per capita income, we must be able to develop a corps of such scientists and physicians as synthesizers and integrators to help guide us in directing our resources toward optimum use.

How few scientific and medical people there really are who have sufficient background in the humanities to effectively place science and technology within the proper perspective of society as a whole.

Can we broaden the development of those rare and gifted individuals who the distinguished editorialist, Dr. Irvine Page, has called polymaths, at a time when scientific and medical training is encountering self-perpetuating, self-accentuating encouragement and more and more limiting specialism? We need such gifted people and must meet the demand for them.

In order to do so, we must develop a more intensive training program for this kind of individual, one who has not disappeared completely but whose ranks have thinned. We must create this kind of fine generalist in the image of a captain of a team, a strong and vigorous leader with intelligent insight and technical know-how in multiple fields; he must be able to overcome the ancient Hippocratic aphorism that the art is getting longer and longer, the brain of the student not bigger and bigger; or the modified version of Benjamin Franklin's "the art is long and the time is short". There must be an upgrading and glamorization of the type of persons who can meet these requirements. He should ultimately be capable of applying science's best technology to society's most acute social needs. Therefore, he must have a firm background in the humanities and a good understanding of the history of science as well as the history of man through the ages. He should be able to demonstrate his creativity, for the truly educated man differs from the trained man only by the extent of his creativity. Because the academic and professional requirements of being perceptive and knowledgeable in more than one narrow sphere of endeavor are difficult to achieve, those who do qualify and who acquire these capabilities should be amply rewarded. The halo of the specialist of the last few decades should be partially refocussed and the glamor of professional acclaim be given to such men because men they would be. In order to generate the process, our educational system in the early years must emphasize respect for the superb generalist and at least a few courses in high school and college should reflect his emphasis. Schools at all levels should give standing to the excellent generalists on the faculty instead of rewarding mainly those concerned with infinitely small. A beginning must be made in reversing this march toward super-specialization and, once begun, will bring forth an increasing number of well-trained generalists who, in turn, will duplicate themselves in their own image. In this way their impact on society, at first minimal, might soon come to be regarded as meaningful.

There is a second mechanism to help in the much greater quantitative and qualitative delivery of effective generalization in science and medicine. If it is difficult to get the knowledge into one person, then the coordinate efforts of many individuals acting in unison is a reasonable substitute. Properly done, the team approach can be both effective and economically feasible, and it has been tried and tested for many years. What is necessary is expansion of the process. In medicine we call it "group practice" and the effort toward expansion of such groups has been advocated recently and strongly by the Surgeon General, Dr. Stewart, and was reaffirmed recently by the President's National Advisory Commission of Health Manpower. This commission reported that only two percent of doctors enter general practice and that the cost of medical care in the United States will jump two and half times in the next eight years unless steps are taken to stop the rise. Group practice will not only help meet the apparent shortage of physicians, but also keep the costs of providing excellent care much lower. Team approaches in other phases of science have been common procedure for many industries and have recently been demonstrated in our national space program, but there are many situations where the surface has been hardly scratched. Much remains to be done, and much should be done.

Pursuit of knowledge by scientists and physicians through their entire career is a *sine qua non* which we all clearly recognize. In recent decades we have seen the pendulum swing significantly too far in the direction of the depth of the super-specialist, rather than the breadth of the generalist. The mandate from the public, and to ourselves in our devotion to a lifetime of learning, suggests that this pendulum swing back sufficiently to provide the harmonious blending of the two.

Is this pursuit of knowledge, synthesis or fragmentation? This is a fundamental issue of our time.

GRAPH THEORY AS A META-LANGUAGE OF COMMUNICABLE KNOWLEDGE*

Edward Kingsley, Felix F. Kopstein and Robert J. Seidel

ABSTRACT

If one is to prescribe to an instructional automaton (a computer in the role of an instructor) how to instruct, it will be necessary to describe the subject matter to be taught largely independent of any particular content. The requirement exists for a meta-language in which to describe communicable knowledge. A strong candidate for this role is the mathematics of nets and graphs, or more generally, the field of topology. Some particularly relevant properties of graphs are reviewed. However, the problem of precisely coordinating formal and empirical structures has not yet been solved satisfactorily. The authors' approach as well as the approaches of French and Russian workers are discussed briefly.

If one is faced, or pretends to be faced, with the problem of specifying to an automaton-instructor, such as a computer, precisely how it must behave so as to instruct a student in some subject matter or other, a number

---

* The research reported in this paper was performed at HumRRO Division No. 1 (System Operations), under Department of the Army contract, with The George Washington University; the contents of this paper do not necessarily reflect official opinions or policies of the Department of the Army. This is a preliminary report subject to revision and expansion. Joint authors are listed in alphabetical order. The authors are with the Human Resources Research Organization (HumRRO), Alexandria, Virginia. Reproduction in whole or in part is permitted for any purpose of the United States Government.

of issues are brought into sharp focus. Computers, contrary to popular expectations, possess no special magic and the mere presence of imposing hardware is, at most, a necessary but not a sufficient condition for effective and efficient learning to occur. A meaningful way of describing the role of the computer in computer-administered instruction (CAI) is to characterize it as a tool - possibly an indispensable tool - for gaining and maintaining a high degree of *facilitative* control over the instructional process. If that tool is to be used effectively, it must be told in complete, precise and totally unambiguous terms what to do. It will not and cannot tolerate obscurity of language in the commands governing its behavior.

1. INSTRUCTION AS THE COMMUNICATION OF KNOWLEDGE

Any definition of instruction (or teaching) must, and in fact does, include the notion of communication of information so as to effect a transfer of knowledge from an instructor to a student. Second, the concept of instruction must not contain anything that is at variance with a rigorous definition of learning. Certain it is that the notion of transfer of knowledge or of a capability for performing or of a response repertoire is entirely consistent with definitions of learning that are generally accepted in scientific psychology (e.g. Kimble, 1961, 1964).

In any formal instruction, as in a classroom or a tutorial situation, at least two aspects of relevant knowledge may be gained by a student (learner). Nothing will be said about irrelevant knowledge. First, he will gain information about the instructor and his characteristics as an instructional agent. Second, it is hoped that he will gain information about some subject matter or other. This subject matter can be viewed - for the moment - as information stored by the instructional agent. The central task for the instructional agent is to transfer a reasonably complete copy of his own course-relevant knowledge to the student. If the instructional agent is an automaton, or if one pretends that he is, two problems present themselves immediately. How much of the information stored by the instructional agent is to be communicated? What is and what is not part of the instructional message? Second, the capacity of the communication channel is such that a simultaneous transmission of the total message is out of the question. A sequential transmission of parts is the only possible one. How is the automaton to decide what is a part, how many there are, what sequences of these parts are permissible under existing constraints (and what are these constraints), what parts and sequences of parts have been transmitted, which of them have been not only transmitted but received and stored, and when transmission of the total instructional message is complete? Allied with this is the problem of specifying the message units and how the parts or pieces or chunks of knowledge may be encoded into these message units to maximize the probability of reception and assimilation.

2. THE SUBJECT-MATTER MAP

Even if one feels a repugnance toward the notion of an automaton in the role of the instructional agent (this is a matter for a different debate), a valuable purpose is served in at least considering the possibility. For if one is to *prescribe* to the automaton how to instruct it will be necessary to *describe* knowledge largely independent of any particular content. If the knowledge were merely a matter of some listing, as in a dictionary, it might be expedient to represent it as a series of numbers. However, simply storing and describing an encyclopedic stock of information does not sufficiently represent what we really seem to mean by knowledge. Without some specified subject-matter organization, there exists no basis on which the hypothetical automaton can decide which facts, principles, procedures, etc., what interrelationships exist among them and which topics may be best presented before or after which other ones. Facts, principles, concepts, and especially their verbal or symbolic representations which are required to render them communicable, exist in contexts and never in absolute isolation (see, e.g., Ackoff, 1962, pp. 16-17). Depending on the particular context in which they occur, they are linked by one or more relations. As a minimum such relations are inevitably demanded by the *inescapable* circularity of definitions (Churchman and Ackoff, 1947).

What has been said hitherto amounts to this: The requirement exists for a meta-language in which to describe communicable knowledge. A strong candidate for this role is the mathematics of nets and graphs, or more generally, the field of topology. The candidate has much to recommend it. Obviously, there is an advantage in choosing a meta-language which is unambiguous, and itself has had an extensive mathematical development. Moreover, it is linked to (i.e. can be restated in terms of) other forms of mathematics - most readily to set theory on the one hand and matrix algebra on the other. It amounts to conceiving of knowledge or at least communicable knowledge, as a space that is structured. This space is represented as a set of points, which are interconnected or interconnectable by a set of lines. To illustrate, an arbitrary net is shown here in Fig. 1. It will be readily apparent that such a representation amounts to a "map" of a knowledge space. Indeed, ordinary roadmaps are technically graphs.

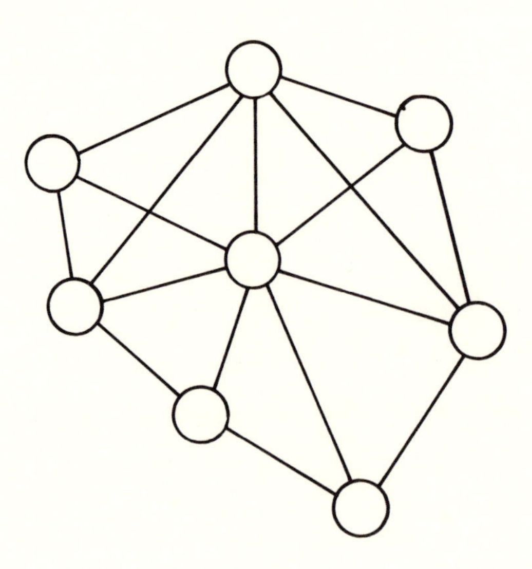

FIG. 1.

Whether the choice of nets or graphs as the meta-language of knowledge is the best possible one, or even a good one, surely will be open to debate for a long time. However some beginning must be made. At the very least it promises new ways of looking at knowledge and thus may lead to new insights. For example, there are two immediate issues in applying graphs to communicable knowledge qua subject matter. What are the boundaries of the knowledge space and what *exactly* do points and lines represent? Before proceeding to discuss these issues, however, it may be well to review the nature of graphs and some of their seemingly relevant properties.

3. GRAPHS AND SOME OF THEIR PROPERTIES

Graph theory is concerned with the systematic study of configurations of points and lines joining certain pairs of these points. More formally, consider the structure consisting of a finite set V, of elements together with a collection, U, of ordered pairs of elements from V.

Such a structure is called a *net* by Harary, Norman and Cartwright (1965) and is denoted by $G = (V, U)$. The elements of V are called *points* (*vertices, nodes*) and are denoted by $\{v_1, v_2, \ldots, v_n\}$. The pairs in U are called *lines* (*arcs, edges*). If v_i and v_j are two elements in V, then the line joining v_i to v_j is written as (v_i, v_j). Lines are said to be *parallel* if a pair (v_i, v_j) is repeated in U. A *loop* is present at a point v_i if the pair (v_i, v_i) occurs in U. A net with no parallel lines is termed a *relation* (Harary et al., 1965). A net with no parallel lines and no loops is called a *directed graph* or *digraph*. (Harary, et al., 1965.)

The above definitions are given in terms of set theory. Other representations can be given pictorially and analytically. The pictorial representation of a graph is extremely useful for depicting structural situations. Their map-like nature will be self-evident. In this representation, points are joined by a line or lines associated with them. The relative position of the points is immaterial as well as the shape of the lines joining the points. Lines may cross each other but such cross-overs are not considered as points. A net is shown in Fig. 2, with parallel lines joining v_1 and v_2, a relation in Fig. 3, with a loop at v_2, and a digraph in Fig. 4.

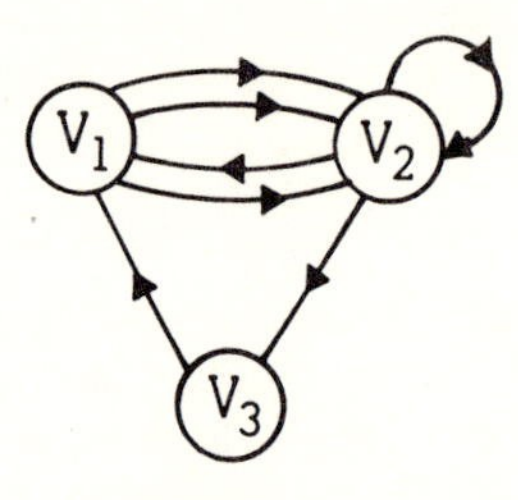

FIG. 2.

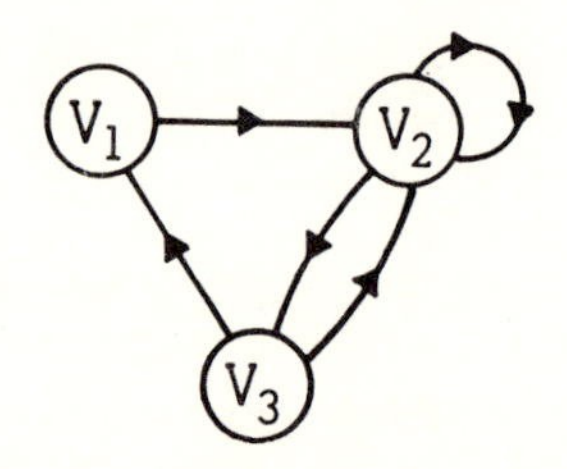

FIG. 3.

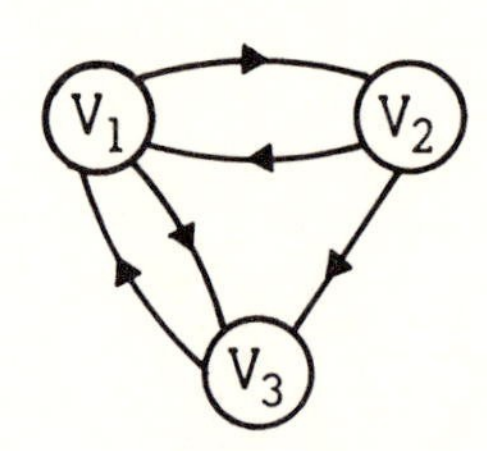

FIG. 4

A *partial digraph* of a digraph G is obtained by using all the points of G and deleting at least one line. A *sub-digraph* of a digraph G is obtained by deleting one or more points from G and their adjacent lines. Possibly partial graphs and sub-graphs might be interpreted as regional or not fully developed maps of knowledge. Fig. 5(a) is one of the partial graphs of the digraph shown in Fig. 4 and Fig. 5(b) is one of its sub-graphs.

(a) (b)

FIG. 5

A *path* is a sequence of lines and points in a digraph such that the terminal point of each line is coincident with the initial point of the succeeding lines. The *length* of a path is the number of lines in the sequence. In the digraph shown in Fig. 6, there are two paths from v_1 to v_3, one path has length 1 and the other path has length 2. Possibly paths and their lengths might describe sequences of concepts or topics (points) and the instructional steps separating them (lines).

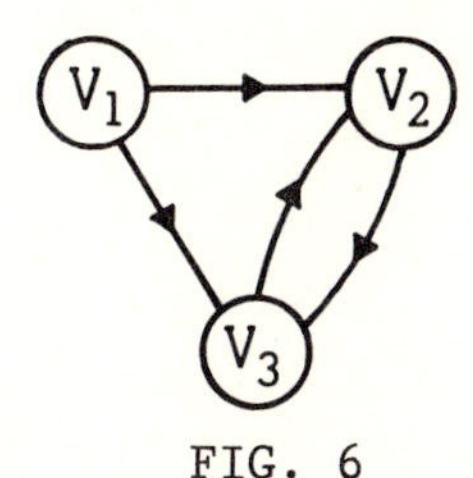

FIG. 6

Two points u and v are *adjacent* if they are joined by at least one line. A digraph is *symmetric* if two adjacent points v_i and v_j are always joined by two oppositely directed lines. A symmetric digraph is shown in Fig. 7.

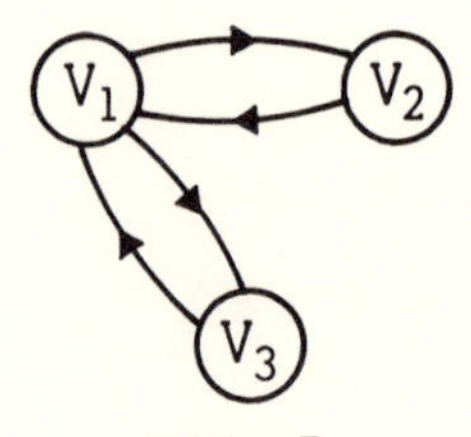

FIG. 7

Two topics in a course of instruction might be considered to represent adjacency if one immediately followed the other along a path leading to maximal end-of-course proficiency. In like manner, symmetry would be represented, if either topic could be taught before the other with equal effectiveness.

A digraph is *complete* if every pair of points is joined by at least one line oriented in one of the two directions. A complete digraph is illustrated by Fig. 8.

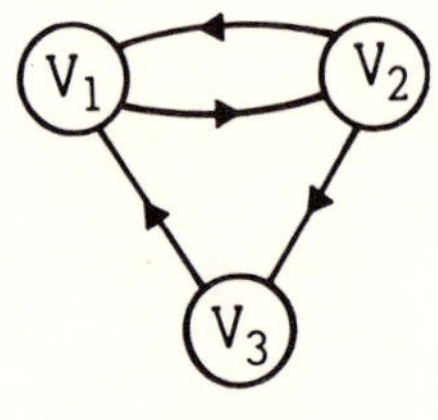

FIG. 8

If a digraph is both symmetric and complete - it is called a complete symmetric digraph. The complete symmetric digraph of order 3 is shown in Fig. 9.

A digraph is *strongly connected* if, for every two points v_i and v_j, there exists a path directed from v_i to v_j. A strongly connected digraph is shown in Fig. 10. Completeness, symmetry and strength might be viewed as characteristics distinguishing inherently highly structured subject matter, such as, mathematics from inherently more unstructured subject matter such as history.

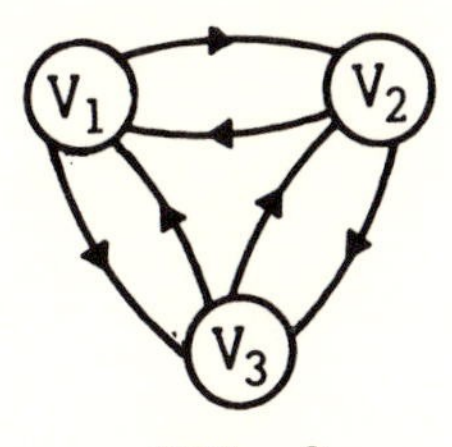

FIG. 9

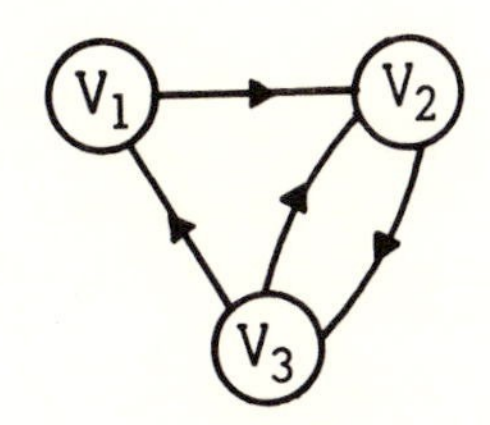

FIG. 10

A method for simplifying a digraph will be described that is useful for obtaining insight into its structural properties. Thus, let $G = (V, U)$ be a digraph and partition its point set V into m subsets $\Pi_1, \Pi_2, \ldots, \Pi_m$, so that

$$\bigcup_{1m} \Pi_i = V$$

$$\Pi_i \cap \Pi_j = \phi, \ i = j; \ i, j = 1, 2, \ldots, m.$$

Then consider the m subsets of the partition as the points of a new digraph, each point, $\pi_i (i = 1, 2, \ldots, m)$, being labeled by the set of elements of G that form Π_i. Finally, a line exists from point Π_i to point Π_j, if and only if there exists at least one line in G from a point of Π_i to a point of Π_j. The digraph, $G^* = (V^*, U^*)$, thus formed is called the *condensation* of G with respect to the partition. The point set of a digraph can be partitioned in many different ways. For this reason, there exists a variety of condensations of a digraph. This is illustrated for the digraph G in Fig. 11. Two different condensations, G_1* and G_2*, are shown for G. The partitions of $\{v_1, v_2, v_3, v_4, v_5\}$, the points of G that form the points of G_1* and G_2* are:

$$\Pi_1 = \{v_1, v_2, v_3\},\ \Pi_2 = \{v_4, v_5\} \text{ for } G_1^*,$$

$$\Pi_1{}^1 = \{v_1, v_2\},\ \Pi_2{}^1 = \{v_3\},\ \Pi_3{}^1 = \{v_4, v_5\} \text{ for } G_2^*.$$

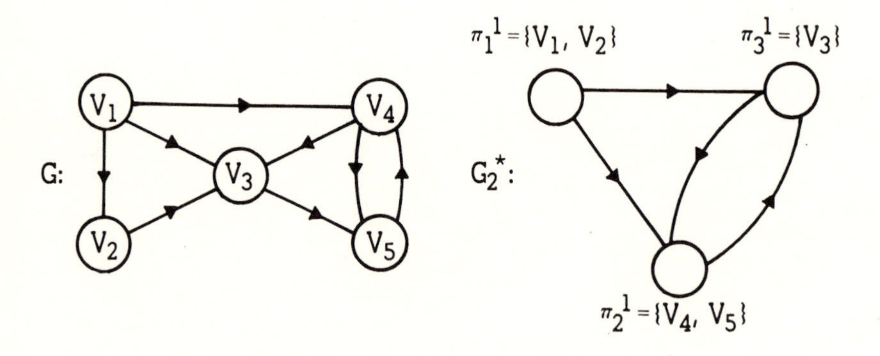

FIG. 11

Complex digraphs can be made progressively simple by repeated use of the condensation method. This technique is particularly useful in studying hierarchical structures. A possible practical application of this technique will be touched upon later. Fig. 12 illustrates two successive condensations. First, the basic digraph G is condensed to G_1* and then G_1* is condensed next to G_2*. The points of G_1* are sets of points of G and the points of G_2* are sets of sets of points of G. The first condensed digraph G_1* is formed by the partition:

$$\Pi_{11} = \{v_1, v_2, v_3\},$$
$$\Pi_{12} = \{v_4, v_5, v_6\},$$
$$\Pi_{13} = \{v_7, v_8, v_9, v_{10}\},$$
$$\Pi_{14} = \{v_{11}, v_{12}, v_{13}\}.$$

The second condensed digraph, G_2*, is formed by partitioning the points of G_1*, $\{\Pi_{11}, \Pi_{12}, \Pi_{13}, \Pi_{14}\}$, by

$$\Pi_{21} = \{\Pi_{11}, \Pi_{14}\} = \{\{v_1, v_2, v_3\}, \{v_{11}, v_{12}, v_{13}\}\},$$
$$\Pi_{22} = \{\Pi_{12}, \Pi_{13}\} = \{\{v_4, v_5, v_6\}, \{v_7, v_8, v_9, v_{10}\}\}.$$

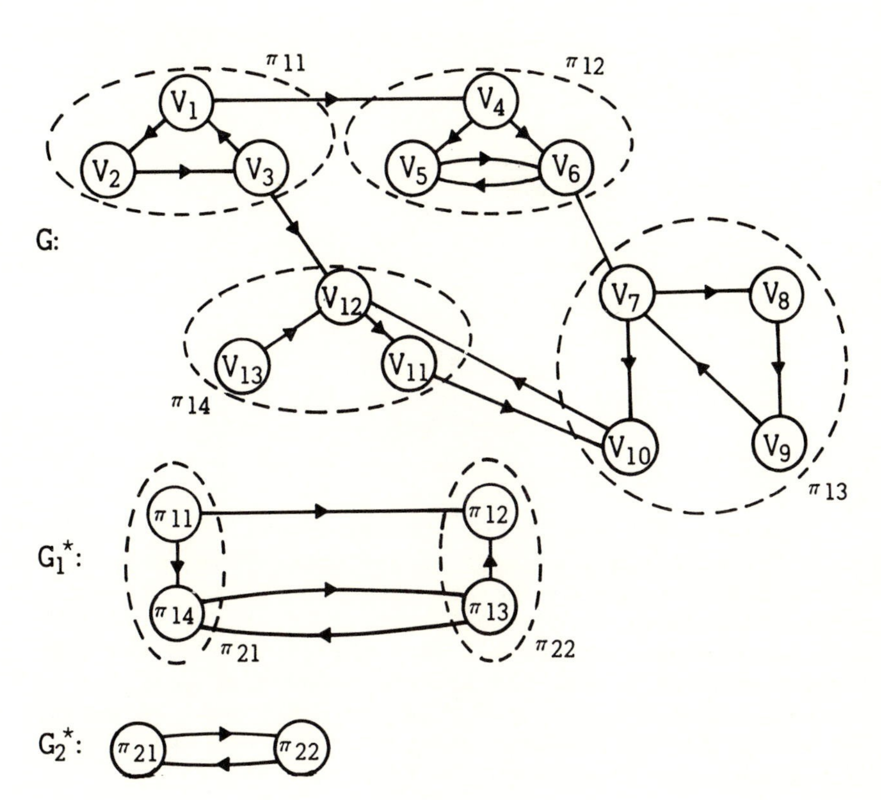

FIG. 12

It will be reasonably self-evident that the condensation procedure is an excellent analog to abstraction. With each successive partitioning and condensation any point in the condensed graph is likely to become more inclusive. At the same time there is a progressive loss of information about the sub-graph(s) that have been condensed. This would seem to suggest that a concept at a given level of abstraction should be presented only after the component subconcepts and their interrelations have been taught.

A digraph G is transitive if, for every three points v_i, v_j, v_k in G, it contains the line $v_i v_k$ whenever it contains both the lines $v_i v_j$ and $v_j v_k$. In other words, if there is a path of length 2 from point v_i to point v_k then there exists a line from v_i to v_k. The digraph shown in Fig. 13 is a transitive digraph.

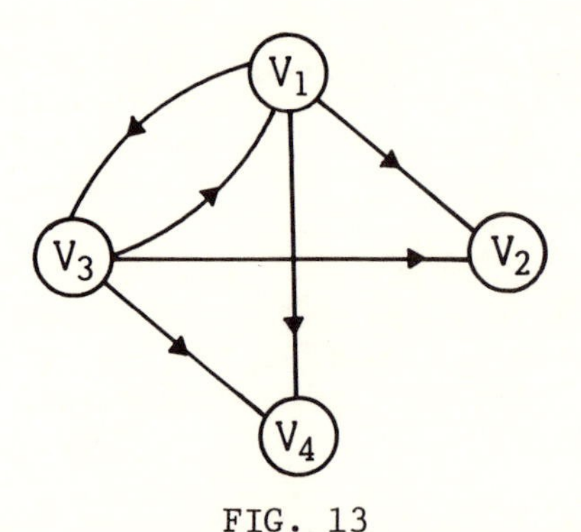

FIG. 13

If a digraph G = (V,U) is not transitive, a new transitive digraph can be obtained from it by constructing the minimal transitive digraph containing G and leaving the same vertices as G. The digraph, $\hat{G}$ = (V, U) thus formed is called the *transitive closure* of G. The digraph $\hat{G}$ has the same point set as G but a different set of arcs. Fig. 14 shows a non-transitive digraph G and its transitive closure $\hat{G}$.

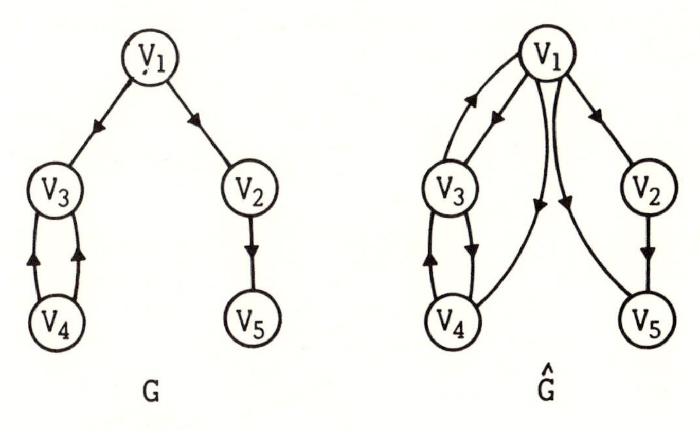

FIG. 14

Provisionally the transitive closure procedures might be seen to have application to such issues as a student developing new "insights" into interrelations among concepts or principles within a subject matter. This will be further explored in the context of relative knowledge space.

If a digraph contains the line v_iv_j but not the line v_jv_i, the digraph is said to be *asymmetric*. In other words, if two points v_i and v_j are connected by a line, there is only one line, either from v_i to v_j or from v_j to v_i.

If a digraph is both transitive and asymmetric it is termed the digraph of a *partial order*. The digraph shown in Fig. 15 is the digraph of a partial order.

It is often useful to be able to arrange the points of a partially ordered digraph in a linear sequence without destroying the partial ordering. That is, the points in the digraph are rearranged so that all the arrows go from top to bottom. The resulting digraph is said to be a *linearly sorted* digraph. The digraph has not changed by this process.

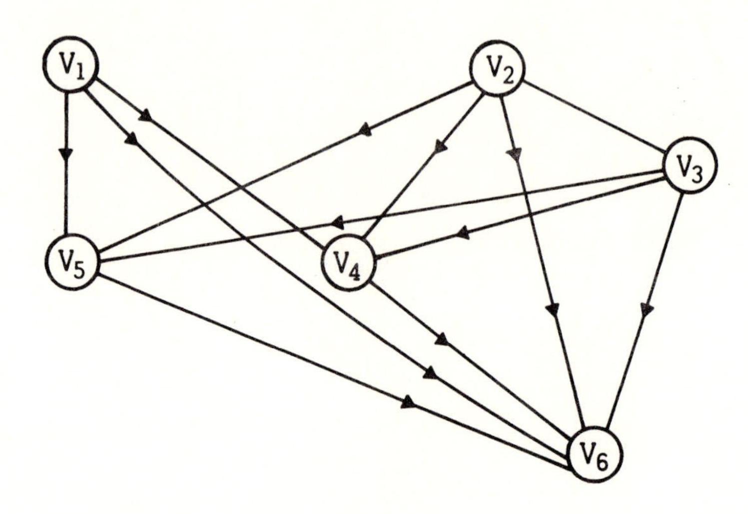

FIG. 15

A particular digraph of a partial order may have several linearly sorted digraphs. This is illustrated in Fig. 16 for the partially ordered digraph G_1 shown there. Three of the many linearly sorted digraphs S_1, S_2, and S_3 are given for G_1.

Linear sorts will be readily identified with specific "routes" through a graph. Where a graph represents, for example, an area of knowledge qua subject matter in a course, the totality of different linear sorts will be equal to the number of different ways in which that subject matter can be presented. The choice among the available routes (linear sorts) might depend on the personal characteristics of an individual student, his within-course history (response patterns), his motivations or preferences, or the preferences or constraints imposed on the instructional agent (teacher, computer program).

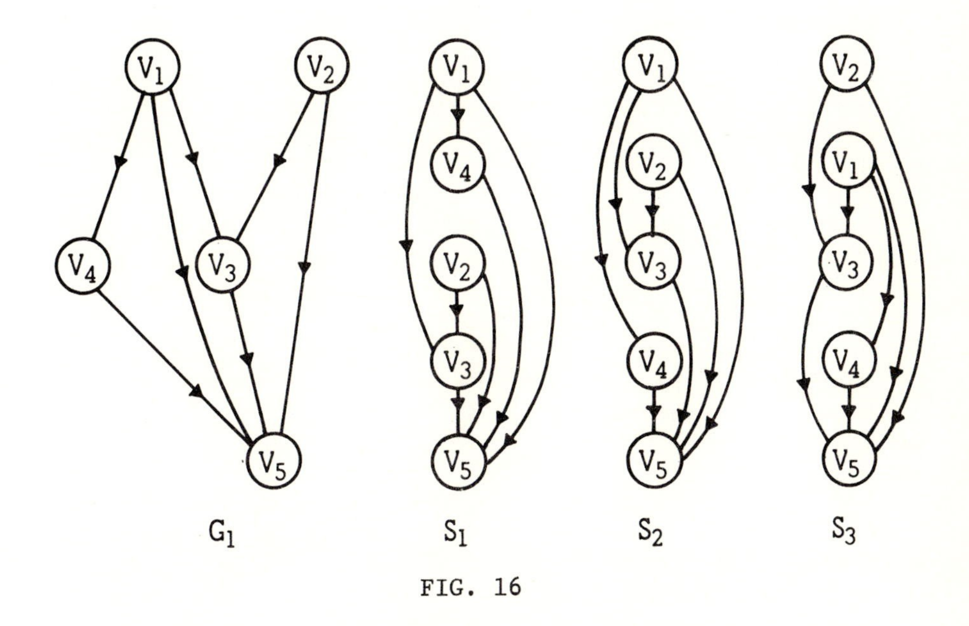

FIG. 16

4. RELATIVE KNOWLEDGE SPACES AND SUBJECT-MATTER GRAPHS

Earlier the question of boundaries on a knowledge space was raised. This problem applies to a total knowledge space, encompassing all communicable human knowledge, as well as to any subject matter in a particular course of instruction.

Operating premises for a pragmatic, conceptual approach to content-independent descriptions of knowledge need to take the following or similar forms: First, any given subject matter is an *arbitrarily delimited portion* of the totality of human knowledge. Second, just as all human knowledge is incomplete and is constantly increasing, so is knowledge of the specific subject matter at hand. Increase in knowledge takes place when new facts are discovered and new concepts are established, or when new relationships among existing facts and/or concepts are discovered. For example, in biochemistry recently established facts about the molecular structure of certain substances (DNA and RNA) are being related to a vast number of other facts and concepts on heredity, evolution of species, cancer, the common cold, and so on. In time, we will become aware of other relationships of the structure of DNA with known and yet-to-be discovered facts in many disciplines.

The DNA example also serves to illustrate that divisions among disciplines and subject-matter areas are arbitrary. Good arguments can be presented for assigning the molecular structure of DNA to the subject matter of physics, of chemistry, of physiology, of genetics and biology at large, or even - perhaps, for example, in the process of remembering qua information storage and retrieval - to behavioral science.

We must distinguish among (1) the total knowledge that is possessed by mankind so far, (2) arbitrarily delimited subject-matter areas within the totality of existing human knowledge, and (3) the knowledge about a given subject-matter area possessed by a specific individual or group of individuals (e.g. an instructor or instructional staff). The knowledge or information about a subject matter possessed by even the most knowledgeable instructor is, in fact, a finite and even quite limited structure. With reference to a *delimited* knowledge structure (e.g. the arbitrary requirements for a course), success in instruction or the communication of knowledge is shown by an inability (in a loose sense) to distinguish between the instructor and the student. That is, the student should have learned all that the instructor knows about the subject matter relevant to the course requirements, and both of them should be able to answer questions or perform tasks pertinent to the requirements of the course equally well.

Although in this sense indistinguishability of instructor and student is an ultimate ideal of instruction, even a close approximation of this state is not readily attained. The knowledge structure of the instructor is communicated to the student over a period of time. Especially in the early stages of instruction, only a small portion of the instructor's knowledge structure will have been transferred to the knowledge space of the *student*. The structure that will have been transferred will be limited in the number of its points (i.e. only part of the facts, concepts, etc.) as well as in its strength (i.e. awareness of interrelations).

The distinctions to be made among the several versions of a subject-matter structure can be viewed as a sequence of partial graphs illustrated in Fig. 17. The points represent the concepts of the knowledge space, and the oriented lines represent the relations among the concepts.

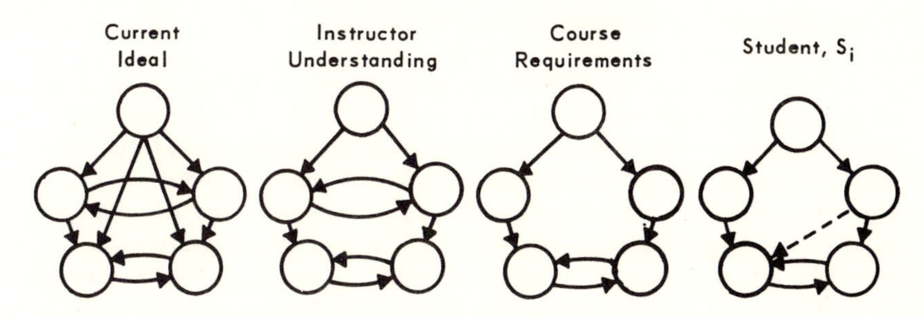

FIG. 17

The "Current Ideal" label reflects the fact that knowledge of any given subject matter is constantly "growing." As for the "Instructor's Understanding": This instructor does not have *all* the knowledge about the subject matter that can be possessed currently, but he possesses enough to serve the requirements of the course; he also has additional relational awareness that is not required to convey the course properly to the students. Finally Student S_i has not only successfully completed the course requirements, but has independently discovered a *new* relation. (He must be a fairly bright student, since it is a relationship not known to his instructor, and not even existing within the "Current Ideal" structure of the subject matter.)

This simple example is intended to illustrate the necessity for considering *relative* states of knowledge about a given subject matter for instructor and student. It accounts for potential individual differences among students. Finally, implicit in it is the notion of generalization - a transfer of potential capability that will engender proficiency in tasks not yet encountered. This capability is important because the course requirements, as indicated, provide but a subset of the potential relations that could be required if a more complete mastery of the subject matter were desired.

5. GRAPHS AND EMPIRICAL REALITY

The question that still remains unanswered concerns the empirical counterparts of points and lines. To be useful, maps need not be simply scaled-down versions of some terrain, but they must retain an isomorphism with some aspects of it. For example, roadmaps can eliminate all cues as to topography and still serve to guide a motorist from town A to town Z. Of course, on arrival at Z the point that represented Z on the regional map will lose its usefulness to the hypothetical motorist. He will need an enlarged city map to find his way about. Note that this familiar characteristic of roadmaps corresponds to the condensation and de-condensation procedures in graphs that were discussed earlier. Note also that what is represented by points and lines on a regional map is *not* absolutely identical with what is represented by them in the city map. Moreover, this would be true even if the enlarged city representation were an integral part of the regional map.

It appears that maps can incorporate a hierarchy of abstractions without demanding complete consistency from level to level. This realization guided one approach to answering the original question. Kopstein and Hanrieder (1966) prepared a pseudo-anthropological account of the culture of a fictitious tribe - the Gruanda. Since it was proposed to use this account as the instructional material in experimental studies of learning as a function of certain structural properties of subject matter, it was necessary to control these structural characteristics. This could be done only by inventing the subject matter to fit the desired structure. Ten *topics* were arbitrarily chosen. These ten topics dealt with various aspects of the Gruanda such as, their territory, their clan system, mythology, rituals, hunting, agriculture, etc. Each topic constituted a paragraph of about 200 words. The topical paragraphs could each stand alone and independent, or they could be related each to one or more of the other topics. In the latter case the paragraph contained some sentences that constituted a cross-reference to one or more other topics. For example, the topic of rituals might contain references to hunting and/or to political structure, and so forth. The resulting maximal structure was as follows. (See diagram.)

Kopstein and Hanrieder identified topics or paragraphs with points and sentences involving cross-references to other paragraphs as lines. They were clearly aware that this isomorphism reflected only one, relatively high level of abstraction and thus referred to as a *macro-structure*. By implication this recognized underlying *micro-structures* - a recognition which entered into the experimental procedures within which this artificial subject-matter was used. Presumably topical paragraphs are supra-organizations of syntactical and grammatical structures (or rather their corresponding behaviors) which, in turn, are supra-organizations of lexical structures, and so forth. Of course, the reference here to the nature of the infra-organization is purely conjectural and intended only as a suggestive illustration. Any serious propositions concerning hierarchical levels of structuring would demand a formally consistent taxonomic array and criteria for assigning given structures (graphs) to a particular level. How far the infra-structures may extend downward is an obvious question, but need not be settled here and now.

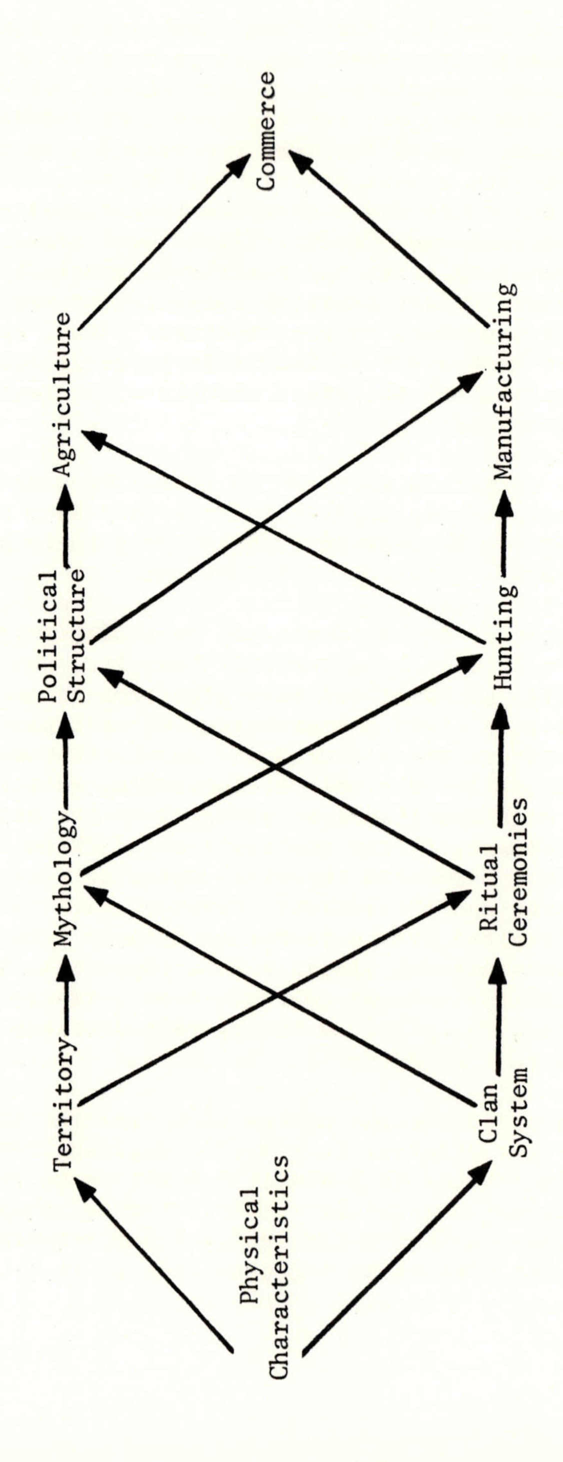
Physical Characteristics
Territory
Clan System
Mythology
Ritual Ceremonies
Political Structure
Hunting
Agriculture
Manufacturing
Commerce

While experimental data from a subsequent larger study by Kopstein and Seidel failed to support hypotheses which essentially predicted that with higher degrees of structuring, learning or, rather, recall of information about the Gruanda would improve, the validity of these hypotheses remains a moot point. Failure to obtain support may be due to the fact that the operational coordinating definitions were faulty. It is also possible that due to the low complexity and restricted scope of the "Gruanda Material" participating subjects imposed their own structure - perhaps micro-structure. Thus, resolution of the issues must await further experimental attempts requiring the mastery of larger amounts of subject-matter of greater complexity.

A quite different approach to achieving congruence between formal and empirical structures has been proposed by Regnier and de Montmollin (1968). They begin by asserting that:

> "All material to be taught can be characterized in precise fashion by a terminal behavior that is specified *a priori* and then partitioned, more or less finely, into elements that we will call units of knowledge which constitute an enumerable set ... we will define the unit of knowledge, whatever the level on which it may be situated in the hierarchy of knowledge, as the putting into relation of at least two terms and therefore being able to give, at least, an 'intelligent' response, i.e. a response characterized by the fact that it puts into play long and indirect circuits and, therefore, is indistinguishable from reflex actions, innate or conditioned, which follow immediately upon the presentation of a stimulus."

Even allowing for the ambiguities deriving from the rendering of the original French into English it would seem that Regnier and de Montmollin assert some questionable mathematical propositions. For example, they assert that a relation is "non-reflexive" and "anti-symmetric," when subsequent discussion suggests that it is "irreflexive" and "asymmetric" (the latter designations are used below). Some definitions (e.g. "simply connected") are not clear. They do propose a unit of knowledge that has a striking resemblance to the familiar S-R in psychology. S and R constitute the elements in the set with a binary relation on the set so defined that, when it is empirically demonstrated, it simultaneously verifies the existence of the unit of knowledge. Note that neither S nor R involve anything that can be pointed to in the physical world except in terms of S being the necessary and sufficient condition for R.* Also, S and R are clearly conceived as being overt and observable.

It was pointed out earlier that graphs are relations defined on sets of points. Regnier and de Montmollin define a relation L on the set of units of knowledge as follows:

> "It is necessary to have acquired the unit of knowledge x_i (i.e. to have correctly responded to the question or to the problem which x_i represents) in order to acquire the unit of knowledge x_j (i.e. in order to respond correctly to the question or to the problem which x_j represents)."

These authors then go on to note that the relation L is non-reflexive, asymmetric and transitive and is therefore a relation of partial order. Since theirs is the problem of trying to determine *a priori* the linear order in which the instructional agent's knowledge structure should be "copied" and transmitted to the student, they are at a loss when L produces a partial order and does not suffice to determine a Hamiltonian path through the resultant graph - i.e. a path traversing every point once.

---

* Scandura (1968) has proposed a set-function language (SFL) that would seem to be pertinent here. In SFL the basic concern is not with S and R per se, but with the rules relating these sets.

Thus a second relation *P* is defined as follows:

> "The acquisition of the unit of knowledge x_i effected immediately before the acquisition of a unit of knowledge x_j facilitates the latter."

Relation *P* is transitive and neither symmetric nor asymmetric. Therefore it is a relation of quasi-order that includes *L*, although *L* takes priority over *P*.

Although their experimentation incorporates some obscure aspects (e.g. the precise differentiation of experimental study materials) and their measure of the structuring of their experimental materials suffers from mathematical defects noted above, Regnier and de Montmollin's experiment represents another initial attempt to investigate a largely unexplored and potentially fruitful domain. Regnier and de Montmollin found preliminarily that "the performance on the final test ... is proportionally inverse to the value of *O*." *O* is the symbol for their measure of structuring. They go on to note that this outcome was "exactly contrary to our hypothesis."

A third approach to representing knowledge in terms of graphs has been outlined by V.N. Pushkin (1967). It would seem that Pushkin's proposals derive from the attempt to describe creative problem-solving by humans as, for example, in certain conditions within chess games. Thus the instructional agent is neither human, nor a simulation but an objective problem situation faced by a person who is in a role analogous to that of the student previously. The thrust of Pushkin's argument is that the position of Newell, Shaw and Simon which equates heuristic computer programs with human behavior is not tenable. First, there is a distinction to be made between an object or situation existing in the real world with given properties and the representation of that object or situation within the problem solver. Second, whereas heuristic programs selectively eliminate unproductive moves toward a goal state in a series of sequential steps, humans preestablish a set of solution-paths qua representations - a solution map - for potential execution. The latter assertion is supported by eye-movement data obtained from chess players. It is the set of problem solutions qua paths to the goal that are represented or representable as graphs.

In Pushkin's approach the points of a graph G would seem to be the possible observable states that the problem object or situation can assume or, at least, a subset thereof that are relevant to the problem. One point represents the initial - the given state and another represents the final or desired state. Given existing constraints there is some finite number of intermediate states that the situation can assume. Lines would seem to be the relational operators that produce the transformation of any given state into an adjacent one. At any time there is a graph G which represents *all* possible valid solution-paths from the initial to the final state. Pushkin seems to argue that out of his representations the human problem-solver forms a problem model consisting of a set of solution-paths which are a subgraph or a partial graph or a partial sub-graph g of graph G.

Mention must also be made of Tesler, Enea and Colby's (1967) attempt to represent belief systems as directed graphs. Whether belief systems such as a psychiatric patient's view of the world, are equivalent to objective knowledge is an open question. Certainly it might be viewed this way, if a psychiatrist sought to describe his patient's belief system to third parties. Tesler et al. identify "concepts" with points and "simple relationships" with lines. Concepts can be *sets*, *individuals* and *propositions*. Directed lines connect concepts whenever any relationship exists. However, lines are more specifically characterized through associated symbols representing the "circumstance" of that line, the type of relation, attitude toward the relation, etc. Details are too complex for recapitulation here. Tesler et al.'s approach suffers from certain defects in its axiomatization, but constitutes still another application of graph theory to represent knowledge.

6. REVIEW

What has been presented here should be viewed as an argument for the *promise* of graph theory as a meta-language of communicable knowledge. In no sense should it be viewed as a complete or final argument. The problem of identifying the points and lines of graphs with some unambiguous empirical reality as, for example, meaningful instructional subject matter remains as the central issue. This is not to be confused with the different although related problem of the *levels* of micro-structures addressed earlier. If we can ever hope to obtain a consistency in application to meaningful material, we must be able to develop a rule or set of rules for defining conceptual units within the instructional content.

The Kopstein and Hanrieder study, and recent studies by Seidel and Kopstein provide good illustrations of this problem. Predicted differences due to varied graph structure were not obtained. Why not? Since the graph structures observed all relevant axioms and theorems, the answer must be in the rules coordinating formal representations with empirical structures. In any empirical application, a legitimate question to be raised is: how does the experimenter decide whether one, two or n-sentences are required for adequate cross-referencing? Secondly, one could also ask: what position in the paragraph would create the best salience for the cross-reference material to transmit the desired message, i.e. what organization will permit the best Gestalt or figure-ground representation? Finally, is there a "natural" or consensual ordering across people appropriate to the topical paragraphs used? In the Kopstein and Hanrieder experimental material, a rough, after-the-fact assessment seemed to support the existence of such an ordering. In fact, these conditions were corrected in the subsequent studies and the preliminary results seem to indicate verification of our suspicions.

This point is raised to emphasize the fact that one cannot simply develop *a priori* a logically tight theory and expect it to be useful without establishing equally tight rules of representation in the coordinating definitions of the theory. It is a difficult chore to accomplish with simple, artificial verbal learning tasks, i.e. isolated words or phrases or short paragraphs. It is particularly difficult - and yet even more important - to achieve with higher order conceptual groupings such as those existing in most classroom instructional materials. The dimensions which must be accounted for in the points, or elementary conceptual units in the latter instance, may well require the application of n-dimensional topological theory rather than simple graph structures. For example, the construction of instructional content itself establishes contextual linkages within the material which linkages potentially may require one or more additional dimensions for adequately specifying what is represented by a point. We are currently pursuing this and other possibilities in characterizing the subject-matter structure of a computer-administered course we are developing to teach computer programming in COBOL, a higher order computer language.

REFERENCES

Ackoff, R.L., *Scientific Method: Optimizing Applied Research Decisions*, New York, John Wiley and Sons (1962).

Churchman, C.W., and Ackoff, R.L., *Psychologistics*, Philadelphia, University of Pennsylvania Faculty Research Fund (1947). (Mimeographed).

Harary, F., Norman, R.Z., and Cartwright, D., *Structural Models: An Introduction to the Theory of Directed Graphs*, New York, John Wiley and Sons (1965).

Kopstein, F.F., and Hanrieder, B.D., "The Macro-Structure of Subject-Matter as a Factor in Instruction," *Research Memorandum* RM-66-25, Princeton, N.J., Educational Testing Service (1966).

Pitcher, G. (ed), *Wittgenstein: The Philosophical Investigations*, Garden City, N.Y., Doubleday, Anchor Books (1966).

Pushkin, V.N., "Psikhologiya Myshleniya i Printsipy Evristicheskogo Programmirovaniya," *Voprosy Psikhologii, 13*(6), 101-117 (1967).

Regnier, J., and de Montmollin, M., *Reconnaissance de l'Organisation, Recherche de l'Ordonnance des Elements et Choix du Mode d'Enseignement de la Matière*, Paris, Société d'Économie et de Mathématiques Appliquées (1968). (Mimeographed).

Tesler, L., Enea, H., and Colby, K.M., *Directed Graph Representation for Computer Simulation of Belief Systems*, Stanford, Calif., Dept. of Computer Science, Stanford University (1967).

Wittgenstein, L., *Tractatus Logico-Philosophicus*, Frankfurt a/M, Suhrkamp Verlag (1960). (Original published in *Annalen der Naturphilosophie*, 1921).

ACADEMIC STRUCTURE AND THE INTEGRATION OF THE SOCIAL SCIENCES

Frederic N. Firestone*

ABSTRACT

It is difficult to design integrative courses in the social sciences because of the structure of the university system. Veblen criticized the structure because of the financial viewpoint of the dominating trustees, and the effect of this view throughout the university. The present neo-Veblenian attacks on the university are concerned more with specific business connections of the university than with the habits of thought of the trustees.

However, the main problem is with the faculty members and their departmental specialization. It is desirable to broaden the introductory courses and bring in specialized material as needed later. Disciplinary allegiance should be de-emphasized in the introductory courses. In economics, in particular, an injection is necessary of a wider range of social phenomena and to include dynamics, or change with time, in order for greater predictive value. While there have been successful interdisciplinary programs at Columbia, Chicago, Wisconsin and elsewhere, they have probably been more characterized by juxtaposition than by integration, to which general systems can make a special contribution.

---

*Frederic N. Firestone is Professor of Economics and Chairman, Economics Department at Illinois State University, Normal, Illinois 61761.

This paper deals with the "specialized deafness" of which Kenneth Boulding wrote, which "means that someone who ought to know something that someone else knows isn't able to find it out for lack of generalized ears"[1]. My especial concern here is the learning of the social sciences.

It is my thesis that certain structural aspects of our college and university system make difficult--and in some cases impossible--the necessary and fruitful synthesis of what we call the academic disciplines of the social sciences.

It is worth noting at the outset that the academic structure has been attacked through the years by various critics who claim it is ill adapted to perform its announced functions. A brief summary of one or two of these complaints, and an inquiry as to its functions are in order.

Thorstein Veblen's *The Higher Learning in America*[2] set the tone of many of the criticisms of this century. One need only recall his intended subtitle--"A Study in Total Depravity"--to assess the extent of his sympathy with academia as he knew it. His most serious criticism concerned the vertical structure of the university. Trustees who oversee the operation set the criteria by which the educational functions are performed. But the trustees, Veblen argues, are not only ill adapted to this function; they are incapacitated for it by their background. Most are products of the business world, in which ability to perceive fact is a negative survival characteristic. Those whose "success" elevates them to academic trusteeship are, he advises us, men with a characteristic adeptness at making a seeming reality out of fiction. The fiction is that of a financial veil which hides the economic realities of production, and subverts the productive process from its nominal aim of generating goods, to the conflicting aim of making profits. Put in charge of a university, Veblen argues, these trustees are still unable to distinguish the pecuniary from the more appropriate goals. Thus they operate the university as a business concern, seeking to maximize the revenue and to minimize the cost. In Veblen's words, "their pecuniary surveillance comes in the main to [be] an interference with the academic work, the merits of which these men of affairs on the governing board are in no special degree qualified to judge"[3].

For Veblen the trustees, though the principal villains, were not the only ones. Administrators and faculty members who sought their own personal gain found it advantageous to act in sympathy with the financial structuring of the university, and thus to institutionalize it as a business firm. In this context he found that the academic world increasingly attracted those willing to maintain the university as a fiscal enterprise, and diverted potential professors and administrators who had other aims.

A recent flurry of attacks on American colleges and universities reflects a portion of the Veblen argument. It has become increasingly fashionable for students--and a few of their elders--to argue that the financial connections of academic institutions set a tone of immorality and hence an organized incapacity to contribute to a better society. This neo-Veblenian argument offers a *selective* revulsion to business ties; it is not the habits of thought of the business world which are seen as detrimental, but rather a connection with specific firms seen to participate in objectionable behavior. Thus firms with military contracts may be seen as objectionable, and their officers considered to be incapacitated to serve on academic boards of trustees. Indeed, in this view, if shares of stock in such firms are held by an academic institution, this further disables the institution in its educational functions.

Now, these attacks on the structure of academia are but two of a large number. They share, however, in the chief ingredient of most criticisms: the *supervision* of the institution is seen as inappropriate to desirable ends and hence the entire process is seen as incapable of attaining those ends.

It is my contention that a more usual--in fact, nearly universal--goals/means conflict has its source not in trustees or (except indirectly) in administrations, but is accountable to faculty members who pursue what they regard as a common interest of the academic community. One effect of their behavior is the fragmentation of knowledge of which a symptom is the "specialized deafness" about which Boulding wrote. The consequent academic separatism is, indeed, of such dimensions that by comparison the French Canadian separatists may be seen as Anglophiles.

A symptom--but not the cause--of the separatism is the departmental structure through which the affairs of a modern American college or university are largely conducted. The separatism itself seems to owe its origins to two phenomena: the inclination to differentiate one's academic product with an eye toward distinguishing oneself within the institution and perhaps within the profession, and secondly the perception that many research problems can be most effectively dealt with at the micro level rather than the macro level. While the motives generated in the first phenomenon may contribute to the accomplishments of the second, it is notable that the first may lead to activities which are less than beneficial to the university and its students.

It is here that the goals of an academic institution become important. While some would argue that the chief purpose should be to create conditions in which students may listen, think and question, and learn through reading and writing and discussion, it may be helpfully added that in any activity with these goals, is is necessary to deal simultaneously with methodology and content. Specific problems and issues must perforce serve as the anvil on which the thinking process is hammered out. One cannot deal with genetics without a detailed discussion of DNA, and one cannot deal with inflation without a detailed discussion of the multiplier effect.

But when one comes down to the nuts and bolts of it, it is the introductory courses which, in the main, affect the orientation of an academic field of study. There is a dual effect: insofar as a student's thinking is affected by an introductory course, it will serve to influence his perception of more advanced material in that field. Also, the introductory course is an important part of the process by which the self-selection of devotees of one departmental interest as against others, is accomplished.

The introductory course ordinarily does not serve as an introduction to techniques of perception and analysis specially relevant in the "subject." Rather, it serves as an introduction to the intermediate courses and in performing this service is amply endowed with the bits and pieces which are known to prove helpful in the intermediate courses ordinarily "taken." These courses in turn serve as introductions to the advanced courses and are organized in their behalf.

The advanced courses, as is well known, in an institution of high reputation serve to prepare students for graduate courses. (Indeed, a standard measure of reputation for a baccalaureate institution is the proportion of its graduates who proceed to post-graduate education.) In graduate courses specialization is the byword. And who is to argue with this? For the graduate schools have the sacred trust of preparing instructors to offer, for undergraduates, introductory courses which will prepare them for narrow intermediate courses which will prepare them for those specialized advanced courses necessary for graduate school and the pinpoint education there which will prepare the next academic generation.

Perhaps this description is unfair. Students do emerge from this system who are capable contributors to our society. But the essential question raised here is not whether our colleges and universities function at all, but whether they function as well as might reasonably be expected.

In achieving such success as is permitted by the specialization described here, our academic institutions become structured in a fashion appropriate to the end.

Bodies of knowledge are placed in individual coffins known as departments.

Departments are jurisdictional units largely related to budgetary matters, which serve, when urgently pressed, to coordinate classes for those interlopers in the academic process, the students.

Each department, in order to preserve its budget and its intellectual integrity, encourages that continual and fearless sifting and winnowing by which alone further distinctions may be found between its jurisdiction and those of competing departments. This of course involves further specialization.

Another aspect of this process is so frequently remarked upon that it will receive only a mention here. Since publication is equated with scholarship, and since the latter is highly valued in the academic community, a substantial portion of academic time is spent on research. This is of course related chiefly to publishable studies and these in turn are disciplinary specialties.

Thus Veblen's complaints, and those of the neo-Veblenians, are seen to be largely unrelated to the current problem. While their focus is on the vertical structure of the academic institution, with impaired values descending from the trustees through the administration to the faculty, the problem discussed here is in the main one of faculty values, and one which is aggravated by *horizontal* structure. Moreover, the current problem is not susceptible to the sort of solution implied by the Veblen-type problems. Wholesale replacement of trustees and, for that matter, wholesale replacement of "objectionable" shares of stock would do nothing to repair the problem.

The problem is made more difficult by its inherent system of positive feedback. The success of the system is evaluated by its practitioners, and not surprisingly these have found the results favorable. Trends toward further specialization are the consequence. And just as introductory courses are the basis of self-selection of students for later courses, the entire academic structure influences the self-selection of those who are to become academic practitioners. Positive feedback again.

This criticism of super-specialization is not intended to suggest that all knowledge is one and should therefore be taught as a single course. But it is intended to underscore the fact that *groups* of "disciplines" have much in common both in methodology and in content, and that it is fruitful, in teaching these areas, to combine the notions they share. It seems of especial importance that the introductory course in each "discipline" be loosely structured in terms of disciplinary allegiance.

One of the difficulties arising in the individual treatment of "disciplines" is that of analyzing change. My own field of economics (if I be permitted to mention a disciplinary interest) for some years virtually limited its notion of change to that of comparative statics. It still suffers difficulties in dealing with the process of change itself, in good part because it remains substantially aloof from other "disciplines."

Here is an example of what's involved: if one bases capital value analysis on present property arrangements (laws, perceptions, attitudes, etc.), then the ensuing analysis may be useful now but is of little predictive value. If one's scope of inquiry, on the other hand, is expanded to include the wide range of social phenomena involved, then it is possible to draw more general conclusions. And if the non-economic information is well-selected, it is possible that a dynamic analysis may be made. There are more variables than this involved, of course, but this illustrates the nature of the problem.

(There are, of course, other perceptions of the make-up of dynamic analysis in economics. John R. Hicks defines economic dynamics as"those parts [of economic theory] where every quantity must be dated"[4]. More generally, dynamic economic analyses are taken to be those in which second derivatives of variables are employed in describing relationships. But Joseph Schumpeter's description comes close to the meaning I find appropriate: "Hence we are led to take into account past and (expected) future values of our variables, lags, sequences, rates of change, cumulative magnitudes, expectations, and so on. The *methods* that aim at doing this constitute economic dynamics"[5].)

A second dimension of the inter-disciplinary approach involves one's perceptions of nature and of the scope of the discoverable. Veblen discussed these matters in 1897[6] and for it was roundly denounced by his fellow economists. They attached to him the most slanderous epithet they could find: they called him a sociologist.

For Veblen, economics could not be studied in ignorance of anthropology, political science, and (even) sociology. Their joint study permits consideration of the time dimension without the stultifying assumption that all changes occur at rates predetermined at the outset of the period considered. The metaphorical synthesis permits an evolutionary overview, in contrast to a static notion that one need but discover the immutable law in relation to the topic studied, in order to know the truth. Cautioning against reliance upon a natural law which "is felt to exercise some sort of coercive surveillance over the sequence of events, and to give spiritual stability and consistence to the causal relation at any given juncture"[4]. Veblen adds,

> The standpoint of the classical economists ... may not inaptly be called the standpoint of ceremonial adequacy. The ultimate laws and principles which they formulated were laws of the normal or the natural, according to a preconception regarding the ends to which, in the nature of things, all things tend[7].

> The evolutionary point of view ... leaves no place for a formulation of natural laws in terms of definitive normality, whether in economics or in any other branch of inquiry[8].

This "evolutionary point of view" may be shared not only within the social sciences, but also with those clusters of sciences to which it owes its first perception. Here, indeed, is an early example of meta-language dialogue.

When one seeks to teach an interdisciplinary course at the introductory level, the critical decision concerns the isomorphs useful to the endeavor. Professor Alfred Kuhn has dealt with this impressively in his text, *The Study of Society: A Unified Approach*[9]. Here a notion developed by John R. Commons in a narrower context,[10], the *transaction*, becomes the pivotal concept of the social sciences. This notion of an exchange--of communications as well as commodities--brings together the operations of society. When to this Kuhn adds the notion of *learning*, all decision-making is encompassed.

If courses such as that which can readily be built around Kuhn's text, were to become common, the immediate issue would be that of intermediate courses left without the usual preparation at the introductory level. The obvious solution is the helpful one: introduce at the intermediate level the more parochial material omitted below. The student, having gained his introduction at the broader level of basic concepts, will be a ready learner of the "missed" material. The impressive improvement with the inter-disciplinary approach is that now intermediate and advanced courses building upon this methodological expertise may be made available.

It is said that the test of the advocacy of a new proposal is the accompanying set of instructions on how to do it. I offer no instructions here on how to break into the departmental system with its incredible feedback arrangement and array of disincentives for effective teaching. There are two possibilities: in the unlikely event that the faculty of a traditional program assents to the advantages of the changes outlined here, the problem solves itself. In the practical case, the introduction of a comprehensive inter-disciplinary general systems program is plausibly expectable in a new university, some of which have arisen in this decade.

There are, of course, famous examples of successful interdisciplinary programs at Columbia, Chicago, Wisconsin, and elsewhere. The difficulty shared by such "traditional" inter-disciplinary approaches is the temptation to use juxtaposition as a substitute for integration. The special contribution a general systems approach can make, is the emphasis upon methodology and the sharing of analytical techniques.

If a more modest approach is to be made to bring the contributions of meta-language dialogue to our academic institutions, one possibility is the introduction of a course in systems analysis for students in the arts and sciences. Such a course is under consideration at the Claremont Colleges. A preliminary study, financed by an NSF grant, suggested the prospect of treating systems analysis non-mathematically in a one-semester course for upper-classmen. After a methodological introduction and some time on the potential of computers (the latter largely to disabuse the students of the notion that computers make decisions), the course could turn to a single broad problem to serve as a testing ground for systems analysis. This might be air transportation or medical care or a subset of the urban problem. While the proposed course would fall short of a meta-language synthetic approach, its benefits should be salutory.

In summary, though American academic structure militates against a general systems approach to the social sciences, such an approach is possible and effective.

REFERENCES

[1] Boulding, K.E., General Systems Theory--The Skeleton of Science, *Management Science*, 2, 198-99 (1956).

[2] Veblen, T., *The Higher Learning in America. A Memorandum on the Conduct of Universities by Business Men*, New York, Viking Press (1923).

[3] Lerner, M., *The Portable Veblen*, New York, Viking Press, 510-11 (1958).

[4] Hicks, J.R., *Value and Capital*, Oxford, The Clarendon Press, 115 (1939).

[5] Schumpeter, J.A., *History of Economic Analysis*, New York, Oxford University Press, 963 (1954).

[6] Lerner, *op. cit.*, 220.

[7] *Ibid.*, 224.

[8] *Ibid.*, 235.

[9] Kuhn, A., *The Study of Society: A Unified Approach*, Homewood, Ill., Richard D. Irwin, Inc., and The Dorsey Press, Inc. (1963).

[10] Commons, J.R., *Institutional Economics*, New York, The Macmillan Co., 2 vols. (1934) and Madison, University of Wisconsin Press, 1 vol. (1959).

A GRADUATE PROGRAM IN CYBERNETICS

T.C. Helvey*

ABSTRACT

A scientific interdisciplinary school is proposed to be concerned with large and ultra-large systems, to be called a Graduate School in Cybernetics, attached to some university. Until outstanding staff can be assembled, the School should be administratively related to one of the existing schools, but not to any one university department, and could begin with a team from other departments of the university. The curriculum would consist of courses specifically included under the rubric of cybernetics as well as courses in mathematics, biology, engineering and electronics. Under cybernetics would be courses such as information theory, bionics, robotics, neurodynamics and engineering cybernetics.

1. INTRODUCTION

Our society becomes more and more aware of the necessity for dealing with large and ultra-large systems, due among other reasons to automation and the strides toward artificial intelligence, which have become a need in the exploration of alien environments such as deep space and deep ocean. The specific knowledge required for the comprehension of complex man-machine interactions, self-organizing systems and their importance for social

---

*T.C. Helvey is Professor of Cybernetics at the University of Tennessee Space Institute, Tullahoma, Tennessee 37388.

problems is now identified. This trend is manifest in the establishment of such instruments as the Cybernetics Forum of the International Union of Railways, The International Society of Cybernetic Medicine, the Cybernetics (Bionics) Laboratory of the U.S. Air Force Systems Command, the Biocybernetic Forum under the auspices of the National Academy of Science, and others. Also the number of professional societies and journals in cybernetics indicates that this branch of the sciences is well established and has become a profession with a broad scope of scientific endeavors and many job opportunities. Therefore, the need for education in this field, and at graduate level, is recognized.

The progress in this area of education, however, is slow because the educational system in most institutes of higher learning in engineering and the sciences is primarily analytically oriented. The only faculty with broad interdisciplinary interest in human affairs, namely philosophy, seldom possesses the necessary background in mathematics, psychology, electronics and technology for the instruction and the study of self-organization, communication and control of complex and large systems. It has to be mentioned that certain aspects of cybernetics are already taught in an analytical way under titles such as Systems Engineering, Information Theory, Decision Theories, Automatic Controls, Computer Technology, Nonstationary Random Processes, Automation, etc. However, there is a great need for expanding the education of specialists in this field. Especially urgent is the preparation of instructors in cybernetics, including, among other things, the electronic simulation of organismic functions, called Bionics, or the design of electronic networks for artificial intelligence, called Intellectronics, or the development of self-organizing, self-optimizing, self-repairing and goal seeking automata, called Robotics, etc. For many years in Russia and its satellites a number of institutes have been devoted to such a task (see their list under Appendix A).

The growing literature in basic and applied cybernetics is the best indicator of the importance of this science, which has a rapidly increasing impact on our total culture. Therefore, it is timely to charter the first Graduate School in Cybernetics. It will promote the state-of-the-art in this new scientific field, and through improvement of the understanding of human interaction dynamics it will hopefully serve the betterment of mankind.

2. SCOPE OF THE SCHOOL

The program of the School will provide:

1) Graduate studies in cybernetics for the specialization in this area and leading to the M.S. and Ph. D. degrees in Cybernetics.

2) A carefully developed curriculum for graduate studies, congruent to the academic standards of the University.

3) Research opportunities in cybernetics and related areas.

4) A dynamic library in cybernetics, which is befitting a graduate school of the University.

Furthermore the School will:

5) Attract outstanding teaching and research staff in cybernetics and related areas,due to the uniqueness of the School. This will increase the prestige of the University as a whole.

6) Foster the study of spin-offs from space exploration to channel its benefits into community development.

7) Service other Colleges and Departments of the University, that wish to incorporate into their curricula special lectures or survey courses in cybernetics and related areas. On the other hand, the School will encourage and seek faculty members of other Departments, whose speciality is related to cybernetics, to give lectures in their field and, maybe, also share research projects.

8) Provide special programs to upgrade the background of teachers in this field, who are specialized in other areas, thus promoting cross-fertilization between various disciplines.

9) Organize short courses and symposia in cybernetics to promote adult education.

10) Develope communication and interaction with those academic institutions, governmental agencies and private enterprises in the USA and abroad, that are interested in the services of the School, and that wish to cooperate in fostering the state-of-the-art in cybernetics and related sciences.

3. MODUS OPERANDI

Because of its interdisciplinary orientation the School should not be an integral part of any one University Department but should have academic administrative relations to one of the graduate schools.

Until outstanding staff can be attracted, which is a slow process, and because of the excellent talent available in various Colleges and Departments of the University systems a team-teaching program should be developed with the extensive utilization of the Telewriter facilities.

A substantial grant should be requested from those funds which aim at such innovations as the proposed Graduate School of Cybernetics. Until such grant becomes a reality, the Colleges and Departments would contribute to the development of the School with some time of the appropriate staff, various facilities useful in cybernetic research and other assistance which a committee will recommend.

Such concerted effort would render the development of the School an all-University project, thus the University would show academic leadership having beneficial effects in many directions. The fact that Russia and its satellites are far advanced in this field has also significant military and socio-economic implications. Therefore, those governmental offices whose task is the maintenance of the technological, economic and cultural leadership of our country, will appreciate the chartering of the Graduate School of Cybernetics.

4. TENTATIVE PROGRAM

Program leading to the M.S. and Ph. D. degree in Cybernetics.

Any person, who is admitted to the Graduate School and fulfills the prerequisites is eligible for admission to the School of Cybernetics. If a candidate does not have all the required prerequisites he can take those courses in any accredited school or can take them by independent study with the consent of the Director and conforming to the University regulations. In special cases and upon the recommendation of the instructor, the Director can waive prerequisites or accept substitutes.

Prerequisites

1) Physical Sciences
   - General Physics
   - General Chemistry
   - Electrical Circuits

2) Life Sciences
   - Zoology
   - Botany
   - Psychology

3) Mathematics
   - Modern Math.
   - Elementary calculus
   - Probability theory

In addition to these, it is highly recommended that the student have a course in Philosophy, in Sociology and Economics.

List of Courses

The following is a tentative list of courses given with a few topics which will be expanded into a course description and syllabus in a separate document. In parentheses after each course title is one of five categories which is used in summarizing the curriculum structure in Table 1.

1) General Systems and Cybernetics Fundamentals (cybernetics)

   Systems philosophy
   Requisite variety
   Decision theories
   Coding and decoding
   The Turing machines
   The heuristic approach

2) Information and Communication (cybernetics)

   Semantic information
   Sensory communication
   Limits of information
   Concept of channel
   Shannon's theorems
   Entropy concept

3) Man-machine Interactions (biology)

   Manual control
   Human transfer functions
   Reaction time
   Information displays

4) Applied Biophysics (biology)

   Anthropometry
   Biostatics
   Biodynamics
   Biokinematics
   Neutral activities
   Sensory modalities

5) Human Anatomy and Physiology (biology)

Man as a system
Integument
Nervous system
Skeletal system
Muscular system
Circulation
Respiration
Digestion
Excretory systems
Endocrine system
Reproductive systems
Embryology and genetics

6) Sensory Physiology (biology)

Vision
Information processing in the eye
Audition and stereophonics
Sensing linear and azimuthal acceleration
Kinesthetic sensation
Mechano-receptors
Tactile sensation
Pain sensing mechanism
Sensory deprivation

7) Servo Systems and Automatic Control (engineering)

Feed-back systems
Nyquist and Bode presentations
Stability analysis
Non-linear systems
Design considerations

8) Probability and Statistics (mathematics)

Sample space
Random variables
Laws of large numbers
Multivariate analysis
Covariance matrices

9) Special Topics in Mathematics (mathematics)

Theory of groups and transformations

Elements of Boolean algebra
Ergodic theory
Laplace, fourier and Z transforms
Elements of complex variables
Distribution functions and H-theorem

10) Linear Active Circuits (engineering)

Transistor theory
Circuit theory
Amplifiers
Complex networks

11) Optimization Techniques (mathematics)

Calculus of variations
Linear programming
Elements of dynamic programming
Markov chains

12) Bionics (cybernetics)

Electronic and computer analogs
Human amplifiers
Elements of intellectronics (artificial intelligence)
Implantable bionic devices
Myoelectric servo control

13) Applied Cybernetics (cybernetics)

Engineering cybernetics
Biocybernetics
Cybernetic medicine
Societal cybernetics
Man in a fully automated environment

14) Human Behavior (biology)

Psychomotor functions
Sensory overload
Emotional parameters
Genetic aspects of behavior
Personality dynamics
Group dynamics
Small group interactions

15) Elements of Robotics (cybernetics)

Pattern recognition
Stereophonic orientation
Learning networks
Vertical sensor and control
Mental control of artifacts

16) Environmental Physiology (biology)

Acceleration (linear and azimuthal)
Vibration (subsonic)
Heat, cold, humidity
Noise and silence
Nuclear radiation
Illumination
Atmospheric physiology

17) Analog Computer Programming (engineering)

Solution of non-linear differential equations
Operational amplifier
Servo-division
Function generation
Time scale change
Magnitude scale change
Forced vibrations of a linear oscillator

18) Control Theories in Physiological Systems (biology)

Application of frequency analysis
The root-locus plot
First- and second-order systems
Steady-state errors in servo and regulator operation
Visual control system
Reflex function of the nervous system
Regulation of the body temperature
Mathematical model of heart rate control

19) Brain Function (biology)

Steady potentials of the brain
Synaptic ultrastructure and organization
Cortical topology

Electroencephalography
Studies on learning

20) Mathematical Sociology (mathematics)

Equilibrium states of social systems
Poisson-type models
Contagious models
Change and response uncertainty
Hierarchization
Local implications
Diffusion of social structures
Probabilistic math. vs. deterministic math.

21) Perceptrons and Neurodynamics (cybernetics)

Classification of perceptrons
Linear and non-linear transmission functions
Code-optimization
Adaptive pre-terminal network
Fixed and conditional response sequences
Choice-mechanisms
Awareness and cognitive systems
Isomorphism of structured information
Memory localization
Mechanism of motivation

22) Biotechnology (biology)

Human decision making
Intended and extracted human output information
Personal equipment design
Comfort engineering
Engineering of micro-environment
Consoles and cockpits
Man-machine task allocation
Test procedures

23) Reliability Theory (engineering)

Wearout failures
Bayes' theorem in reliability
Component reliability measurement
Confidence limits
Failures and system stress

Series and parallel systems

24) Elective in Sociology, or Philosophy

The candidate is free to elect any graduate course.

25) Elective in Economics or Business Administration

The candidate is free to elect any graduate course.

26) Elective in Chemistry or Biochemistry

Student Term Papers

The Legal System in 2068 AD
Communications and the Computer
Automated Democracy
The Rule of Three
Future Applications of Computers
"2068"
Predictions
A Cybernetic Look at the V/STOL Wind tunnel
Systems Approach to Supersonic Transport
Cybernetic View of Future Transportation Development
Cybernetics in Engineering Education
Interaction of Spacecraft with Man
Unique Biological Systems in Lower Life Forms
Automation and the Future
Man and his Machine - Flying
Theology in the Cyberculture
Cybernetics and My Homeostatic Niche
Toward a Unified Theory of the States of Matter
Man-Machine Interface in Light Aircraft
The Future of Work
The Scope of Our Society in 2068

Student Class Lectures

The Nyquist Criterion
Human Engineering

Toward the Cybernetic Factory
Teaching Machines
Psychology of Language in Human Interactions
Probabilistic Aspects of Information Theory
Logic Elements
Negative Rodex Number System
The Neuristor
Basis of Boolean Algebra
Information Sequence in Language to Describe environment
Reliability and Maintainability
Physical Analogs of Biological Models
Numerical Control
Focus on Behavior
Cybernetics of the Vascular System
Man-Machine Symbiosis
Probability of English Letters
Homeostosis in Man
2nd and 3rd Generation Computers
Adaptive Controls
Noise
Information Displays

Master Degree Curriculum

For the Master Degree the following curriculum is recommended:

1st Quarter: General Systems and Cybernetic Fundamentals (1)
Information and Communication (2)
Human Anatomy and Physiology (5)

2nd Quarter: Applied Biophysics (4)
Sensory Physiology (6)
Servo Systems and Automatic Control (7)

3rd Quarter: Probability and Statistics (8)
Linear Active Circuits (10)
Human Behavior (14)

4th Quarter: Man-Machine Interactions (3)

Bionics (12)
Applied Cybernetics (13)

Doctor Degree Curriculum

For the Doctor Degree the following additional courses are recommended:

1st Quarter: Special Topics in Mathematics (9) or Optimization Techniques (11)
Brain Function (19) or Biotechnology (22)

2nd Quarter: Environmental Physiology (16) or Control Theories in Physiological Systems (18)
Analog Computer Programming (17) or Reliability Theory (23)

3rd Quarter: Elements of Robotics (15) or Perceptrons and Neurodynamics (21)
Elective in Biochemistry (26)

4th Quarter: Mathematical Sociology (20)
Elective in Economics (25)

5th Quarter: Elective in Philosophy (24)

All courses carry 3 credits.

Curriculum Summary

For the summary of the above curriculum structure see Table 1.

For the Master's degree the required research carries 9 credit hours and the candidate should have a total of 45 credits to graduate.

For the Ph.D. degree the thesis carries 36 credit hours and the requirement is a total of 108 credits.

TABLE 1. Curriculum Summary

(a) for the Master Degree:

	Cybernetics	Mathematics	Biology	Engineering	Electronics	TOTAL
1. Quarter 1, 2, 5	2		1			3
2. Quarter 4, 6, 7			2	1		3
3. Quarter 8, 10, 14		1	1	1		3
4. Quarter 3, 12, 13	2		1			3
	4	1	5	2		12

(b) for the Doctor Degree:

	Cybernetics	Mathematics	Biology	Engineering	Electronics (and others)	TOTAL
1. Quarter 9 or 11: 19 or 23		1	1			2
2. Quarter 16 or 18: 17 or 23			1	1		2
3. Quarter 15 or 21: 26	1				1	2
4. Quarter 20, 25		1			1	2
5. Quarter 24					1	1
TOTAL	1	2	2	1	3	9
Total Curriculum	5	3	7	3	3	21

Research Project Titles

1) Experiments with man on form-recognition
2) Cybernetic study of comprehensible questions
3) Modeling of automated mill for furnace-welding of pipes
4) Investigation in the area of navigation and orientation
5) Cybernetic study of fetal distress
6) Investigation in the area of aerohydrodynamics
7) Cybernetic investigation of human temperature dynamics
8) Investigation in the area of man-machine interface
9) Cybernetic study of cardiac functions
10) Mathematical approach to endocrinological diagnosis
11) Feedback in endocrinology
12) Amplitude-Time classification of images
13) Calculating protein binding of drugs with computers
14) Mathematical study of the excitation of nerve tissue
15) Cybernetic regulation of cell numbers in the epidermis
16) Implantable electronic organs and components
17) Electro-hysterographic research

18) Respiratory system as a biological feedback regulator

19) Swallowable intestinal transmitter

20) Cybernetic therapy of deafness

21) Cybernetics of the nervous control of muscles

22) Monitoring, Prognosis and Control of the State of Man-Machine Systems

23) Universal Analyser of a Multichannel Diagnostic System

24) Simulation of a Monitoring and Decision Unit for Ship Navigation Systems

25) Bioelectrical Control of Technical Objects

26) Cutano-Galvanic Stimulation as Information Generator

27) Study of the Formation of Visual Images

28) Experiments in the Descrimination of Compace Set of Images

29) Procedure for Conversion of Images into Sound

30) Simulating Human Hearing

31) Functional Simulation of Nerve Elements

32) Cyclic Processes in the Biosphere caused by Cosmic Factors

33) Bionic Model of Color Vision of Man

34) Use of Computers to Process Information from Organismic Loops

APPENDIX A

Institute of Automation and Electrometry Novosibirsk, Siberia

Institute of Automation and Remote Control, Moscow

Mathematics Institute, Moscow

Labor Red Banner Mechanics Institute, Leningrad

Institute of Automatics and Mechanics Riga, Latvia

Institute of Electronics and Computing Technology Riga, Latvia

Institute of Mathematics Kiev, Ukraina

Institute of Cybernetics Kiev, Ukraina

Institute of Electrical Engineering Kiev, Ukraina

Institute of Cybernetics Tbilisi, Georgia

Institute of Electronics, Automation and Remote Control Tbilisi, Georgia

State University Stalin Tbilisi, Georgia

Institute of Mathematics and Computer Technology Minsk, Belorussia

Institute of Cybernetics Tallin, Estonia

Institute of Engineering Cybernetics, Sophia, Bulgaria

Institute of Cybernetics, Techn.Univ. Budapest, Hungary

SECTION III

SOCIAL SYSTEMS IN THE GENERAL SYSTEMS SPECTRUM

In the series of papers constituting this section, we first discuss the nature of systems, particularly as relating to social systems. Not everything is a system, unless one wants to define it so. There are degrees of systemicity, or systemness. Certainly there must be some stability of relationships over time for an entity to fulfill the common notion of system. Even if there are apparently wild oscillations and rapid motions, there are system laws, at least in natural systems. All of these papers express in varying degrees the search for law or invariance in the systems considered. Some system laws, or at least concepts, relate to natural as well as social systems. Some of the papers emphasize the commonality of system concepts across all fields; others emphasize the differences.

In Kuhn's paper, he makes a strong effort to identify what is and what is not a system, and to institute a classification of systems. His particular objective, of course, is to relate social systems in the overall systems spectrum. It is necessary to concern ourselves with the different types of systems we encounter in different fields. Fundamental to general systems is the concept that there is system theory that is applicable across many fields. However, this does not mean that a particular concept may apply to all fields. Some concepts may apply to very few fields. Kuhn provides a useful service in examining the nature of system concepts and differences between various kinds of system models and their transferability across different fields.

Berrien first defines a social system. The definition is applied to biological system with suitable interpretation of the terms of the definition and on this basis he discusses the analogy between an organization and an organism. Maintenance inputs to the system are those which supply sustenance, and the external sources are prestige, regard, reputation and considerations emitted by superiors. Group-needs satisfaction (G.N.S.) is a feedback from the social interaction of the group to the maintenance input. Signal inputs are stimuli to the internal processes of the system to perform those functions made possible by the structure of the system. The outputs of the system are F.A. (Formal Achievement), which is used by the suprasystem in which the system is embedded, and waste. He discusses the proposition that the amount of F.A. is some function of the maintenance input. The model is applicable primarily to non-coerced social systems, and he applies it to several types, and relates his models to those of other theorists.

Schossberger directs his attention to the individual human being in the social system and his relations to other humans, and in particular, the messages and symbols by which meaning is transmitted. He discusses the nature of messages and information flow, and the relationship of primitive art to written language and of this to control of the hand and of aggressive impulsions. Inner subsystems have a language or information processing that is different from one of which they are part, that is electrical or chemical. However the systems must be mutually regulated. In fact, of course, human language operates only in part on an intracranial component, the outer system being all existing recorded information. The author tries to show how all this hangs together through tetradic linkages between symbols, objects and their makers. He finds that his subject matter makes it necessary for him to be poetic and metaphysical as well as scientific, and subjective as well as objective.

In the next paper, the Maccias discuss the application of their model to the study of a particular educational system as a social system. Their model (SIGGS) employs set theory, information theory and general systems theory as a basis of analysis. Most interesting to the reader will be their listing of system properties, such as openness, adaptiveness, wholeness, centralness, etc. Although they describe how they use set and graph theory for relating these concepts, the particular example can be understood in their discussion relating these concepts in everyday English.

Marien also looks at education from a systems point of view. However, he considers the whole national education "complex". An overall picture, which he attempts to construct, examining the internal relations among the parts of the education complex, and the ramifications of this complex in its environment, is in itself a valuable systems contribution. He blends both systems and social science language in his discussion of education as a social system.

House contributes the only paper that is directly concerned with the urban system (a city). There are many subsystems of different types in an urban system, and in order to examine their interrelationships in a meaningful way a multidisciplinary approach is necessary. One of the difficulties in doing this, House says, is the lack of a general urban theory in which to incorporate multidisciplinary approaches. His way of approaching the problem is to examine and quantify to a reasonable extent the elements of the problem so that the "system" can be embodied in a computer program, with the exception of human decision-makers. Games are played with data being presented to human beings from the computer, with the humans making decisions which are fed back into the computer, each game being played out with the city developing along a different course in each game, depending on the inputs and decisions during play of the game.

Milton D. Rubin

TYPES OF SOCIAL SYSTEMS AND SYSTEM CONTROLS

Alfred Kuhn*

ABSTRACT

The term "system" is currently used so loosely that the statement, "*X* is a system" conveys little or no information about *X*. First, we must stop calling things systems that are not systems at all. Second, we must identify separate types of systems so that distinctive generalizations can be stated about each (a proper basis for scientific classification), and the following main categories are suggested.

Certain generalizations apply to all *acting* systems; none of them to any *pattern* (nonacting) system. Among acting systems some generalizations apply to both *controlled* (cybernetic) and *uncontrolled* systems; others to one type only. *Formal organizations* are controlled and *informal organizations* uncontrolled systems of multiple humans.

Human organizations must be distinguished from *nonhuman systems*. For the former, communications are semantic and transactions are value-based. Behavior of the formal organization as a unit is subject to systems analysis. Its interactions with other organizations, and interactions of its component subsystems, are subject to communicational and transactional, but not systems, analy-

*Alfred Kuhn is Professor of Economics and Senior Associate in Psychiatry, University of Cincinnati, Cincinnati, O. 45220.

sis. Behaviors of systems must never be considered systems.

Types of system controls are implicit in the definitions of the various types of social systems.

1. INTRODUCTION

Every science classifies and categorizes the phenomena within it. The goal of a classification system is a condition in which the act of identifying some phenomenon as belonging to a certain category will tell you something about it - if you already know something about that category. Such knowledge is possible under certain conditions. First, all items within the category must be similar in certain respects, those similarities being the things one can know from the act of categorizing. Second, different categories must be reasonably discrete. Biological categories, for example, are kept nicely discrete by the fact that cross fertilization occurs across only a very narrow species spectrum. Were this not so, we would long ago have arrived at the point where every individual would, in effect, be its own species - as, for example, if we could cross fleas with elephants and then cross breed their progeny with the offspring of dogs and salmon.

If you tell me something is a mammal or a lever, I immediately know something about it. But if you tell me something is a system, I am not sure there is anything I can know from your statement that I did not know before, however well versed I might be in systems analysis. While this situation may have been inevitable in the early days of a science, it seems about time we try to do something about it. In that spirit, the purpose of this paper is to suggest some sub-categories of the genus 'system' so that to identify something as a particular kind of system will provide information. There are undoubtedly loopholes in this proposal, some of which reflect my orientation in the social sciences. We have nevertheless got to start somewhere, and this is a tentative trial.

2. THE CLASSIFICATION SCHEME

Since a classification scheme is clearest in outline form, an outline of this one is appended. I have included in it examples of each type of system, but with sharp reservations, knowing that examples, like analogies, are apt to be more controversial than the generalizations they illustrate. The difficulty is that no real system is a precise counterpart of the pure analytical construct.

Before dealing with the divisions shown in the outline I would like to distinguish even more broadly between systems and non-systems, to counter the tendency nowadays to refer to practically anything as a system. An automobile crankshaft, for example, is not a system when viewed solely as a component of the engine. No two parts or characteristics of it change relative to one another. The crankshaft, of course, changes position relative to the engine block and the pistons. Hence, the crankshaft plus those other components do constitute a system. But the crankshaft alone is a non-system. The term 'mountain system' is also common. While I will not try to change the layman's usage, my point is simply that various parts of a mountain 'system' do not interact in any sense relevant to the typical user of the term. For most purposes we could level the whole of the Appalachians north of Mason-Dixon and precisely nothing would happen to the part south of the line. However, if a geologist is dealing with some developments in which the parts *do* interact, he may legitimately refer to the mountain chain as a system. In this vein I would like to require every system analyst to make a long list of things that are *not* systems and paste it over his desk.

Acting vs. Nonacting (or pattern) Systems. Now to the outline. The first division is between (I) action systems and (II) pattern, or non-action systems. Action systems *do* something. By contrast, pattern systems have things *done to* them, but do not themselves do anything. I see this as a fundamental distinction, and feel sure that no significant analysis that applies to one will also apply to the other.

To deal first with pattern systems I would subclassify these into real pattern systems and analytical or conceptual ones. The distinction is the reasonably familiar one between information level and matter-energy level. In this connection I would note Peter Caws' presidential address to this Society two years ago when he observed that theories about systems are themselves also systems. (Caws 1968, p.3). This he considered a serious ambiguity. I would strengthen his point by saying that only chaos can result if we do not sharply distinguish real systems from the pattern systems which are their analytical counterparts. There is an important sense, of course, in which statements about an analytical model are also true of its real counterpart, as we shall see below. But there is also a fundamental difference, as you would quickly discover if you thought you had been put in charge of General Motors and were then told you were only in charge of its analytical model. A map of North America is not North America, and the schematic of your TV set will not bring you Huntley-Brinkley.

There are many real pattern systems. For example, the skeletal shape of a skyscraper and the structural characteristics of steel beams are related in a systematic way. Yet the shape does not *do* anything to the characteristics of steel beams, which are the same whether or not they are put into the skyscraper, and whether or not skyscrapers are ever built. Conversely, the characteristics of steel beams do not themselves *do* anything to the shapes of buildings. The only *doing* that occurs is done by the human being. Once he has learned the constraints placed upon him by the traits of steel he then shapes his building according to his own criteria of safety, appearance, cubic capacity, land values, and the like. Although there is a "systematic" relationship between steel strength and building shape, neither *acts* on the other. Only the human being acts in this case.

Similarly, the parts of a language are interrelated in highly systematic fashion, and a sentence holds together only if its parts follow the system. But if by accident or ignorance one half of a sentence is utterly inconsistent with the other, neither half *does* anything to rectify the discrepancy. Only if some human intervenes will a change be made. Again, the human acts; the language does not. The same general logic also applies to the pattern system which is mathematics.

As of the moment I am extremely skeptical that systems analysis, as such, has, or ever will have, anything to contribute to understanding pattern systems. If anything we learn from studying language systems is applicable to our understanding of steel-skyscraper systems, this result, I think, will be purely fortuitous. I certainly have no objection if someone can find some common traits among all kinds of pattern systems, but I think it unlikely. A Gregorian chant is a pattern system and so is the theoretical structure of neoclassical economics. Again, I doubt if any generalizations about one will be applicable to the other. If linguistic and mathematical analysis have some things in common I suspect the reason is that mathematics is in an important respect also a language.

If an analytical system is created as a model of an acting system, then the logic of the acting system may also be said to be valid for this corresponding pattern system. This is only so because the analytical system *is* a model of an acting system, and the logic is that of the acting system. For example, if I use a mathematical formula to compute the traits of a particular transistor the result is determined by the laws of electrons, not of mathematics. Only to the extent the analytical mathematical model happens to fit the electronic reality can the former be used for the purpose at all. By the same token if several models (pattern systems) of different acting systems show similarities, the similarities are those of the acting systems they represent, not of any logic of pattern systems themselves.

By contrast, my strong suspicion is that any systems science that can be developed will apply solely to acting systems. Here I think we can already say that as soon as something is identified as an acting system we *know* something about it. The things we know are those already widely discussed by systems analysts, and include:

1. Under certain conditions the system can reach and maintain equilibrium. The best known condition is negative feedback, in which a deviation from some point sets in motion an opposite action which pushes the system back toward that point. Less widely discussed is positive feedback when one or more variables reach a limit. A fire, for example, is subject to positive feedback in that the hotter it gets the faster it burns and the faster it burns the hotter it gets. The system may nevertheless reach an equilibrium if fuel or oxygen is supplied at only a limited rate, or if a temperature is reached at which heat is dissipated as fast as it is generated. Social analysis has many such positive feedback equilibria. For example, the more accurately people communicate the more similar their ideas become and the more similar their ideas the more accurately they communicate. Here we have an equlibrium at the asymptotic limit of identity.

2. A system will be subject to explosion or shrinkage under positive feedback without any limit to its variables.

These two points give us some information. As to the first, unless a particular system is subject to some equilibrium it is not likely to last long enough to be worth investigating as a system. Hence merely to identify something as a continuing system implies that within it can be found either some negative feedback relation, or positive feedback with at least one variable limited. Hence the basic understanding of any continuing system lies in identifying its negative feedback elements or the asymptotically limited positive ones. I cannot recall where I saw a generalization to the effect that the first law of systems is that they tend to maintain themselves. This is an interesting and basically correct statement, but to my mind it reverses the logic of things. The proper statement, it seems to me, is that only those things that tend to maintain themselves are worth classifying and studying as systems. The other version has a mildly anthropomorphic quality, and seems to overlook something important about the basic nature of science.

3. If the system is closed it will be subject to entropy, which in its broadest sense may be construed as loss of differentiation. By contrast, if the system is open and in some way subject to positive feedback, it may

undergo evolutionary change into greater differentiation. Stated in reverse, if the system *does* evolve, then we know it is open and subject to positive feedback (Maruyama 1963).

4. The final state of a closed system, or any intermediate state, is determined by those forces and processes already within it at the moment of closure.

5. The final state of an open system, or any intermediate state, is determined by both the forces within the system at the moment it became open and the forces of the environment.

I do not know how many extant generalizations are applicable to all *action* systems, or how many more may eventually be developed. So far as I can see, however, none of these applies to pattern systems, as I have defined them.

Controlled (*cybernetic*) *vs. Uncontrolled* (*noncybernetic*) *Systems*. Action systems can again be subdivided into controlled or cybernetic systems and uncontrolled or noncybernetic. The former has some kind of goal and behavior as a unit, and maintains at least one variable within some specified limits. Whenever the variable moves beyond those limits, the system reacts to bring it back. The household thermostat is the classic example.

By contrast, an uncontrolled system will presumably reach some equilibrium since if it does not it is probably not worth analyzing - as indicated above. But it has no "preference" for any *particular* level of equilibrium. A river system, for example, at any given moment will reach a level which equilibrates such factors as amount of rainfall, area of watershed and width, depth, and slope of river channel. But if rainfall increases or the channel becomes clogged and the river, therefore, rises, the new equilibrium is "accepted" just as readily as the old one. The river system does nothing to restore its previous level. Similarly in microeconomic theory the equilibrium price of a commodity is that which clears the market. If supply increases, as through lower cost of production, a new equilibrium will prevail. The new equilibrium is just as "right" as the old one. The system will not respond to restore the previous equilibrium, which would now be "wrong" under the increased supply.

Note that the word 'controlled' here refers only to internal system controls, not external. For example, the path of a purely ballistic missile can be externally controlled when one sets its angle of fire and propulsion energy. But no internal mechanisms rectify its course if it goes astray, in marked contrast to the controlled guided missile, which detects and corrects such deviations.

The obvious reason for distinguishing controlled and uncontrolled systems is the difference in applicable analysis. Although both are subject to the five generalizations stated above, and possible others, that particular sub-species of system analysis called 'cybernetics' is applicable solely to controlled systems. This distinction between controlled and uncontrolled systems will be elaborated below in connection with human organizations. Pattern systems obviously cannot be subdivided into controlled and uncontrolled. The distinction is relevant only to systems that act, and pattern systems do not.

Organizations (of humans) vs. Non-organizations. Both controlled and uncontrolled systems must, I think, be subdivided between those whose elements consist of multiple interacting human beings and those which do not, to reflect fundamental differences in the nature of system interactions. To give them a name, we will call all systems consisting of multiple humans 'organizations'. Although system analysts widely use 'organization' and 'system' interchangeable, much other usage is already in accord with the present one.

The distinguishing differences between organizations and non-human systems are indicated in the outline. First, communications between humans and their organizations are mainly linguistic or semiotic. They are based on (pattern) systems of signs and referents in semantic communications. Now there is plenty of communication going on inside an amoeba or the genes of human beings, and much of it is amenable to rigorous information and communications analysis. But it is not amenable to linguistic analysis. By contrast, in the linguistic communication so typically human, communication theory may tell us something about the quantity of information as measured by numbers of signs. But it can tell us nothing about the quantity of information which constitutes the referent of any particular sign in the head of any individual.

The second distinction is the transactional. Systems other than humans engage in many transfers (inputs and outputs) of matter and energy. These exchanges are based on certain technical characteristics of the systems, and can be described in terms of principles of biology, chemistry, or physics. By contrast, the transactional exchanges between humans are based on the *values* of the things exchanged. Except in some primitive, analogic sense the kinds of transactions dealt with in bargaining power theory and supply and demand analysis, for example, occur only among humans. No other systems similarly engage in negotiations and establish terms of trade based on utility functions of the interacting systems. Human systems are therefore distinguished because the interactions between them are subject to utterly different analysis than are those between nonhuman systems.

Formal vs. Informal Organizations. Since this paper is directed primarily at social systems I will deal in further detail only with systems of humans. Here I suggest calling controlled systems 'formal organization' and uncontrolled ones 'informal organization'. The main distinctions are:

1. The formal organization has some kind of preference as a unit; the informal does not.

2. The formal organization makes and executes some

kinds of decisions as a whole unit - possibly of a sort that will optimize something - while the informal does not.

3. The formal organization to some degree determines its own internal structure; the informal does not. By this we mean that it specifies its own subsystems and the role each is to play in the whole system. How well the controls operate will depend on some of the control mechanisms to be indicated below.

4. The analysis of formal organization (like that of any controlled system) can be focused on its detector, selector and effector functions - of which more later. By contrast the uncontrolled system cannot be said to have any detector, selector or effector functions as a unit.

The biological eco-system can be viewed as a lower prototype of informal organization. All its significant subsystems are themselves controlled systems, of which each seeks its own goals in its own ways. The subsystems also interact in numerous ways, but only in pursuit of their own separate goals. The total set of components and interactions constitute a system in a very full and clear sense. But the system engages in no behavior as a unit, and has no goal beyond those of its separate subsystems. So far as the whole system is concerned, any equilibrium it reaches is as good as any other and it will institute no behavior as a unit to move toward a different one. If and when controls are instituted, as in the Tennessee River system and the American economy during the 1930's, by definition the system then moves into the controlled category.

I mentioned earlier that my examples may be arguable and will try to clarify several. For example, I have listed a laissez-faire economy as one particular kind of social system at the level of the whole society. One might argue, for example, "But the American economy is a laissez-faire system, and yet it has lots of controls!" This is the typical problem of classifying complicated real phenomena into pure analytic categories. By a laissez-faire economy I mean a laissez-faire economy - that is, one with *no* controls. If an economy that is primarily laissez-faire nevertheless has some controls, then to that extent it must be categorized as a controlled system, not uncontrolled. We must then apply the analysis of formal organization to the controlled part and of informal organization to the uncontrolled part. The fact that much of reality is very messy does not mean that our analytical system must also be. Elementary physics tells us nice things about perfectly hard, perfectly spherical objects rolling down perfectly hard, perfectly plane surfaces. But most phenomena of the real world do not fit the neat categories of the physicist - like the rubber headed mallet bouncing down your cellar steps. Although physics is enviable in that many pieces of its reality are, or can be made, very similar to analytical models, for most other sciences (and many physical phenomena) we can make precise statements only about the analytical models, not about reality. We can all regret this, but there is not much we can do about it.

Similarly I classify a cocktail party as informal on the assumption its sole aim is amusement of its individual members. But if its members are brought together with some collective intent, such as to found a new chapter of Alcoholic Anonymous, then that cocktail party is to that extent a formal organization. To understand its behavior may then require both kinds of analysis.

In the outline I have classified the social system as informal. More strictly it probably justifies an intermediate category, which elsewhere I call the semi-formal. It is informal insofar as it has no central controlling mechanism and no single goal. But it does contain some formal elements in that many behaviors of many individuals are directed toward their views of the total good, not solely their own. The social organization of tribal societies lies closer to the formal end of the spectrum. They are small enough so that any reasonably perceptive person has some sense of his own effect on the total society. If additional controls are exerted by individuals on their peers, even without controls by a chief, these provide some degree of formality in being directed toward a preferred state of the whole society. Again, however, the validity of the overall classification scheme does not depend on how precisely we can categorize particular real entities.

Interactions within Organizations (and between humans). The next question is why the above references to interactions among human beings and organizations have been described as communications and transactions. This formulation arises from work I have been doing elsewhere in connection with "unified social science", and is related to system analysis as follows. Human beings are controlled systems behaving in an environment. Any such adaptive behavior seems to involve a logically irreducible list of three ingredients: (1) The environment, (2) the system, and (3) the adaptive response.

Now a system does not respond to its total environment, but only to those aspects which impinge upon it, or are "known" to it in some sense. It is only to the extent that the system has itself been modified by its environment that it can respond to it, and in the strict sense the system responds only to these modifications, not to the environment as such. When the system has been thus modified, we will say that it "contains information" about the environment. The process by which a system acquires such information can be referred to as its *detector* process.

But all systems do not respond alike to a given environmental situation. Due to differences in their internal logic or goal structure they "select" different responses. These differences may characterize a whole species of systems, or they may differ from individual to individual within a species, depending upon their built-in and learned differences. The tendency of a system to respond in a certain way will be construed as its goal or value structure. This is the sense in which a system selects behavior, and the function of doing so will be referred to as its *selector* function.

Having selected a response a system then effectuates

it, and the processes or structures of doing so will be referred to as its *effector* function. Couched in this language we then say that, given its structure, the behavior of any controlled system can be diagnosed by examining the state of its detector, selector and effector processes. For systems such as human beings we can say that these three things represent respectively the concept-perception function which processes information about the environment, the goal or value function of the individual, and the ability of the individual through muscular strength and coordination or other mechanisms actually to carry out the selected behavior. The effector process normally involves a transformation of the environment or of the individual's relation to it. For the present paper we need not be concerned whether that environment consists of nature or other human beings. The same three functions - detector, selector and effector - very closely correspond respectively to the Stimulus → Organism → Response relationship of behaviorist psychology. They also correspond to the three functions of watching, comparing, and responding or adjusting as used by Philip Chase in his discussion of armaments races (see p. 468 ff.) and the correspondence is not coincidental. The same trio is widely found elsewhere including the head, the heart, and the hands of the Campfire Girls. Further, the detector and selector functions correspond to the distinction between factual or scientific judgments and value or preference judgments which date from at least the Fifth Century B.C.

For purposes of unified social science I have been suggesting a parallel trio in the relations between humans or organizations. Parallel to the detector, which processes information, is the *communication* which transfers it from one person to another. Parallel to the selector, which deals with values or preferences, is the *transaction* which transfers valued things from one person to another. More strictly, communications and transactions are transfers analyzed with respect to their information and value contents, respectively. The third intrasystem function is the effector, and the corresponding intersystem behavior is *organization*, which is the joint effectuation of some behavioral reponse or transformation of environment.

For reasons which need not be spelled out here, but which are probably intuitively obvious, organization necessarily fulfills the definition of a higher level system, whose components are the two or more individuals comprising it. To the extent the behavior of the organization is analyzed as a unit we turn back to the tools of intrasystem analysis - namely detector, selector and effector. This then leaves the communication and transaction as the only distinctively intersystem types of behavior, and they can occur between either persons or organizations.

In this connection we should refer to the earlier distinction between acting systems and pattern systems, and the related suggestion that only chaos will result unless we sharply distinguish the two, and apply systems analysis solely to the former. In a similar spirit we must equally sharply distinguish between system action and systems. Again only chaos will result if we attempt to deal with actions of systems as if they are themselves systems. It is perfectly true that various actions of a given system are interrelated, and those who yearn to use the term may refer to these relationships as a system. But if so, it should be noted that this is a pattern system, to which (I think) systems analysis does not apply.

In this same vein I also have some reservations about calling a particular collection of things a system on the basis of a particular kind of interaction among them. I cannot state this as clearly as I would like, but will try to illustrate. General Motors is a system. It consists of many subsystems, which communicate in certain ways. It is common to list or diagram these parts and the information flows among them and to refer to the whole as a 'communications system'. Certainly this is not a subsystem of General Motors in the same sense that a vice president, the Buick division, or a research department is a subsystem. The latter three *act* in a sense that the 'communications system' does not, since communication is itself already an action of a system. If one insists on calling it a system I would view it as a pattern system. Given free choice I would probably avoid the problem entirely and call it simply a communication net or pattern.

I would similarly talk of the *pattern* of power within some organization, formal or informal, but not of the *system* of power. Here as elsewhere, if we want to develop a body of tight generalizations about systems we may have to be a bit more discriminating about the kinds of things we will call systems.

CONCLUSIONS

The title of this paper refers to "system controls", a matter I have not dealt with directly. Differences in types of control, however, are implicit in the classification system itself, and are, in fact, part of the definition of some types of systems. For example, the formal organization is one which specifies - i.e., controls - its own subsystems. In present language this means that it specifies the detector, selector and effector functions of each subsystem, and the kinds of communications and transactions they engage in, either within the main system or across its boundaries. The internal controls of the formal organization are themselves assumed to take place by communications and transactions. For example, a system does not respond to its environment as such, but only to its information about it. To the extent such information depends on the communications received, an organization's control over the information flowing to its subsystems controls their behavior. In addition, within a formal organization the control exercised by transactions is essentially that of authority, which I have defined transactionally as the ability to grant or withhold rewards (or punishments) in return for the performance or non-performance of the organization's instructions. Furthermore, whenever an organization requires a group decision, as contrasted to decision by a single executive, the decisions themselves are made by communications and transactions which are roughly the same as persuasion and payoff, respectively. If these fail the decision is made by a dominant coalition, a group which holds the largest bloc of the particular type of power relevant to the organization. The power of the dominant coalition is itself based on its internal communications, transactions and possible subcoalitions.

By contrast, the structure of informal organization is not specified or controlled by the main system. Furthermore, its work is done exclusively by each subsystem's pursuit of its own goals by its own techniques, not on instructions from the main system. Interactions among the subsystems of informal organizations consist of communications, transactions, and suborganizations. Should the whole society prescribe some particular types of communications, transactions, or organizations, the informal organization to that extent becomes formal. *Government* itself, of course, is a formal organization.

I note in passing that a system might be defined, or at least identified, as any situation to which the *fallacy of composition* could apply - namely, that the whole is something distinct from the sum of its parts.

In summary, as I see it, each different kind of system I have suggested requires a different kind of analysis, yet (except for pattern systems) an analysis which is common for all systems within each type. That is, different propositions apply to one type of system than to the other types. All in all, I know no better scientific basis on which to build a classification system - a particular type of pattern system to which systems analysis does not apply!

REFERENCES

Caws, P., Science and System: On the Unity and Diversity of Scientific Theory, *General Systems, XIII,* 3, 1968.

Maruyama, M., The Second Cybernetics: Deviation-Amplifying Mutual Causal Processes, *General Systems, VIII,* 233, 1963.

Systems. A Tentative Classification Scheme with Special Reference to the Behavioral Sciences

I. *Action Systems* (Acting, or behaving systems): Mutual cause-effect relations between at least two elements, *A* and *B*. A change in each element, by movement of matter-energy or information, induces a change in the other(s). All action systems are real systems.

A. *Controlled* (*Cybernetic*) *Systems*: Has at least one goal of the system as a unit, and engages in some co-ordinated activity as a unit to achieve it. (Maintains at least one variable within some limited range.)

1. *Formal Organization*: (Involves multiple humans) A particular case of controlled interactions of controlled subsystems.

a. Traits (defining characteristics);

(1) All lowest level components are human beings (though some social behavior of animals partially parallels that of the formal organization)

(2) Communications between formal organizations and among their subsystems are semantic

(3) Transactions (exchanges) between formal organizations and among their subsystems are value-based, and involve mutually contingent decisions.

b. Levels and types (examples from Western societies)

(1) Whole-society level: governments & controlled economies

(2) Parts-of-society level: firms, churches, charities, clubs, unions, professional

associations, etc.

2. *Biological and Nonliving Systems*: Includes single human and subhuman systems (The latter may incorporate humans as components at some points, but in a technical, not social, role)

a. Traits

(1) Components may be subsystems, human or nonhuman, or non-system elements

(2) Communications are nonsemantic (except perhaps with computer components)

(3) Exchanges of matter-energy are not value-based

b. Examples: Organisms, organs, thermostatic controls, a plane operated jointly by pilot and instruments, computerized inventory control, TVA water-level controls

B. Uncontrolled (Noncybernetic) Systems: Has no goal and no behavior as an entity; hence does not interact as an entity with any other system. All variables of the main system fall where they will.

1. *Informal Organization*: (Involves multiple humans) The solely human parallel of the ecosystem, q.v.

a. Traits

(1) All lowest level components are human beings

(2) Communications among subsystems are semantic

(3) Transactions among subsystems are value-based

Note: There are no communications or transactions between informal organizations, as such, since they engage in no behavior as entities.

b. Levels and types

(1) Whole society level
(a) A laissez-faire economy
(b) The social system - i.e., the part, other than governmental and economic systems, studied by sociologists and anthropologists. (In primitive societies the social system in this sense approximates the controlled, formal organization.)

(2) Parts-of-society level: groups of persons who interact through communications and/or transactions, but who hold and execute no joint goal. E.g., a cocktail party, informal groups in firms.

2. *Ecosystem*: A system produced by the uncontrolled interactions of its controlled subsystems, each seeking its own, not main system goals

a. Traits

(1) At least two components are organisms or species, one of which may or may not be human. Conceivably, but not probably, ecological analysis might be applied to uncontrolled interactions of controlled, nonliving systems.

(2) Communications among components are nonsemantic

(3) Transactions among components are not value-based

b. Examples: The total interactions of all organisms in a pond or forest, the interaction between a human society and its natural environment, a predator-prey relation, the "nitrogen system"

3. *Wholly Nonliving Systems*

a. Traits

(1) No components are living systems, though some may be controlled. Some components, and all lowest level components within the relevant analysis, are nonsystems

(2) Communications are nonsemantic

(3) Exchanges of matter-energy are not value-based

b. Examples: Noncybernetic man-made machines, the solar system, a natural river system, a meteorological system

II. *Pattern Systems* (Nonaction systems): Relations between at least two elements, *A* and *B*, such that *A* and *B* are consistent with one another within the criteria and constraints of some action system(s), *C*. *A* does not "act on" *B*, or vice versa; only *C* acts on either. Pattern systems do not *do* anything.

A. *Real Pattern Systems*

1. Traits: these systems take the form of spatial or temporal patterns in matter-energy, including the behavior of action systems.

2. Examples: the relation between the shape and structural materials of a skyscraper or pottery bowl, a family structure, a style of music or painting, the pattern of any one composition or painting, the correlation between income and voting behavior, the relation between

technology and economic structure, the total pattern of a personality or a culture.

B. Analytical or Conceptual Pattern Systems

1. Traits: these are conceptual constructs in the heads of humans (and possible other animals) or in computers. They may or may not be constructed to represent action systems or real pattern systems. Whatever their origin, they may at times be coded into external form, such as the Tinkertoy model of a molecule, the printed drawings and propositions of Euclidean geometry, or the projective behavior of a person suffering delusions. Although in such cases they give rise to real counterparts, their essence remains that of the conceptual form.

2. Examples: logical and mathematical systems, languages, classification systems of science, the world view of a particular society or individual, capitalism (not the capitalist bloc, which is an acting system), the relation between a particular theology and its associated moral code, the total conceptual set of an individual.

Note: To the extent we focus on the conceptual set as such, the conceptual content of a culture or an individual is classified as a conceptual system. But to the extent we view it as part of the behavioral pattern of an actual person or culture it is classified as a real pattern system.

A GENERAL SYSTEMS APPROACH TO HUMAN GROUPS

F.K. Berrien*

ABSTRACT

The paper presents rigorous definitions of system, boundaries, components, inputs and outputs applicable to any system including human groups. Several propositions pertinent to human groups and based on the definitions lead to a model having adaptive characteristics. Some general anecdotal observations, but very little experimental data, are available supporting the model for understandable reasons. A final section offers some speculations about the model as it applies to the national system and the potential utility of the proposed national data bank. The model also suggests an unrecognized source of imbalance (between GNP and GNS**) in an expanding economy.

1. INTRODUCTION

I should like to consider the ways in which human groups are like machines on the one hand and biochemical systems on the other. It is the boast of general systems theory that some common principles of operation apply across all such organizations. To meet this challenge, I must lay down some assumptions and definitions, follow

---

* F. Kenneth Berrien is Professor of Psychology at Rutgers, The State University, New Brunswick, New Jersey, 08903.

** GNS = group-needs satisfactions.

these with the draft of a model, and then apply the model to typical small face-to-face groups. Finally, with some brashness, I will also present some observations about the national domestic scene as seen within the framework of the social systems model.

I am well aware that these notions are at least audacious, perhaps unfruitful, and surely in need of further development. I am encouraged however by the knowledge that what follows bears some correspondence with the concepts offered by Sells (1966), Parsons, Bales, & Shils (1953), and Katz & Kahn (1966), and in one sense may merely be extrapolations or higher level abstractions of what has already been offered. I shall have to start with a set of rather pedantic definitions.

2. DEFINITIONS

A social system is a set of interdependent role components surrounded by a boundary of social norms which screen certain signal inputs from other systems and produce some service or product acceptable to another social system (terms to be defined later).

I should like to emphasize that the elements of the definition just proposed are not limited to human beings, groups or organizations. Professor André Lwoff of the Pasteur Institute, Paris, and a Nobel Laureate in physiology, began his Nobel lecture with these words (1965): "An organism is an integrated system of interdependent structures and functions. An organism is constituted of cells and a cell consists of molecules which work in harmony. Each molecule must know what the others are doing. Each one must be capable of receiving messages and must be sufficiently disciplined to obey."

If one were to re-read this statement substituting for *organism*, organization; for *cell*, group; for *molecule*, person, the statement would still have the ring of truth. In other words, if one takes a general systems view the language of the biologist is not foreign to the social psychologist when either is talking about the fundamental processes in each field.

Let me now return to the so far undefined terms in the basic definition.

The *boundary* of a system is the screen or filter through which inputs must pass to enter the system and outputs must pass to be discharged. Light waves are transformed into nerve impulses by the boundary retina. Beef protein molecules are transformed by the digestive processes into human protein molecules before they are assimilated by human cells. The perceptual system of an individual is selective, screening in, and screening out some stimuli. Spectator behavior acceptable at an athletic contest is frowned upon in a church, for example. For a social system, approved and established behaviors (norms) appropriate to a given system are its boundary. For the individual the full range of possible social behaviors of which he is capable is constrained by the role behaviors he is compelled to adopt, and these change as he moves from one social system to another.

The roles the system permits are only some among many.

Under some circumstances the boundary may not select properly. Microbes and viruses in addition to appropriate nutrients, after all, do penetrate biological systems and cause mischief. However, as Walter Cannon and others before him emphasized, the healthy biological systems immediately react to neutralize, combat, or expel the noxious intruders. In like fashion, the norms and mores of a group may not always *prevent* a group member from behaving in ways at variance with those norms, but if he does, the group treats him in various ways that have the effect of neutralizing and insulating his influence, or in extreme instances, physically excluding him from the group's territory (Schachter,S., 1951; Merei, F., 1949). The norms, like the semi-permeable walls of a cell, screen inputs according to criteria intrinsic in the components of the social or biological system itself. Should the boundary fail in its function, the rest of the system mobilizes its defenses, protecting itself from destruction. Yet even these defenses may be overwhelmed and the system--either social or biological--may disintegrate.

We have spoken of inputs. I should like to make a distinction between two kinds of inputs; maintenance and signal. These are also screened by the boundary. Maintenance inputs are those which sustain the system. A computer requires electrical power before it will do any work. Food, air, water are needed merely to sustain living biological systems. The sleekest of airplanes or autos require fuel before they will move. What about social systems?

Let me propose, following Homans (1950), Newcomb (1950), Katz & Kahn (1966), Thibaut & Kelley (1959), and a host of others, that a social system requires a minimum level of social interactions which feedback to provide maintenance which I find convenient to call group-needs satisfactions, or simply G.N.S. By this I mean that a group is maintained once the social interaction has started by the satisfactions its members experience as a consequence of group membership. We set it as a proposition that no human group will *voluntarily* hold together and be maintained unless its members find a minimal level of satisfaction from their interpersonal interactions. Moreover as we shall see later, there are several other auxiliary sources of G.N.S. This does not exclude the possibility of coerced groups such as a chain-gang of prisoners, or a disgruntled work group under economic pressures accepting employment at high wages (a substitute for or supplement to G.N.S.). It does say that as these coercive conditions *diminish*, G.N.S. must increase merely to make possible the system's responsiveness to signal inputs; that is, to do something other than just exist.

The external sources of maintenance for a social system are different for different types of systems. A work group in an organization would receive maintenance from what Fleishman and Harris (1962) call considerations emitted by supervisors. A committee of a church, a fraternal club, or a university would receive maintenance partly from whatever prestige accrues to the members from being appointed and from the good regard with which its output is held by the larger organization. Or, if one wishes to consider a total church, a business enterprise, a university, a hospital or a charity like the Red Cross, its external sources of maintenance come from its public reputation as perceived by its components.

The problem of measuring G.N.S. is formidable but not insurmountable. Gross (1966) has proposed a number of indirect or surrogate measures for systems as large as nations. The pending proposal to establish a national data bank, although fraught with "invasion of privacy" issues, nevertheless may provide a basis for adequate statistical measures which conceivably could develop satisfaction indicators. Gross suggested that national expenditures for education, leisure, recreation, parks, travel, artistic and cultural affairs, although not measures of satisfactions, per se, may be calibrated in some rational way as indicators of human satisfactions. Furthermore, if such a national data bank existed it would be feasible to break down the data on regional, ethnic, socioeconomic, or other lines, thus measuring G.N.S. in segments of population. The implications of such developments I wish to reserve for the moment. Yet in spite of the difficulties in dealing with their measurement, it is evident that as Gross remarks, "The performance of any family, group, or organization--no matter how it may be ranked--is inevitably associated with the satisfaction of certain human interests." (Gross 1966, p.216)

Signal inputs to a system are those messages or stimuli which trigger the internal processes of the system to perform those functions of which it is capable. This definition requires us to make explicit an assumption which lies behind much of what has already been said, namely that the structure of a system, the attributes of its materials and components are determiners of its functions. A telephone system will not do the work of a cake mixer. Or one person with his repertory of skills may not do the work of another with a different skill pattern. A bank cannot manufacture automobiles. Each of these systems is built differently and consequently performs differently. Moreover the signal inputs acceptable to each are different because of the difference in the system's structure. A savings deposit mailed to the automobile manufacturer would cause some consternation, exceeded perhaps only by the disturbances in a bank that received an order for a car. I have used extreme examples only to make evident the fundamental principle that *each kind of system, whether sense organ or organization, will accept only certain kinds of signals as process starters and the processes themselves depend upon the structure of the system.*

The *outputs* of a system like the inputs may be divided into two kinds: (a) the outputs for which the system was designed, and (b) wastes or entropy. The first of these we label Formal Achievement (F.A.) when speaking of groups or organizations but other descriptive terms apply to other systems. However, to understand what we mean by Formal Achievement we must deal with two or more assumptions. First, throughout the presentation so far we have assumed that all the systems with which we deal are *open* rather than *closed*. That is, the systems are nested in such a way that they receive inputs from, and discharge outputs to collateral, sub- and supra-systems. Moreover, each of the supra-, or sub-, or collateral systems with which a particular system is related possesses the properties we have been describing. This being the case, our second assumption is that *some of a system's output must be acceptable* to some other system as an input. We have already said that inputs are filtered; hence for a collection of systems to function as a supra-system it is necessary that the F.A. produced by the components be of a limited, special kind. Moreover, since a given structure can perform several functions and thus produce several different outputs it is the "receiving" system which selects the outputs useful and acceptable to itself. Those outputs which are not useful to some other system or to the system for its own use are wastes. But even the outputs rejected by one system may in some cases be useful to another. Those not used by any system add to entropy. It is by this process of producing and selecting appropriate outputs that small systems collaborate to form larger systems, whether they are individuals forming a group, groups forming organizations, or organizations forming community.

What is considered waste by all other systems may not be waste to the system which produced it. A sugar mill produces sugar for commercial sale, but it also produces bagasse--the dried refuse of the sugar cane--that in years past was discarded. It is now burned in the furnaces that propel the mill itself. Or a manuscript rejected by several editors may nevertheless have served to increase the author's intellectual grasp and clarified his thinking.

I have gone into these prosaic illustrations to make an additional point, namely that *portions of the outputs considered waste by some supra-system may feedback into the system to help as maintenance inputs*! A group may take pride in its formal achievements and this may contribute to its G.N.S. even though the consumer of the group's F.A. may consider such pride irrelevant. The labor union whose F.A. is the protection and enhancement of its members' wages, working conditions and rights--sometimes at considerable cost to its members--or the winning football team, are illustrations of the proposition that Formal Achievement may in some cases be contributor to the maintenance of the system.

We have introduced another concept which needs some attention--feedback. This merely refers to the arrangement found in nearly all stable systems for controlling the input rate as a function of the output. Ecologists, biologists, and physiologists are currently searching for the "natural" feedback channels in their various systems, which go a long way toward understanding how a system is able to survive within larger systems that themselves vary both with respect to the input resources and output demands. Social psychologists, like the engineers, however, work with manmade systems but unlike the engineers we have been less interested and skillful either in deliberately building feedback channels or in discovering precisely what they are in social systems.

We have talked about adaptation and equilibrium without a clear understanding of how the system was adapting, or what was in equilibrium. The social

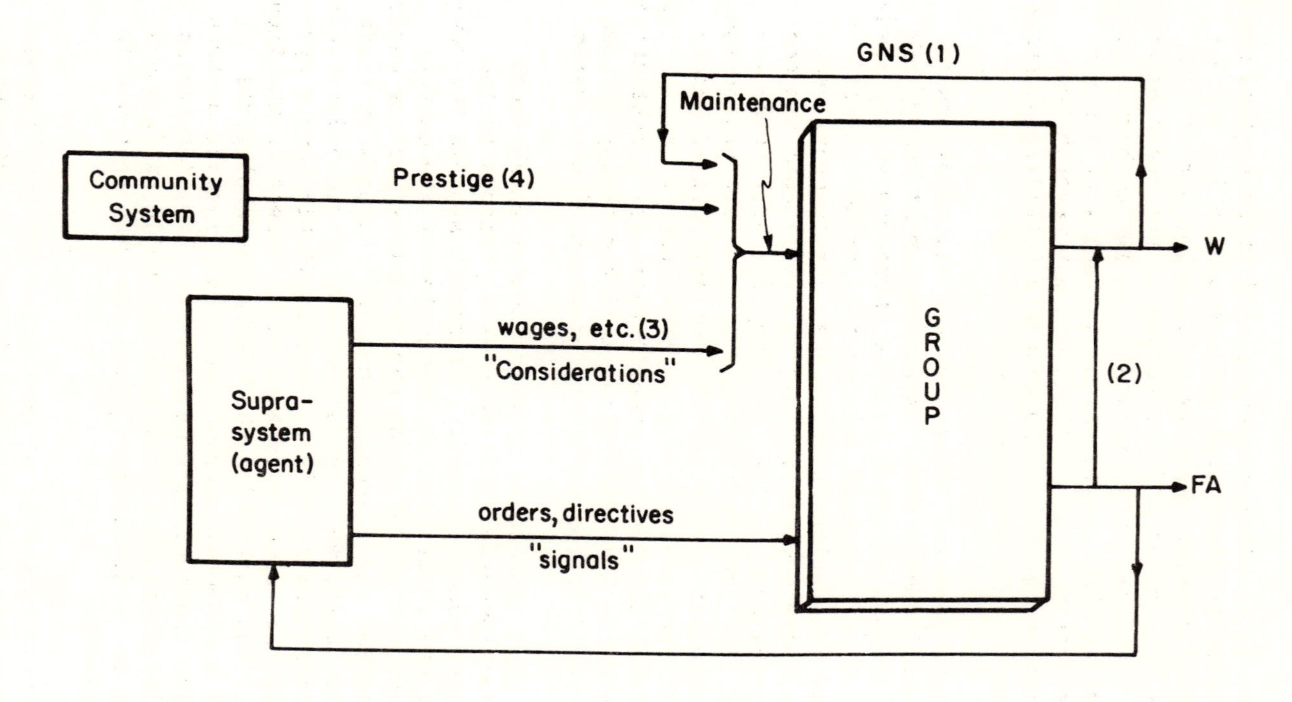
GNS (1)
Maintenance
Community System
Prestige (4)
W
G R O U P
wages, etc. (3)
"Considerations"
(2)
Supra-system (agent)
FA
orders, directives
"signals"

Fig. 1

engineers of the future must pay more attention to feedback loops, deliberately building them and measuring the flow of information along them.

3. A SYSTEMS MODEL

I should therefore like to offer a model of a general social system, which I believe may be applicable to organizations, groups and the social psychological aspects of individuals. With this I believe it is possible to conceive of adaptation as the consequence of feedback from output to input on both the maintenance and signal side. Figure 1 is an oversimplified diagramatic presentation, but like all such simplified maps it cannot possibly portray all that is implied.

It will be noted first that we have not specified the components of the system nor the nature of the relations between the components. This is a black box. We can say parenthetically that all the relationships between the components (or between systems) can be described as some combination of mathematical relations. That is, their inputs and outputs either add to, subtract from, or compare (equal to, greater than, less than) with each other. Furthermore, and this is more important, the relationships between the components may change from time to time and the detailed description of all the existing relationships among the components is called the *state* of the system. A system of N components with R possible relations may have any one of $(R)^{N(N-1)}$ possible states at any particular point in time. With as few as 4 components and 5 possible relationships the number of states is $(5)^{12}$. This is a very large number. Since the relation between components may be modified at the next moment in time, a different but not completely new state becomes possible. Some aspects of a state are relatively stable and others are easily changed merely as a consequence of processing inputs. Evidence for this lies in the fatigue of systems as a consequence of work; the phenomena of learning and forgetting; of growth and decay. The difficulties in describing all possible states of the components at any point in time is therefore formidable and furthermore, the complete description of any single state may have only transitory and temporary validity. There is no reason for ignoring the state, but especially with respect to living social systems we must take into account its labile nature.

4. ADAPTATION IN SOCIAL SYSTEMS

Let us return to the problem of adaptation, in the light of the model simplified in Figure 1. We have earlier assumed that a certain level of maintenance (GNS) is necessary before a non-coerced social system can respond to signal inputs. I should like to build a proposition on that assumption, namely that *the amount of Formal Achievement of a social system is some function of the amount of maintenance input*. The most parsimonious function to assume is that of a straight line, until empirical data provide a basis for some other relationship.

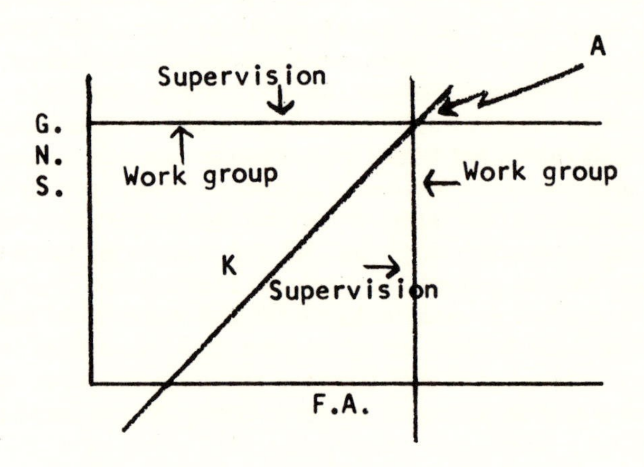

Fig. 2

We may chart this proposition very simply as it might apply to a work group in an industrial organization. The figure illustrates a situation in which the work group functions in such a way as to increase its G.N.S., but this level is balanced by the supervision or supra-system which serves as a damper or suppressor on G.N.S.* At the same time the supra-system's principal pressures are toward increasing the F.A. while the work group provides a balancing resistance to these F.A. pressures. In such a system both outputs are controlled by a balancing of conditions. The *upper* limit of F.A. is determined by the physiological capacities of the workers, their norms, and methods of operation. The *lower* limit of F.A. is established by the supra-system which demands a minimal level of output on pain of reprimand, or in extreme cases, separation from the supra-system. The *lower limit* of G.N.S. is in the hands of the work group itself which will disband when satisfactions reach some unacceptable level. Finally the *upper* G.N.S. limit is controlled by the supra-system which imposes restraints on what it construes as too much socializing, coffee time, or other interpersonal interaction not contributing directly to F.A.

Although we have used a typical work group to illustrate the adaptation principle, it can be applied to organizations such as a church, an academic department, social clubs, political parties, or even temporary systems such as ad hoc committees (M.B. Miles, 1963). Such a model borrows heavily from two sources; first Lewin's concept (1947) of driving and resisting forces whose confrontation results in a dynamic equilibrium, and second, Walter Cannon's (1945) propositions concerning homeostasis in which he held that any disturbance to a system will be met by some mechanism which will dampen or oppose the disturbance. The point of resolution between levels of G.N.S. and F.A. we propose should fall along or close to the line K if the system is to be considered adaptive. The greater the departures from K brought about either by an excess of

---

* This is not a contradiction to the earlier assertion that some maintenance may come from supervision. Here we are asserting that *external* control over some sources of external maintenance comes from supervision or the agent of the supra-system.

G.N.S. or F.A. represent conditions of reduced adaptability. For example, groups in which there is a great deal of sociability, frequent parties, but low in productivity because supervision is either lax, or incompetent, are groups that we would expect to be poorly adaptive to emergencies or fluctuations in signal inputs. These are the socializing goldbrickers in industry. Contrariwise, groups in which the pressure for F.A. is excessive and G.N.S. is low would likewise be considered by this model as poorly adaptive. These are the disgruntled, complaining, grievance-sensitive employees who may nevertheless be productive to the limits coerced by their supervisors, yet relatively unresponsive to changes in signal inputs such as routines, work methods or assignments.

A biological system which is poorly nourished, living on the bare edge of subsistence is not one which can accept a major shift in its signal inputs either positive or negative. We are arguing that the same holds true for social systems. In like fashion biological and social systems which are excessively "fat" (great maintenance inputs) are likewise lethargic to shifts in stimuli which would provoke "leaner" systems to outputs useful for some other systems. The crux of the adaptation concept is that some system variable (body temperature, blood composition, metabolic rate) is maintained relatively constant in the face of external variations. Such constancies are possible only by appropriate system changes in other variables. It is paradoxical that suprasystem variations excite changes that result in some constancy.

Figure 2 shows K with a slope of 1. This is purely hypothetical and is not meant to be fixed. Any other positive slope may be characteristic of a set of groups. The specific slope for a given case is a function (linear or log) of the scaling of the two measures which has not been specified. However, it is likely that the intercept of K, when G.N.S. is zero, will be some positive value of F.A. Exogenous conditions (economic pressures) sometimes compel individuals to be productive even when G.N.S. is at low or zero levels. Prison camp groups working under the pain of punishment or death would be examples.

The resolution of forces at point A cannot be considered a sudden or immediate consequence which comes about quickly. For those groups which have existed for some time, point A is "norm determined." That is, the norms of both the sub- and supra-systems play an important part in bringing about the resolution of their respective needs. It should be clear from our earlier discussion of norms that these, in turn, may vary from one group to another. Within one firm having a given reputation, kind of work, and personal policy, the resolution at point A may be quite different than in another kind of firm. The Michigan studies, for example, of a large insurance company revealed that in work groups having "employee-centered" supervisors the norms were such as to lead employees to expect a moderately high degree of personal considerations. This finding was not confirmed by the same team of investigators when studying railroad section gangs (Katz, Maccoby, Gurin and Floor, 1951, pages 33-34). The railroad workers, whether in high or low productive gangs, evidently were operating within norms that did not require supervisors to give them as much consideration as was true of the office workers. Both studies, however, are silent on the question of adaptability.

The model further leads one to expect that if some exogenous or other supra-system conditions operate in such a manner as to force F.A. westward to some low value without simultaneously reducing G.N.S., the sub-system will tend to counteract the F.A. changes partly because of the G.N.S. link to F.A. and partly because of the positive feedback from interactions among the components. This is essentially what has happened in those organizations in which their F.A. becomes irrelevant or no longer useful to the supra-system. Consider the National Foundation for Infantile Paralysis. After the vaccines were discovered and distributed to large masses of the population, the original mission of the Foundation was largely accomplished. The supra-system thereafter had no use for the F.A. of the Foundation. But it did not disband. It modified its name and found a new target for its F.A. This has been the history of several other similar organizations. Garden clubs often evolve out of organizations originally formed to protest a real estate developer's failure to complete his portion of a contract. If an organization accomplishes its mission (supra-system no longer requires its F.A.), the organization will disintegrate unless it finds another F.A. that is acceptable. In the absence of an acceptable F.A. the supra-system operates to suppress the F.A. level which is opposite to the situation when F.A. is "demanded" and the sub-system is to some extent limited in its F.A. output. In terms of the model, if the supra-system forces augment--rather than counteract--the sub-system forces, the sub-system is in grave danger of collapsing.

Adaptation is that process whereby survival is ensured. The model suggests how this is accomplished in various ways by social systems.

5. LIMITED EMPIRICAL SUPPORT

Is there empirical evidence to support the dynamics of such a model as we have presented? I can report that some exists, but it is both indirect and inconclusive. For example, taking measures of G.N.S. in an industrial plant four months before, and about a week after, an announcement that the plant would be permanently closed showed no marked drop in G.N.S. (Berrien & Angoff, 1960), indicating the reverberating positive feedback from interaction alone. In another situation G.N.S. was extremely low. Here slight changes in work routines created major disturbances in F.A. and further depressed G.N.S. (Indik & Tyler, 1961). The studies by Bales & Strodtbeck (1951) of problem solving groups in which they observed an alternation between supportive and non-supportive interactions tend to fit with the proposition that G.N.S. and F.A. are interdependent and mutually change adaptively. Much of the supporting data comes from case studies and anecdotal reports which do not meet the highest standards of empiricism. Moreover such data as do exist have to be reinterpreted into systems terminology. One reason for this state of affairs lies in the fact that General Systems approaches are simply too recent to have generated the necessary experimental tests.

A second reason for the paucity of empirical support stems from the fact that the requisite research designs involve frequent, or better, continuous measured observations over an extended period on bona fide systems. Most of the social experiments conducted in laboratories are on groups with no group history and a short life using cross-sectional rather than longitudinal measures. The variables are generally taken on comparable groups before and after some experimental manipulation. It is therefore not surprising that a theory which attempts to address the essence of *process* should at this point have little quantitative data on which to rely. The justification for presenting it in its empirical nakedness now is the justification for any theory--the possibility that it may stimulate research to prove its naiveté. We need a major change in the kind of data we collect.

6. THE NATIONAL SYSTEM

You will pardon me, I hope, if I now make a broad jump from the sterile bones of the model to some applications of the model in our present domestic situation. The 1968 national elections have clearly demonstrated the deep cleavages in our society. Although the disturbances during both national conventions exposed our sores before huge TV audiences, they were minor compared with the loss of life and property in the riots of the previous two summers from Los Angeles to Newark. It is patently obvious that large segments of the national population are dissatisfied whether they are in the disadvantaged minorities, among the middle class affluent who cry for law and order, or those who would emphasize correctives for social injustices. But simultaneously, this country's Gross National Product per capita soars to historic heights. As a nation we are economically rich but psychologically poor. The "Formal Achievements" of this national system have clearly overbalanced the collective Group Satisfactions. One is reminded of Dickens' "It was the best of times; it was the worst of times." Today we have a growing number of feedback loops largely unrecognized as such in the form of social statistics but nevertheless serving no adaptive functions. One notices the current efforts to cool off the economy--primarily for reasons of monetary stability, but I submit that a slower gain in GNP may be necessary to permit GNS to catch up to an appropriate level.

The national need today is said to be unity. Translated to the model this means a proper balance of F.A. and G.N.S. in all segments of the population. Unity is a by-product of such a balance. Like happiness it cannot be achieved by directly striving for it. Earlier I referred to the projected National Data Bank and its possibilities of providing reflections of human satisfactions, in the manner suggested originally by Gross. Although this project will not help us to achieve unity during the next few years, a total systems understanding will enable us to find creative ways to utilize the data bank for feedback controls. Moreover, the possibilities of breaking out various sub-populations and sub-systems from the total, offer great promise for maintaining adaptive balances throughout the segments of the national system.

As Ken Boulding has so well pointed out, Malthus' vision of an under-employment equilibrium had to wait a hundred years before the requisite national economic statistics were available to permit Keynes to revive and elaborate the Malthusian theory.

I view our society as a dinosaur--huge in its productive capacity, small in its intelligent self-corrective controls.

With the possibilities just across the horizon of obtaining the requisite measures of human satisfactions, we are on the verge of being able to construct the feedback loops for a man-made dinosaur society. A model like the one here presented is based on cybernetic principles which characterize self-corrective adaptive systems of nature. These have guided the design of man-made electro-mechanical systems and hold some promise for suggesting ways of managing our man-made organizations, whether they be small face-to-face groups, or the enormously more complex national systems.

7. SUMMARY

In line with the general systems approach to social groups, I have tried to present the essentials of a systems theory in terms which frankly simplify, perhaps over-simplify, many complex details. This however is the function of any theory--to find the uniformities in any phenomena that at first blush seem unrelated. I next sketched a model of adaptive groups in terms of the input, output, feedback concepts which found their first expression in electro-mechanical systems but which appear to be useful to an understanding of social systems. Finally the last part of the paper presents some limited empirical support for the model but, more importantly, indicates the need for continuous tracking of social variables in much the same manner as electro-cardiograms record heart actions, or security prices are recorded in stock exchanges. We talk about dynamic social processes but generally record data too infrequently to represent the processes. The general systems approach to social systems holds the promise of generating new kinds of process data sorely needed.

REFERENCES

Bales, R.F., The Equilibrium Problem in Small Groups, in T. Parsons, R.R. Bales and E.A. Shils (eds.), *Working Papers in the Theory of Action*, New York, The Free Press, 111-161 (1953).

Bales, R.F. and F.L. Strodbeck, Phases in Group Problem Solving, *Journal of Abnormal & Social Psychology, 46*, 485-495 (1951).

Berrien, F.K. and W.H. Angoff, Homeostasis Theory of Small Groups, IV: Light Manufacturing Personnel, *Technical Report No. 6*, New Brunswick, N.J., Rutgers, The State University Press (1960).

Boulding, K.E., *The Impact of Social Science*, New Brunswick, N.J., Rutgers, The State University Press, Chapter 1 (1966).

Cannon, W.B., *The Way of an Investigator*, New York, W.W. Norton (1945).

Fleishman, E.A. and E.F. Harris, Patterns of Leadership Behavior Related to Employee Grievances and Turnover, *Personnel Psychology, 15*, 45-53 (1962).

Gross, B.M., The State of the Nation: Social Systems Accounting, in R.A. Bauer (ed.), *Social Indicators*, Cambridge, The M.I.T. Press (1966).

Homans, G.C., *The Human Group*, New York: Harcourt Brace (1950).

Indik, B.P. and J.M. Tyler, A Technique for the Longitudinal Study of Group Stability and its Application to Group Homeostasis, *Technical Report No. 9*, New Brunswick, N.J., Rutgers, The State University Press (1961).

Katz, D. and R. Kahn, *The Social Psychology of Organizations*, New York, John Wiley and Sons (1966).

Katz, D., N. Maccoby, G. Gurin, and L.G. Floor, *Productivity, Supervision and Morale Among Railroad Workers*, Ann Arbor, Survey Research Center, University of Michigan (1951).

Lewin, K., Frontiers in Group Dynamics, *Human Relations, 1*, 2-38 (1947).

Lwoff, A., Nobel Lecture (1965), Stockholm, Sweden, The Nobel Foundation (1966).

Merei, F., Group Leadership and Institutionalization, *Human Relations, 2*,23-39 (1949).

Miles, M.B., On Temporary Systems, *Innovation in Education,* New York, Teachers College (1964).

Newcomb, T.M., *Social Psychology,* New York, Dryden (1950).

Schachter, S., Deviation, Rejection and Communication, *Journal of Abnormal & Social Psychology, 46,* 190-207 (1951).

Sells, S.B., Ecology and the Science of Psychology, *Multivariate Behavioral Research, 1,* 131-144 (1966).

Thibaut, J.W. and H.H. Kelley, *The Social Psychology of Groups,* New York, John Wiley and Sons, Chapter 9 (1959).

THE INDIVIDUAL AS A COMPLEX OPEN SYSTEM

Janos A. Schossberger*

ABSTRACT

For the purpose of understanding the complex interrelations by which individuals interact, biosemiotic symbol theory contributes certain general principles. The creation of concrete objective symbols derives from and depends upon the faculty of pictorial representation. From this, the role of writing in enhancing hemispheric dominance by channelling the unified strands of written, read, heard, spoken or typed, etc. language is readily traced. Human beings are constantly exposed to each

---

* Janos A. Schossberger is Directing Psychiatrist, Kfar Shaul Work Village, Jerusalem, Israel, and at time of presentation of paper was Visiting Professor of Psychiatry, The John Hopkins University, School of Medicine, Department of Psychiatry and Behavioral Sciences.

This investigation has been aided by a grant from the Foundations Fund for Research in Psychiatry.

Thanks are due to Professor Joel Elkes, Psychiatrist-in-Chief, The John Hopkins Hospital, for enabling the author to take time out for recollection and thought and to the Ministry of Health, Jerusalem, Israel, for granting the necessary leave of absence.

This paper is a condensation of a longer one with the same title.

other's meaningful communications. The products of this activity codetermine, by unceasing repetitions, what the individuals in a social group will be like.

The individual, however, is not merely the respective receiver of these influences from the informational field that surrounds him. Alternating with this receptivity, each individual contributes to the informational field messages of his own creation. These messages about any definite meaning may be expressed in several ways, so that the term equiclamatory expression suggests itself for the investigation of the problem whether any messages can be translatable in terms of each other.

Individuals being both the source of information and the consequence of their impact, there is a certain resemblance between a theory of an informational field in which the individual appears as a "singularity," a region where the general field rule is discontinuously modified, and between the gravitational field theory, where the trajectories of material objects may be viewed as determined by the objects, that themselves follow such trajectories.

Not more than a hint can be given about the effect of the tremendous multiplication of intermediate message transmitters (e.g. "the media," books, but also money, treaties, laws). These separate as well as unite the individuals to which they apply. The elaboration of these problems is left for a more comprehensive study.

But from the point of view of general systems theory, the present essay defines how, within the social system, links between individuals must be understood as "tetradic" structures, comprising 1) both participants, 2) the "designates" of their intercommunication, and 3) the messages and symbols by which the meaning is transmitted. Many failures in the social communication process -- "misunderstandings" -- are due to neglect of one of the four tetradic elements: 1) observing subject, 2) observed subject, 3) their object, 4) their message about it all.

I. The subject of this presentation shall be some aspects of the *individual as a complex open system.* There is no question that individuals should not be studied in isolation for the purpose of understanding their functioning. Even in psychoanalysis, where considerable constraints are at work in this direction, curative effects can only be assessed by observing how the patients become newly or differently activated in their environment. This paper represents an attempt to relate the study of the individual to the social system of which he is a part.

Because of reasons of space, an introduction to general system theory cannot be presented. However, the paper can be read without such background preparation. Such material can be found in the book, "An Outline of General Systems Theory" by Ludwig von Bertalanffy [1].

II. It takes a different order of magnification to focus down from the social system to the individual human element. This problem links up with the concept of "meaning" and leads to the core of my subject matter. At this level, individuals appear as complex systems in their own right, interacting with their similar and yet uniquely different counterparts in constant mutual interplay. By their similarities, individuals conform to the general rules valid within the system. Simultaneously, they display their individual character by incessant meaningful differences. Therefore, one must not merely appraise the human element from the viewpoint of generally valid systemic rules, as feasible and customary in general systems theory. In addition to this, the role played by the individual actors requires for its assessment a number of explanatory concepts not otherwise needed in describing non-human systems. I shall turn to these concepts now.

III. C.S. Peirce introduced the concept of *triads*, and mathematical methods based on multiple triadic relations. He noted that the world picture of reality results from interpreting stimulus clusters from which the sensual percepts are collated into concepts about things and events. Peirce's "thirdness," "secondness," and "firstness" correspond roughly to the coding levels of concept, percept and stimulus cluster. Experiencing and awareness entail a tripartite observational process, using "symbols" through which the world is interpreted by way of "indices," from the "ikon" of actuality.

Let me stress that Peirce's concept of triads did not cover the problem of meaning actively conveyed by persons and animated beings. He confined himself to elucidating the impact of the physical inanimate reality upon the "interpretant": he refrained, thereby, from exploring communication proper, but defined it as the problem of "meaning" signalled through his "utterance" by an "utterer." [2] This however is precisely the problem to be solved in order to describe what individuals do. I shall show that for the intercommunication between individuals, a concept of tetradic relations is required, because in contrast to the ikon of pragmaticist semiotic, "utterances" are ready-made, precoded symbols in their own right. In keeping with Peirce's usage, the interpreting or decoding of pre-coded configurations would have to be termed "fourthness."

IV. F.S. Rothschild was led to apply semiotic, the science of signs and symbols, to biology. Proceeding from comparative neuroanatomy, he investigated the functional and the structural aspects of living communication systems, and their interrelations with the conative strivings displayed by their hosts. Moles, birds or fish live in dissimilar worlds. Their intentional strivings are manifestly different. So is the neuroanatomical structure of their central nervous systems. However, the structural scaffolding of nervous systems could now be correlated both to the patterning of inner stimulus functions taking their course in them, and also to the expressive value conveyed by these activities. The structural, the functional, and the intentional aspects of biological communication systems could, in this way, be treated as mutually symbolic of each other. This biosemiotic method could explain how nervous systems had come by their actual architecture. Repositories of "aeons of evolutions," they had become a symbol, a logogram, as it were, of the course taken by their own evolutionary history [3,4].

V. In order to deal with symbols and signalling processes, the concept of transcendency was now introduced. Symbols operate as messages, "*from* someone, *to* someone, *about* something." This led to investigating the denotative functions of language, in which symbols were used by communicating individuals in order to transmit information. Any exact description of communication processes between "utterer" and "interpretant" through the link of "utterances" or messages was incomplete as long as it disregarded the denotative link to the content of the conversation between the two. This content of communication may be physically remote, temporospatially inaccessible, and indeed completely unobservable during the communication process. Nevertheless, this "denotated" content was whatever the communication process was about.

Thus, irrespective of whether the bee's dance is viewed as a language proper or merely as a complicated trigger system for releasing alternatively preset activities, it reflects immediately effective circumstances beyond the hive: directions, distances, flowers, and whatever else is conveyed by the performance. A full description of information transfer had always to reckon with the ecosystem in which the transactions occurred. Messages between "utterer" and "interpretant" "stand for" something else, and moreover, whatever the messages stand for, that is what it is all about, so to speak.

VI. This illustrates a new type of multiple intercorrelations between the constituent elements of organismic assemblies and their ecology. Beyond physical processes based upon the commerce of energy and material, the individual elements were interrelated through the fluctuations of information exchange. The distinction between mere patterned flow of matter and energy and communication flow derives from the fact that message patterns have been precoded by the shared idiom of common language, by its denotative use. Thereby, the ecologically relevant objective environment enters the awareness of some of the participants indirectly, through the artifice of symbol transfer. Through it, the designated or denotated content about their inner and outer environment is transmitted from the awareness of one communicant to that of the other. As with bees, in grouped living systems the common shared language is firmly inscribed in the organic structure of their constituent individuals, with the notable exception of human languages. In these, the intraorganismic language functions, located in the various speech centers of the dominant cerebral hemisphere, are supplemented by an external network of objectively real symbols.

VII. Before dealing with this most recent development in biological communications, I shall turn now to the hierarchic coordination of the increasingly complex biological communication systems, expounded in Rothschild's biosemiotic. Higher organisms must be viewed not merely as unilingual assemblies, but rather as compounded, hierarchically stratified unities. In contrast to computers, where older models are superseded by their improved successors, in the living systems many language networks operate simultaneously, because the oldest idioms with their original morphological structures have been preserved through the ages.

Thus, at the cellular level, transcellular integrative effects of signal transmission between cells and cell-groups led gradually to the formation of cell assemblies. By their greater motility and efficiency, these display a completely new way of interacting with the environment. Multicellular organisms, for instance jellyfish, override the independent amoebic existence of their isolated elements. These subordinate their admittedly limited functional range to a newly acquired group identity.

Already in the morular and gastrular cell-clusters, the negotiations of organismic signal exchange require two idioms. An *outer language* deals with the respective environment, including any other organisms of the same kind. Second, an *inner language* for broadcasting information into the organismic interior, for the various coding, activating and suppressing operations that concern it in its entirety.

This led to clarifying the interrelations between the various superposed systems: the unifying language pools at any level send signals to the next higher network, and reciprocally receive steering and informing signals from it in turn. The inward language of superposed systems conforms to the outward idiom of the subordinated ones. As a leading unit of organismic unity, the superposed system exerts inhibiting, dominating, and decision-implementing functions. To this *dominance* of the respectively most recent, highest agency, a *polarity* is joined: outer languages conform to the environmental "otherness," while the inner networks convey the inward "selfness." The principle of dominance and that of polarity dovetail with the third principle, that of organismic *unity*: the highest, latest systems operate by remaining firmly based upon the viable function of all their archaic antecedents within the organism. The inhibiting effects of dominance are countervailed by the nurturing aspects of unity [5,6].

VIII. As noted, human languages are the highest and newest acquisition in the series. However, they are distinguished by a striking difference from all the preceding levels. Only the inner part of human languages is located in brain centers and operates, so to speak, on an intracranial fundament. The outer language, no longer reposited in anatomical structures or physiological processes, has become consolidated in the environment as the sum total of recorded social information. Thereby human ingenuity, and learning, creativity and memory, experience, intention and endeavor, all are liberated from the shackles of somatic constraint. This newest information network ranges in coherent transcendence over considerable spans of time, and has extended the external memory of the race, through documents, treaties, and the written tradition. By contrast, the oral transmission of tradition limits the amount of information for its beneficiaries in each generation to the amount conveyed "in real time," by mouth-to-ear communication.

The human languages must be newly acquired in infancy: there is nothing instinctually innate and structurally pre-set to determine their grammar and syntax "a priori."

IX. These innovations derive from a completely new use of the brain, traceable to the ability to create *pictorial records* of the inward imagination in wall paintings, rock paintings, and carvings. These achievements are due to a fundamental change in the use of the hands, and a corresponding change in the visual performance: both tool-wielding and club-swinging rely on inserting the implementing hand passively in the field of vision, subservient to visual control. This arrangement is suitable for discharge-like performances of the hit-or-miss type, needed for instance to catch a flying ball or a fleeing prey. However, for the ability to draw, the dominant leading hand must be guided by inhibitory controls, illustrated by the synergistic - antagonistic bridle effect. This restraint inserts a pause between the visual impulsions and their manual discharge. Pausing to "look and see" leads to a "stop and think" effect and to the phenomena of "ideational motility," the retracing of events in mid-air. Plainly, pictorial imagery reflects the inward imagination and preserves it through the times.

The inhibitory control of the leading hand coincides with the internalized mastery over aggressive impulsions. These are prevalent prior to the acquisition of drawing and writing, both in the palaeoanthropological evolutionary, and the ontological, developmental series. By the newly acquired maneuver of pointing towards an object rather than grabbing at it, the leading hand could in due course be held in the field of vision in the familiar, "painters' gesture" of pointing and sizing up. Then, by swivelling about the vertical axis of the spine while keeping his bearings constant, a projective reproduction of imagery could now be accomplished by early man.

X. Pictorial representation, from which writing finally derives, marked a turning point both in evolution and ontogenesis. It modified and enriched the environment which now contained not merely people and things, but also real images of both. With increased proficiency, stereotyped ideograms and ultimately the alphabet created new ways to symbolize *word images* objectively. An external linguistic store now mirrored the internal language system. Furthermore, within the inner system, a new intercorrelation now connected the oral and auditive with the visual and manual-graphic verbal expressive channels. Current simultaneous transcoding in reciprocal correspondence had been introduced, and was now established in the central nervous system. How this affected the group process may be illustrated by the fantastic achievement of integrating a complicated musical score into a concerted process of concurrent, mutually supplementary unanimous orchestral display of the composer's intended message.

The inward communication processes devolve simultaneously in mutually translatable contextual strands, connected in accordance with compatibility rules of frequency distribution and patterning. In analogy to the concept of equifinality, the term "*equiclamatory message transmission*" suggests itself here, in order to denote the internal symbol transfer processes by which written, spoken, heard, and read sequences are transcoded in terms of each other: a veritable, currently proceeding, internal, cerebral automatic language translation. The sound and the aspect, the utterance and the manual graphic performance of the same word or the same statement may be regarded as conveying the same thing. The written, spoken, heard, and read "thank you," all have a like message to bring, respectively to dispense, they express the self-same meaning in various "equiclamatory" ways and modalities.

XI. In order to tackle the problem of meaning, one must first consider a fundamental change in twentieth century science. The intrusion of the subject-object relationship in the area of physical measurement led to renewed controversies such as whether the activity of thinking and the prevalence of "thought as an active factor in the real world" (Peirce) ought to be acknowledged as truly real. Concepts regarding the nature of reality and the physical world are closely connected with the precursors of general systems theory. As an illustration, one may cite the concept of celestial bodies in the gravitational field of planetary systems; then, later, the concept of molecular aggregates subjected to thermal motion, to be regarded as closed systems; lastly, electrically charged subatomic particles travelling in electromagnetic fields in closest interconnection with the wave field around them. In the world view of modern physics, nature is likened to ephemeral oscillatory events with which our methods of observation happen to be in tune. Physicists were led to realize that events could not be described apart from their observers, whose presence in the physical system co-determined the qualities, the quantities, and the very nature of observed events.

Thus one of the first statements about thermodynamically closed systems referred to their entropic decay. But the growth of organic forms and the emergence of organismic units has led to the concept of "open systems." This was plainly applicable to human organisms also. Their activity of thinking and the reality of their thoughts could only become objectivated in the world of real events through recognizing their effects, as interpretable, pre-coded, intentionally symbolic expressions of *meaning*. Other peoples' thinking could be observed through their active *communications about it*. This is why the problem of meaning was bound up with the fact that physical reality in its evolution had produced human beings, animated creatures with intentions similar to those of their observers. Their thoughts beyond and prior to their utterances could only become exteriorized by *symbols about those very thoughts*. With these insights, both the extremes of primitive and infantile animism and those of an ill conceived materialism could be corrected. Lifeless objects and symbols, felt in earlier stages to be animated and spontaneously active, could now be viewed, more objectively, as physically determined. But on the other hand, persons need not be reduced by science to their mechanical, physically determined processes: plainly, as opposed to the inanimate objects of mechanics and dynamics, people at least were certainly animated: they consisted of stably patterned substance, but manifested the expressions of having a mind of their own.

XII. This introduced the problem of *interpretation as a scientific method.* As noted, even physical data had to be interpreted as bearing the traces of the confluent processes by which they had come about. This led to attempts to explain the presence of interpretants in the systemic aggregate, from the growth of organisms to the deployment of symbolic structures, including the self-expressive and symbol-making activities of human individuals. The human emergence now appeared as the culmination of cumulative patterning processes symmetrically opposed to the entropy function. They may be illustrated by the Ising effect and other self-organizing phenomena [7,8].

Accordingly, within their psycho-social, socioeconomic, ecologic and physical systems, individuals had to be viewed as open systems in their own right, and from two viewpoints: first, they followed the rules of permanently patterned physico-chemical processes. The commerce of matter and energy flow maintained their structural pattern, by "burning them up while they lasted." But, secondly, they were interconnected by communication processes. Thereby, they became "open systems" with regard to the commerce of symbol-transfer. They produced, spontaneously, real stores of information. This connected them among each other, and co-determined their intercorrelations with the systemic unity. Their picturing activities transcended the range of their lifetime. By distinguishing patterns of matter and energy transport (for instance, pulsar emissions) from patterns of precoded symbol transfer (for instance, expressive activity in animals or distant sources), the difference between the animated and the lifeless condition could now be defined. Individuals could be spotted by their current emission of symbols denotative of their environment and of their own intended meaning.

XIII. By their expressive activities, individuals suffuse their system with strands of linguistic, induction-like influences. Now strands of message transmission constitute the *communication flow* extensively analyzed in the mathematics of information theory. The concepts of signal-noise ratio, redundancy, channel capacity and the quantification of information by binary items were used. But it was tacitly assumed [9,10] that, once faultlessly decoded, the signals would also, as a matter of course, be intelligible to recipients, who were considered to have agreed upon the choice of a common language. Thereby both the problem of meaning and that of the role played by a connecting syntax became specifically excluded from the theory.

By contrast, whenever real persons communicate, they decide first what language to use and second what they choose to say. Furthermore, any genuinely new information must be constantly repeated in order to gain general acceptance. This introduces a new type of redundancy, applicable not to binary items but rather to signal groupings with specific denotative significance, such as for instance Einstein's formula linking mass with energy. Once accepted, such groupings recede into an oft-repeated, hence relatively redundant, syntactic structural background. Conversely, one may then regard messages proper as conveyed together with the communication processes that flow in the coherent contextual channel. In a multilingual situation, any one among several languages current at the moment may be selected. This shows how the context of any message operates as a group identifying signal: the in-group participants stay tuned to their group idiom; within that context, they convey any specific message to each other. Relevant passages turn into stereotypes by repetitions, while constantly new relevance is created through communications about unprecedented experiences. This outline must suffice here to show that whenever communication occurs, it consists of two simultaneously present streams of information flow: first, the relatively redundant current of *contextual background*. It signals the onset of information transfer by confirming the established connection, and it also proclaims the group identity. Secondly, the more significant passages contain *messages of relevance*, forming improbable inclusions, regions of minimum redundancy within the more repetitive background information. As noted, for example, mathematical statements exemplify highly significant, low redundancy methods of conveying meaning. But of course, they, too, are embedded in the ever-expanding language field of calculus and symbolic logic, growing by repetition and multiple application of its significant findings.

XIV. The stream of information flow engages the participants in perpetually changing roles of becoming respectively emitters or receivers of information. Consequently, within the language pool, *meaning is propagated* by alternate reciprocal emissions and receptions of the group signal -- a signature tune of high repetitiveness -- furthermore, by specific relevant messages enclosed in that background flow, as terse formulations of high significance, low redundancy, and low probability. In this sense one may envision a *language field* with oscillatory features in which individuals, including their observers, trade information. In this sign making process, meaning at any time achieved by the participants is conveyed among them and becomes respectively completed and supplemented. Gradually the low redundancy relevant messages merge by frequent reiteration into the high redundancy explanatory context.

XV. As to the individuals within the observable environment, they are now exposed both as complex and also as multiple open systems. In each of their stratified networks, incessant emissions, transmissions and receptions, both of metabolic commerce and of information exchange, take place. Thus already at the level of electrolyte and water flow and energy transport, cohesive phenomena of patterning ensure the formation of equifinal states, through the homoeostatic balance of anabolism and catabolism. However individuals are more than mere metabolic vortices of pattern-preservation. Theirs is a message carrying pattern, operatively effective in structuring the ever-expanding field of significantly formulated information. In other words: the individuals structure their ecologic realm by adding to it a constant supply of symbols about it. These have become an external independently structured network of sign systems. Incidentally, the individual torchbearers of message transmission are less hardy than the field that surrounds them. The messages produced by them in works of art and written records preserve the creations of their imagination across the oceans of time.

The step from the level of complexity of electrolytes to that of macro-molecules points, for instance, to virology: individuals are open to infections; they respond with immunities, but spread whatever afflicts them. Next, at the level of being physical bodies, the human frame is subject to contusions, concussions, and fractures; conditions of crowding and the behavior of packed humanity during rush hours allow of statistic appraisal. At the next higher level of central nervous structure, analogies to computers emerge in the reverberating circuits and channelling or shunting operations. Information retrieval, multiple interferometric (holographic) and transmodal signalling are all standard procedures in the central nervous system, so to speak. Now at the highest systemic level, individuality emerges, by leaving permanent traces on the stage of actuality. It involves a total meaningfully relevant reciprocal engagement through communications, and it also provides the specific impact of human individuals upon their human observers.

XVI. We have now listed the elements of the human environmental system: first, objects, objective events and configurations; second, subjects and animated beings; third, objectivated symbols referring to and denotative of all the other elements. The object world and its rules belong to physics. But already the temporo-spatial conceptualizations, for instance of wave mechanics, entail statements about meetings between the observing subject and the observed passage of object-like transient eventuations. Proceeding further, one arrives at configurational concepts and symbolic connections. Here the demarcation between subject and object is tenuous indeed. This is so, because purely configurational relationships, for instance the geodesics that define the trajectories of orbiting celestial bodies, are not at all actually to be found anywhere in reality. Much rather, they are descriptive and predictive reports about regularities experienced throughout observation periods of widely varying duration. This explains the striking congruence between the inner syntax of mathematical methods and their supposedly objective observational counterparts. Actually, the external sign system of mathematics and symbolic logic are merely precise and increasingly compact restatements of human experiencing. This is so, because in contradistinction to objects and lifeless configurations, the symbols given to us by our predecessors and reposited in the environment convey the recorded expression of their creators' experiences and intentions. Mathematical treatises, operation manuals and the like belong to this category of symbolic extrajects.

Now, as regards animated beings, they too convey encoded messages about their condition, even by their very presence. The decoding of such messages (e.g. purring, or tail-wagging) entails acquiring an additional group idiom, so to speak. As to the human species, a cardinal conceptual difficulty arises here: observers, of course, owe their own physical constitution to their being human. They start out from being completely attuned to the shared idioms of posture, motor activity, gesture and features. This includes autonomous activities such as blushing and paling, tremor and perspiration. The very shape of human beings conveys significance. This is best shown by considering well-known features, e.g. Nefertiti's head or Michelangelo's Moses. Merely by facing them, the viewer understands what these sculptures convey, although they lack movement displayed by people even when at rest. In contrast to speech, this part of human expression need not be acquired by "learning" it. The onlookers identify with the inward intentions of their counterpart precisely because they too have been shaped by processes of unconscious identification throughout uncounted generations. One's counterparts are intelligible expressions of themselves, just as one feels oneself to be an expression of one's own personality, its history, present character and future intentions. From their features, appearances, behavior and attitude, we "read" the persons whose presence we share. In this sense too, they are *symbols of their own life*.

XVII. By bearing the image of human sameness, individuals partake of a shared redundant context: their fundamental traits are those of the Old Adam. Also, as with the group signalling functions of context, they carry the imprint of their locality and times. Nevertheless, out of this sameness, any single person is directly recognizable as an incomparable, unique singularity, the like of which can only be found in imaginary stories of the Dorian Gray type. Through the uniqueness of their respectively relevant present instant, at any moment, whatever else they express, individuals convey to their observer that he is not alone. As a sentient subject, he shares with them a common universe of relevance [11].

XVIII. From here one is led to formulate the concept of a constantly evolving *nootic field* with its specific rules of organization. Embedded in this field, the participants appear as compound systemic elements. In the unfolding transpersonal aggregate, they act as emitters of information flow, or as receivers of information supply. Alternatingly, they generate an informational field that oscillates or fluctuates between the participants. These, like all other biological organisms, carry the traces of their past origins. Human organisms, and before them, the evolutional avenue of increasingly lower and simpler aggregates, all point backwards in time, to an immense stream of formative processes. Of these, the specific structuring of the human environment through a symbolic field system is a comparatively recent development.

Viewing individuals as the source of their nootic field, one is led to regard them as local singularities not unlike the massive bodies that create the gravitational field to which they are subject. Incidentally, communication processes consume time while they take place; one must therefore always reckon with four elements and their specific role in the communications that connect them. Information from subject, to subject, about objective events, in terms of messages, amounts to a growing progression of four, not three interacting elements. The *informational tetrad* of *subject, message. denotate, and fellow subject* represents a dimensional expansion in comparison to the concept of triads. One may visualize these four elements as the four corners of a triangular pyramid. Any observation of multiple tetradic interactions, must reckon with the time span required to perform it. This is why data about actually progressing information processes can never apply simultaneously to all the four tetradic elements [12].

XIX. As noted, the meaning created by individuals may be explored from the point of view of frequency analysis. The various psychosocial processes range in a gradation of redundancies, from predictable, stereotyped behavior patterns, for instance manners, customs and protocol, to exceptional circumstances such as unique ceremonial occasions peak performances or rare moments of insight. All these events are capable of being ordered into staged arrays of *frequency distribution,* combining at any moment the two extremes of statistical probability and uniqueness [13]. In retrospect, individuals always act predictably in the frame of their time and place. But concurrently they can only be conceived as the compounded interference pattern of the information load carried by them at any moment to the next further instant of significant field patterning. The view of individuals as compounded of innumerable periodicities is not intended here as a loose analogy. Respiration-rate, heartbeat, the sleep wakefulness cycle, are samples from a list that can be extended to annual, seasonal, menstrual and circadian cycles, down to the minute oscillatory phenomena of scanning rhythms and the compounded operation of the various "internal clocks" [14]. An imaginary observer from a distant planet, unacquainted with the forms of life on earth, would have first to concentrate on such recurring regularities. The better his survey, the sooner he would notice the slight but *recurring irregularities* by which the welter of synchronic and diachronic redundancies becomes punctuated: sudden dilatation of pupils, changes in heart-rate or breathing, and many more. He could view such phenomena as negligible rarities, but he could also perhaps, *link them with environmental contingencies*. At that precise point he would have discovered their denotative function and their character of messages, peremptorily and objectively indicating the presence of life in the system. From there to the complete decoding of the cultural texture and the full understanding of whatever is meaningful for those focal centers of activity, he would merely have to progress ever onward in carefully attempting to break the code without damaging the inhabitants. Finally, with the intimate knowledge of familiarity, he could participate not merely by recording the accepted habits, but also by partaking in the current of constantly created new meaning. He would be comparing notes with them, thus making the shared nootic field grow, let us say, by one bit.

REFERENCES

[1] Bertalanffy, L.V. An Outline of General Systems Theory, *The British Journal for the Philosophy of Science*, Vol. I, 134-165 (1950); or Bertalanffy L.V. *General System Theory*, New York, George Braziller (1968).

[2] Peirce, C.S. *Values in a Universe of Chance*, N.Y. P. Wiener ed., Doubleday, Garden City (1958).

[3] Rothschild, F.S. *Symbolik d. Hirnbaus*, Berlin, S. Karger (1935).

[4] Lorenz, K. Kant's Lehre vom Apriorischen im Lichte gegenwaertiger Biologie, *Blaetter f. Deutsche Philosophie 15*, 94 (1941).

[5] Rothschild, F.S. Laws of Symbolic Mediation in the Dynamics of Self and Personality, *Annals N.Y. Acad. Sciences 96*, 774-784 (1962).

[6] Rothschild, F.S. Concepts and Methods of Biosemiotic, *Scripta Hierosolymitana*, Vol. 20, Jerusalem, The Magnes Press, 163-194 (1968).

[7] Buckley, W. ed. *Modern Systems Research for the Behavioral Scientist*, Chicago, Aldine, 39-44 (1968); Purcell, E. *Parts and Wholes in Physics*.

[8] Fox, S. Self-organizing Phenomena and the First Life, *General Systems Yearbook*, Vol. 5, p. 57, (1960).

[9] Shannon, C.E., Weaver, W. Mathematical Theory of Communication, Urbana,Ill., Univ. of Illinois Press, p.8 and p.75 (1964).

[10] Luce, R.D., Bush, R.R., Galanter, ed., *Handbook of Math. Psychology*, N.Y., Wiley (1963).

[11] Maruyamah, M. Metaorganization of Information, *Gen. Syst. Yearb. 11*, 55 (1966).

[12] Caianiello, E.R., ed., *Proc. School on Neural Networks*, Ravello (1967). Mathematics for a Junket to Mars, W.S. McCullogh and R. Moreno-Diaz.

[13] Planck, M. in *Maxwell's Influence on Theoretical Physics in Germany*, in Maxwell, J.C. A Commemorative Volume, Cambridge U.P., p.63 (1931),

[14] Richter, C. *Internal Clocks*, Johns Hopkins Press.

USE OF SIGGS THEORY MODEL TO CHARACTERIZE EDUCATIONAL SYSTEMS AS SOCIAL SYSTEMS

Elizabeth Steiner Maccia and George S. Maccia*

ABSTRACT

Education is one organization among those that can be called 'social organizations'. The purpose of this paper is to show how general systems theory can lead to understanding education as such an organization. In order to accomplish this purpose, aspects of a given educational organization are explained by means of a given development of general systems theory. The educational organization selected was the Ocean Hill-Brownsville School District of New York City. The development of general systems theory utilized was the SIGGS Theory, a development resulting from the integration by the authors of set theory (S), information theory (I), graph theory (G), and general systems theory (GS).**

1. PURPOSE OF THIS PAPER

Education is one organization among those that can be called 'social organizations'. It is the purpose of

---

* Elizabeth Steiner Maccia and George S. Maccia are Professors in the Department of History and Philosophy of Education, Indiana University, Bloomington, Indiana. 04740.

** Siggs represents set theory (S), information theory (I), graph theory (G) and general systems theory (GS), this paper to show how general systems theory can lead to understanding education as such an organization. In order to accomplish this purpose, aspects of a given educational organization, the Ocean Hill-Brownsville School District, are explained by means of a given development of general systems theory, the SIGGS Theory.

2. EDUCATION AS A SOCIAL ORGANIZATION

A social organization is a group which has structure and culture. To say that a group has structure is to say that there is a "specific relational system of interaction among individuals and collectivities" (Kroeber and Parsons, p. 538). Culture of a group, however, is "transmitted and created content and patterns of values, ideas, and other symbolic-meaningful systems" (Kroeber and Parsons, p. 538). Structure gives rise to status or positioning of individuals and collectivities in the group, while culture gives rise to role or functioning of individuals and of collectivities in the group. Since structure is the network of social relations, it tells how one member stands in reference to the other members of the group (for example, to whom a member is subordinate) and so gives rise to status. Since culture is the shared beliefs and orientations of the group, it provides guides for the conduct of members of the group (for example, whether a member should use a command as his basis of choice) and so gives rise to role. An example would be the association between the status involved in subordination, and the role involved in reliance on command as a basis of choice.

Social organizations emerge whenever and wherever individuals live together, i.e. whenever and wherever individuals are related and share values and norms. Men and women living together with their offspring, for instance, evolved the social organization called 'the family'. Sometimes individuals set up a specific social organization in order to achieve certain goals. The formal social organization comes into being. An industrial firm is a formal social organization, because it is a social organization deliberately shaped to provide profit to its owners.

Education is not simply learning. At the very least, education involves a possibility of learning, something to be learned, and a deliberate attempt to bring about learning. In other words, education at least involves a student, subject-matter, and a teacher. Except in the case of self-education where the teacher and the student are one and the same individual, education, therefore, is a group of at least two individuals, the teacher and the student. Even if the group be only two, it is a social organization. There is both structure and culture. The very terms 'teacher' and 'student' designate status and role. The student is subordinate to the teacher, and the student is expected to behave as a student (base his academic choices on teacher commands) while the teacher is expected to behave as a teacher. Status and role arise from structure and culture respectively.

Education as a social organization consisting of only two individuals who are related and who share values and norms so that one is a teacher and one is a student is not education formalized. Education formalized is a trifle more complex. Besides technical employees, such as teachers, for adaptation and goal attainment, managers are required to mediate between various parts of the organization and to coordinate their efforts, and directors to maintain the organization in the larger society. (Parsons, pp. 16-96).

3. THE SIGGS THEORY

The SIGGS Theory consists of a set of propositions about education which were devised from the SIGGS Model in order to explain education as a social organization. A group of related terms which are characterizations of a system in general constitutes the SIGGS Model. The characterizations of the SIGGS Model are summarized in Table 1 and Table 2. Table 1 is a list of primitive (undefined) and defined terms which do not directly characterize general systems but which are needed to do so, while Table 2 is a list of defined terms (most of which are properties) which directly characterize general systems.

TABLE 1.

(19, 21, 24, and 25 are negasystem properties.)

Indirect System Characterizations*

Primitive

1. universe of discourse, $\mathcal{U}$
2. component, s
4. characterization, CH
10. condition, $\underline{F}$
15. value, V

Defined

3. group, S
5. information, I
5-1. selective information, I_S
5-1-1. nonconditional selective information, I_S^N
5-1-2. conditional selective information, I_S^C
6. transmission of selective information, $\vec{I}(I_{S_1}, I_{S_2}, \ldots, I_{S_i}, \ldots I_{S_n})$
7. affect relation, R_A
7-1. directed affect relation, R_{DA}
7-1-1. direct directed affect relation, R_{DA}^D
7-1-2. indirect directed affect relation, R_{DA}^I
9. negasystem, $\bar{\$}$
12. negasystem state, $ST_{\bar{\$}}$
14. negasystem property, $P_{\bar{\$}}$
17. negasystem property state, $ST_{P_{\bar{\$}}}$
19. negasystem environmentness, $E_{\bar{\$}}$
21. negasystem environmental changeness, $EC_{\bar{\$}}$
24. fromputness, FP
25. outputness, OP

* Maccia and Maccia, p. 68. The number before the term indicates the order of presentation

TABLE 2.

Direct System Characterizations*

Non-Properties

8. system, $\bar{S}$
11. system state, $ST_{\bar{S}}$
13. system property, $P_{\bar{S}}$
16. system property state, $ST_{P_{\bar{S}}}$

Properties

18. system environmentness, $E_{\bar{s}}$
20. system environmental changeness, $EC_{\bar{S}}$
22. toputness, TP
23. inputness, IP
26. storeputness, SP
27. feedinness, FI
28. feedoutness, FO
29. feedthroughness, FT
30. feedbackness, FB
31. filtrationness, FL
32. spillageness, SL
33. regulationess, RG
34. compatiblenss, CP
35. openness, O
36. adaptiveness, AD
37. efficientness, EF
38. complete connectionness, CC
39. strongness, SR
40. unilateralness, U
41. weakness, WE
42. disconnectionness, DC
43. vulnerableness, VN
44. passive dependentness, D_P
48. interdependentness, ID
49. wholeness, W
50. integrationness, IG
51. hierarchically orderness, HO
52. flexibleness, F
53. homomorphismness, HM
54. isomorphismness, IM
55. automorphismness, AM
56. compactness, CO
57. centralness, CE
58. sizeness, SZ
59. complexness, CX
60. selective informationness, SI
61. size growthness, ZG
62. complexity growthness, XG
63. selective information growthness, TG
64. size degenerationness, ZD
65. complexity degenerationness, XD
66. selective information degenerationness,ID
67. stableness, SB
68. state steadiness, SS
SS

TABLE 2. (continued)

Direct System Characterizations*

Properties	
45. active dependentness, D_A	69. state determinationness, SD
46. independentness, I	70. equifinalness, EL
47. segregationness, SG	71. homeostasisness, HS
	72. stressness, SE
	73. strainness, SA

'SIGGS' draws attention to the development of the characterizations from set theory (S), information theory (I), graph theory (G), and general systems theory (GS). Set theory allows explication of a system as a group of components with connections between them marked off from its surroundings, the not-system or negasystem. Information theory allows explication of categorization of components and connections of a system and its negasystem and exchanges taking place between components. Graph theory allows explication of kinds of connections between components. Using set theory, information theory, and graph theory, therefore, permitted greater development in the characterizations of a system. Illustrations should make this greater development more apparent.

The basic system properties and negasystem properties developed from information theory are presented in Fig. 1. The set characterization complement gives meaning to the negasystem, $\not{S}$. With respect to the components and connections selected for consideration (within the universe of discourse, $\mathcal{U}$, selected), the components and connections which are not the system, $\overline{S}$. are $\not{S}$.

---

* Maccia and Maccia, p. 69. The number before the term indicates the order of presentation.

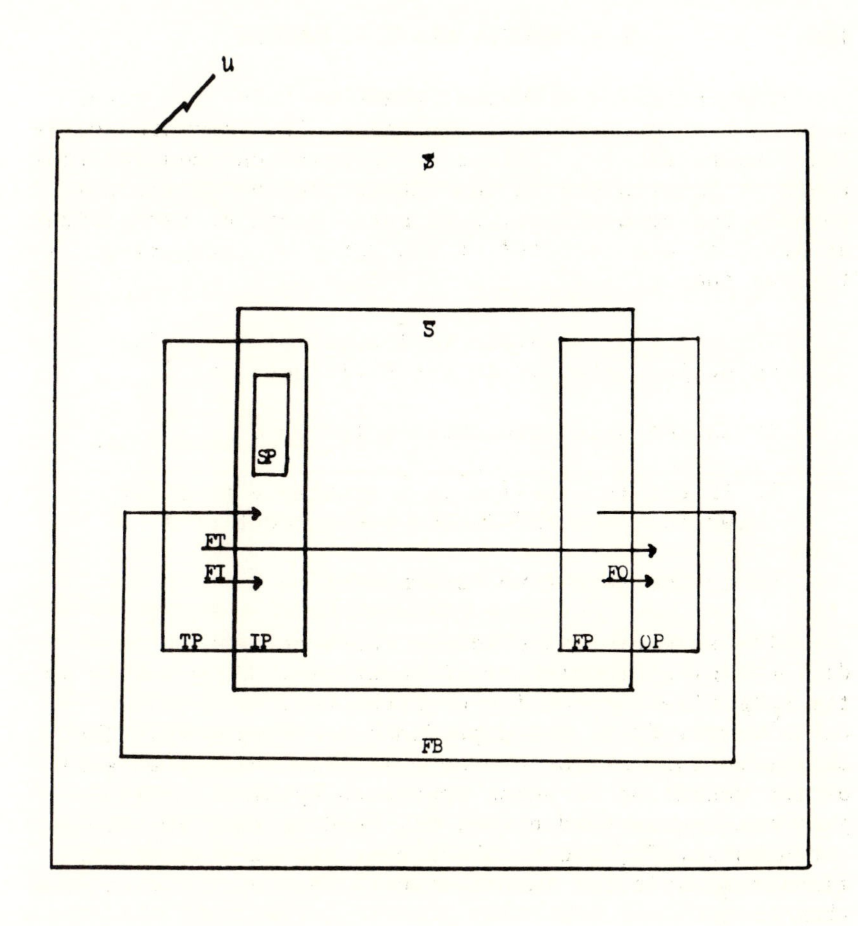

'U' - universe of discourse
'S̄' - system
'$̄' - negasystem
'SP' - storeputness
'FT' - feedthroughness
'FI' - feedinness
'TP' - toputness
'IP' - inputness
'FO' - feedoutness
'FP' - fromputness
'OP' - outputness
'FB' - feedbackness

Figure 1

The information characterization selective gives meaning to toputness, TP, inputness, IP, fromputness, FP, and outputness, OP. Selective information requires that there is uncertainty in the categorization of the components and connections, i.e. not all can be categorized as being in one category of the array of categories. It follows that

> TP is the information of $\overline{\not{S}}$ available to $\overline{S}$ for selection, i.e. TP is the environment of $\overline{S}$
>
> IP is the selective information on $\overline{S}$
>
> FP is the information on $\overline{S}$ available to $\overline{\not{S}}$ for selection, i.e. FP is the environment of $\overline{\not{S}}$
>
> OP is the selective information on $\overline{\not{S}}$

The selective information characterization conditional gives meaning to storeputness, SP, since SP is the selective information of $\overline{S}$ not available to $\overline{\not{S}}$. In other words, SP is the dependency of IP upon FP. The selective information characterization sharing as well as the condition of being separated by time intervals give meaning to feedinness, FI, feedoutness, FO, feedthroughness, FT, and feedbackness, FB, because a transmission of selected information is involved. It follows that

> FI, a transmission of selective information from $\overline{\not{S}}$ to S, is the shared information between TP and IP where TP is at a time prior to IP
>
> FO, a transmission of selective information from $\overline{S}$ to $\overline{\not{S}}$, is the shared information between FP and OP where FP is at a time prior to OP
>
> FT, a transmission of selective information from $\overline{\not{S}}$ through $\overline{S}$ to $\overline{\not{S}}$, is the shared information between TP, IP, FP, and OP where TP is prior to IP, IP to FP, and FP to OP
>
> FB, a transmission of selective information from $\overline{S}$

through $\overline{\$}$ to $\overline{S}$, is the shared information between FP, OP, TP, and IP where FP is prior to OP, OP to TP, and TP to IP

Although the seven system properties (22, 23, and 26 through 30, Table 2) and the two negasystem properties (24 and 25, Table 1) are not all of the information theoretic properties (19 and 21 from Table 1 and all of 18 through 37 as well as 60, 63, and 66 from Table 2 are also such properties), they do represent an expansion of the usual trio of input, output, and feedback. The greater development of system characterizations, thus, is patent. Nevertheless, a greater development through graph theory too will be indicated. This indication, moreover, will serve a second purpose, namely to reveal the treatment of a system's structure. A system has structural properties as well as activity ones, and information theory is not adequate to explicate the former.

Graph theory provides the perspective for setting forth structural characterizations of a system. An exemplification of the use of the perspective would be the devising of complete connectionness, CC, strongness, SR, unilateralness, U, weakness, WE, and disconnectedness, DC, from graph theory. The graph characterization, connection between every two components, gives meaning to CC, SR, U, and WE, since only a disconnected system does not have connections between every two components. Another graph characterization, directedness, also gives meaning to CC, SR, and U, for connections with directedness are channels between components and completely connected, strong, and unilateral systems have channels. Only completely connected and strong systems have two-way channels and so bi-directedness gives meaning to CC and SR. Finally, the graph characterization, direct, gives meaning to CC, because only completely connected systems have two-way channels that do not go through other components. To summarize

CC is two-way direct channels between every two components

SR is not CC, since one or more of the two-way channels between every two components is not direct

U is not CC or SR, since one or more of the channels between every two components is only one-way

WE is not CC or SR or U, since one or more of the connections between every two components is not a channel

DC is not CC or SR or U or WE, since one or more pairs of components do not have connections

See Fig. 2 for a diagrammatic summary. Again not all of the graph theoretic properties were presented; from the 20 in Table 2 (38 through 52, 56, 57, 59, 62, and 65) only 5 (38 through 42) were.

Now that the SIGGS Model has been discussed, attention can be shifted to the SIGGS Theory. The characterizations of a system in general, which as you recall constitute the Model, were interpreted in the light of an educational organization. For example, toputness became the factors external to the educational organization that place demands upon it. To state the matter differently, a system of concepts about educational systems was developed (Maccia and Maccia, pp. 126-133). Then relationships between these concepts were derived from the Model in the light of what is known about educational organizations. These proposed relationships constitute the SIGGS Theory. To illustrate

$$TP\uparrow \wedge FP\uparrow \Rightarrow FT\uparrow$$

is one proposition from the 201 making up the SIGGS Theory (Maccia and Maccia, pp. 138-167). Given the readoff as

If the toputness increases and the fromputness increases, then the feedthroughness increases

the proposition is as follows

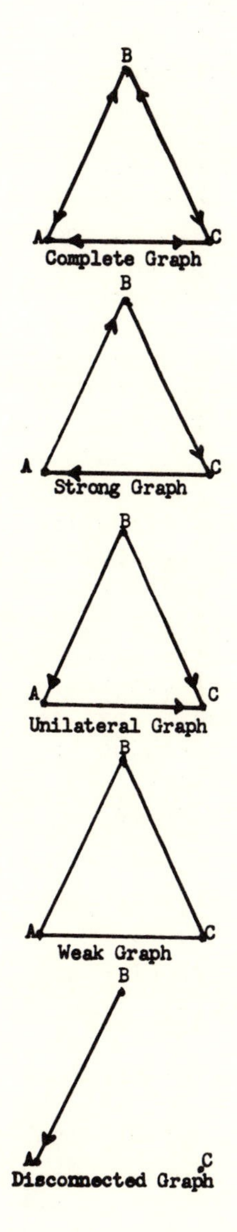

Figure 2. Types of graphs

> If the demands upon an educational organization increase and what it can offer (its supply) increases, then transferring of demands out of it increases.

One illustration will suffice, for it is not the purpose of this paper to set forth the entire SIGGS Theory. Rather the purpose is to show how general systems theory leads to an understanding of education as a social organization. Already discussion of the SIGGS Model and Theory has indicated how they lead to delineating and relating the activity and structural properties of an educational group with respect to its culture and structure. Given that the SIGGS Theory is a development of general systems theory, that understanding is being able to delineate properties (to characterize) and relate properties (relate characterizations), and that a social organization is a group with culture and structure, then the purpose has been attended to. Attention with a greater sense of accomplishment, however, would be forthcoming if one could use the SIGGS Theory to explain aspects of the Ocean Hill-Brownsville School District. Understanding should lead to explanation, because understanding provides relationships or regularities which make sense of our happenings. To explain is to appeal to regularities, i.e. to appeal to theory.

4. THE OCEAN HILL-BROWNSVILLE SCHOOL DISTRICT CASE

Although the New York City School Board responded to the Supreme Court decision, segregation continued. For instance, IS 201 in Harlem was, according to the Board of Education, to be integrated when it opened in September of 1967. It opened as a segregated school. Moreover, there was discontent with the schools, particularly with those in ghetto neighborhoods. Consequently, there were efforts by civic, educational, and community leaders to decentralize the New York City School System. One result of these efforts was the Ocean Hill-Brownsville School District.

The Ocean Hill-Brownsville section was selected

by the Board of Education as one of the ghetto sites for community run demonstration school districts. Another demonstration district was to center in IS 201 in Harlem, and yet another in the Two Bridges area between the Manhattan and Brooklyn Bridges. These demonstration districts were supported in part by Ford Foundation money. The Ocean Hill-Brownsville Demonstration School District was established in July of 1967 with $45,000 from the Ford Foundation and subsequently received more than $128,000 for development of its programs.

The structure of the District is depicted in Fig. 3. A governing board of 19 members was formed as follows: 7 parents elected by the community, 5 community leaders selected by the elected parents, 2 principals and 4 teachers chosen by their colleagues in the 8 district schools, and a professor of education who acted as a consultant to the Board before his election to it. In the proposal it was "assumed that the Board is indeed to govern; if it is to govern, then it will have the power to select and appoint personnel, initiate and approve programs, request budget appropriations, and make budget allocations" (McCoy, p.1). In spite of the fact that the Governing Board did appoint its unit administrator, Rhody McCoy, it became clear that the powers the Board took as requirements to govern resided with the Superintendent of Schools, the Administrator of the New York City Board of Education. The Unit Administrator and the Governing Board were taken as subordinate to the Superintendent of Schools. They were not free to govern, through their principals, the eight schools of the district, each of which averaged 1,100 students and 70 teachers.

Due to clarity about such a lack of freedom as evidenced in instances like the State Supreme Court's illegality ruling with respect to the Governing Board's selection, without regard for the Board of Education's civil service eligibility lists, of four principals, the Governing Board attempted to gain freedom by assuming it.

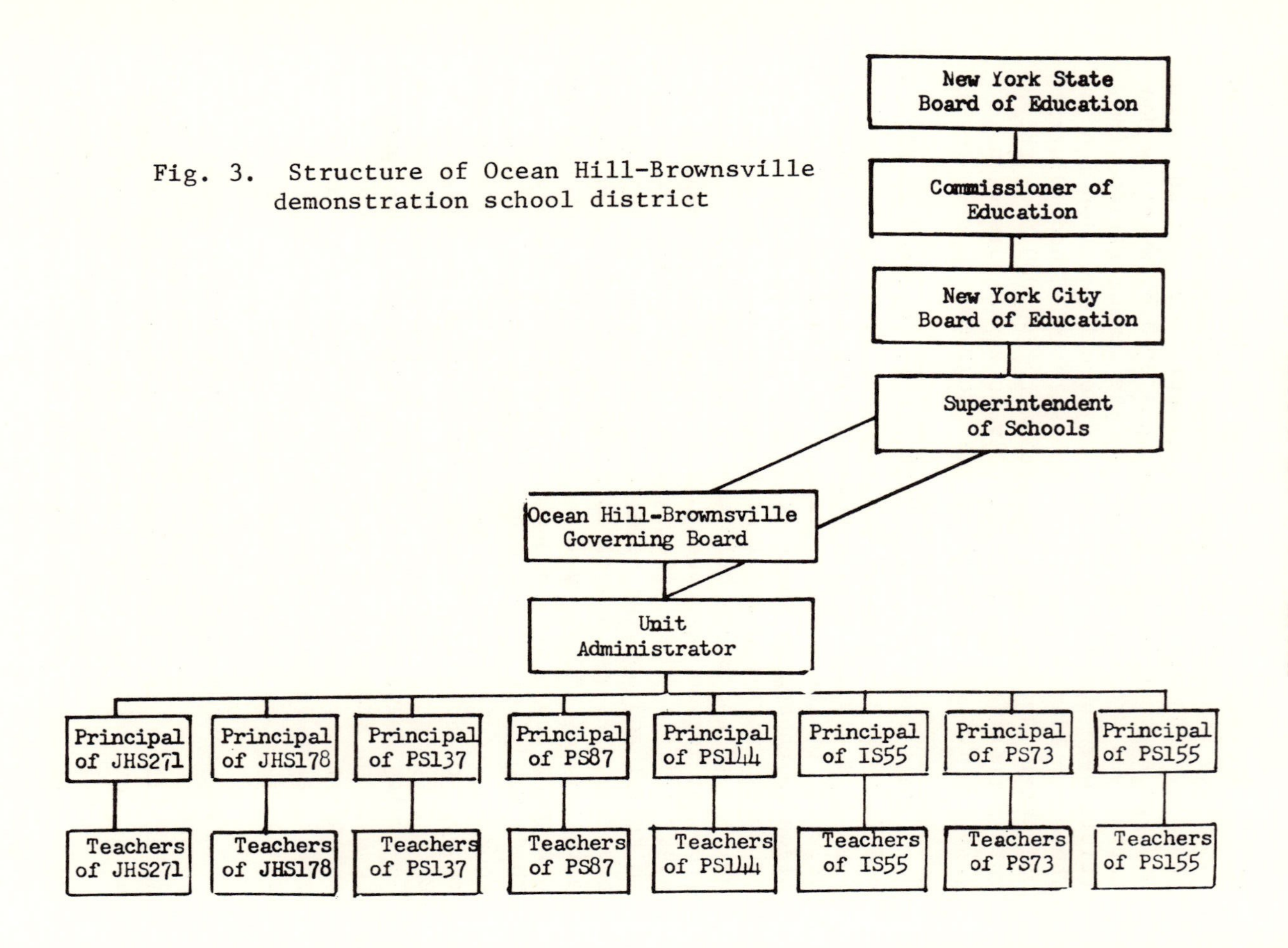

Fig. 3. Structure of Ocean Hill-Brownsville demonstration school district

On May 9, 1968 the Board through its unit administrator dismissed 1 principal, 5 assistant principals, and 13 teachers. 350 of the 556 teachers in the district walked out in support of the 19. This caused turmoil in the district, since there were only 200 teachers for 9,000 students.

The Superintendent insisted that the Unit Administrator bring formal charges against the 19 dismissed. Even though the charges brought against 10 out of the 19 who wanted to be reinstated (the principal, 5 assistant principals, and 2 teachers requested transfer and 1 teacher was reinstated) were declared inadequate by Former Judge Rivers, trial examiner, the Board would not reinstate the 10. Also the Board hired 350 teachers for September of 1968. These teachers were young and enthusiastic for involvement in the District. With the hiring of these teachers, 200 of the 350 teachers who walked out and did not transfer could not be reinstated. Adding this 200 to the earlier 10, now 210 were in need of reinstatement. The 210 finally reduced to 79. After 3 city-wide teachers' strikes called by the United Federation of Teachers, the 79 were reinstated. During these strikes the Ocean Hill-Brownsville School District continued their classes largely without incident, but as of December 1, 1968, there is turmoil due to suspension of the Governing Board, the four principals appointed by them, and three teachers and due to the reinstatement of teachers. The District, however, has been placed directly under the State Commissioner.

5. ASPECTS OF THE CASE EXPLAINED

Since the Ocean Hill-Brownsville School District was a result of efforts to reduce centralization in the New York City School System, the understanding of centralization should be a good beginning in explaining some of what happened. Centralization, or centralness, CE, as we call it, is a concentration of channels in a system. When a social organization such as the New York City School System has much centralization, there are parts of the system where there are relatively few channels and so, for instance, relatively few lines of communication. In the words of McCoy:

> In Ocean Hill-Brownsville, the need for decentralisation has a simple genesis; in turn, what decentralization offers in the fulfillment of a simple need. There are people here who feel themselves out of sight of other people, groping in the dark. The city takes no notice of them. In the midst of a crowd, at church, or in the market place, these people are about as obscure as they would be if locked somewhere in the cellar. It is not that they are censured or reproached, they are simply not seen - the invisible people. To be wholly overlooked and to know it is intolerable.
>
> The District 17 School Board is a typical example of the obscurity these people experience. Ocean Hill-Brownsville has not been represented on the Board since its inception two-and-a-half years ago. Nor has the community been recognized as having anything worthwhile to contribute to alleviating the problems that confront the schools here. (p.1)

Obviously centralization has its effects whether it is increased or decreased as in decentralization. The SIGGS Theory sets forth these effects. See Fig. 4. Knowing effects, of course, puts one in a position to explain as well as to control.

If centralness of an educational organization increases, then toputness of that organization decreases, i.e.

$$CE\uparrow \Rightarrow TP\downarrow$$

Not everything in the surroundings of an educational organization necessarily is attended to; not all of the surroundings of an educational organization is necessarily its environment. What will be the educational organization's environment is what can place demands upon it, that is toputness. Not much in the surroundings of obscure parts of the New York City System can be environ-

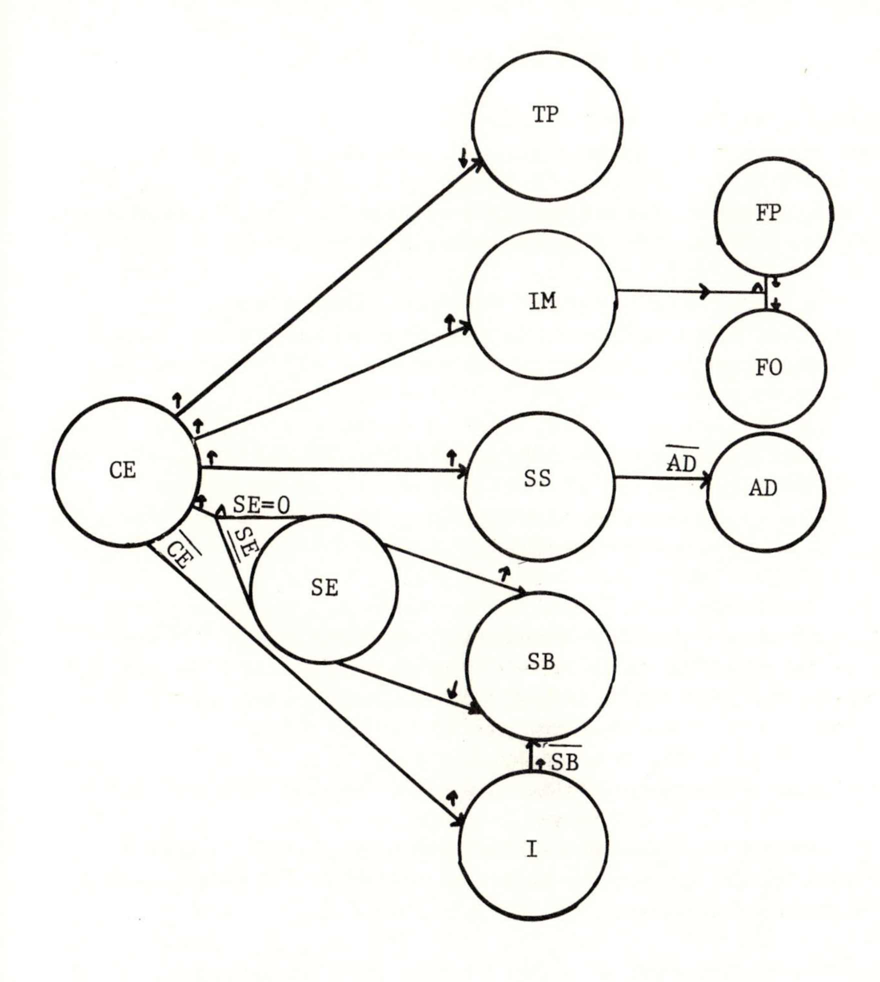

'CE' - centralization
'TP' - environmental demand on organization
'IM' - uniformity
'FP' - supply
'FO' - supply transmission
'SS' - stability under environmental change
'AD' - adaptability
'SE' - severe environmental change
'SB' - stability
'I' - independence
'↑' - increases
'↓' - decreases
'∧' - and
'_' - is greater than some value
'¯' - is less than some value
'⟹' - if ... then

Fig. 4. Effects of centralization

ment of that system, because such parts have few channels to make the needs of their surroundings demands. The schools in ghetto neighborhoods are glaring examples. Unless there is decentralization, ghetto people cannot look forward to their schools attending to their needs.

Uniformity is another outcome one can expect from centralization. If centralness of an educational organization increases, then isomorphismness of that organization increases, i.e.

$$CE\uparrow \Rightarrow IM\uparrow$$

When a city has a standardized curriculum, as evidenced by practices like city-wide reading lists, then there is isomorphismness in the curriculum. The structure of the curriculum in corresponding parts of the school system is the same. Centralness which reduces expression in parts of the system leads to uniformity. Mrs. Evelyn Stener, a teacher at one of the Ocean Hill-Brownsville schools, states it well:

> "This [new program known as 'Follow-Up Kindergarten]," she said, "is the sort of thing we can do in decentralized schools like the ones in the demonstration districts. Bring in innovations and get some challenges for the children.
>
> "The old system worked against such initiative," she asserted. "It was so rigid. Every time you started an idea like this, they made you stop it." (Farber)

Uniformity too has its effects, and thus centralization has yet other effects. For example, if isomorphismness of an educational organization increases, then fromputness and feedoutness of that organization decreases, i.e.

$$IM\uparrow \Rightarrow FP\downarrow \wedge FO\downarrow$$

Fromputness is the supply the educational organization

has for its surroundings, while feedoutness is that supply actually transmitted. Student achievement would be a kind of fromputness, while employment of students after graduation would be a kind of feedoutness. When the curriculum displays uniformity, then the needs of all students cannot be met and low performance results. Again, it is the ghetto schools of New York City that suffer.

> Even before the city's Board of Education disclosed the great extent of the disparity in pupil achievement between schools in disadvantaged areas and those in better communities, concerned slum parents were well aware that the majority of their children were severely retarded in reading and that relatively few would be graduated from high school, fewer still with an academic, or college-preparatory diploma. (Buder)

The New York State Board of Regents, a board superordinate to the State Board of Education, also recognized the need to increase educational achievement and that to do so its cause, increasing centralization, would have to be removed. Consider the following excerpt from their proposals for decentralization of the city school system:

> Achievement tests administered by both the state and the city, and other measures of progress, reveal that shockingly low performance levels still exist in a large number of the schools of the city. It is obvious that remedies more basic than any previously proposed must be sought.
>
> In seeking these remedies it cannot be too strongly emphasized that the central issue is how best to raise and maintain the level of educational achievement of all boys and girls, with special attention to those who, for whatever reason, are failing to acquire the most fundamental tools of elementary education .....
>
> Recognizing this as the only justification for change, careful consideration has been given by the

> department to many proposals, plans and points of view, leading to the increasing conviction that the most promising remedy for the ills of the city's school system lies in decentralization. ("Text of Regents' Proposals for Decentralization of City School System")

The lack of innovation in the New York City School System not only reflects uniformity but also stability even though the environment changes. The more stable a system is the less it changes. Increases in centralization brings about this system rigidity in the face of changes in the system's surroundings. If centralness in an educational organization increases, then state steadiness in that organization increases, i.e.

$$CE\uparrow \Rightarrow SS\uparrow$$

Up to a point this state steadiness does not adversely affect the adaptability of the educational organization. In other words, too much rigidity in the educational organization in the face of changes in the communities served by that organization conduces to substandard satisfaction of the demands of these communities. If state steadiness of an organization is greater than some value, then adaptiveness of that organization is less than some value, i.e.

$$\underline{SS} \Rightarrow \overline{AD}$$

The basis for the demand to decentralize the New York City Schools is just such over-stability which results in under-adaptation

> (1) The huge system, with 1.1 million pupils, a staff of 60,000 and 900 schools, has failed to respond to the expectations of the parents and to many national educational experts' demands for change.
>
> (2) The system's strong commitment to civil-service procedures, with special examinations and

> eligibility lists, designed at an earlier era to eliminate patronage and graft, today largely reinforces conservative personnel policies. It makes the selection process too standardized to feed into the system rapidly enough the kind of unconventional talent needed to cope with the urban slum schools.
>
> (3) These issues, though long a matter of concern, have become critical under the angry pressure by the Negro communities who see the present system's rigidity as a major reason for their children's failure. (Hechinger)

The angry pressure by the Negro communities which became lockouts and boycotts points to the crisis facing the schools of New York City.

> Yet the crisis is grave indeed. A system already grown rigid in its negative powers has been called upon to meet the unexpected challenge of an extraordinary immigration of impoverished citizens whose children have special needs for the very best our schools can offer, and the system has not effectively met this challenge. The new needs of large numbers of Negro and Puerto Rican students from low-income families may be the most dramatic, but they are not the only group which now needs better schools. (Bundy, et al.)

A crisis means that the stress on the schools is not 0 but rather is great enough so that increased centralization can only lead to a decrease in stability. The following relationship does not hold in New York: if centralness of an educational organization increases and stressness is equal to 0, then stableness of that organization increases, i.e.

$$CE\uparrow \wedge SE=0 \Rightarrow SB\uparrow$$

Rather the following one does: if centralness of an educational organization increases and stressness is greater than some value, then stableness decreases, i.e.

$$CS\uparrow \wedge \underline{SE} \Rightarrow SB\downarrow$$

The decrease in stability evidenced itself in the formation of the Ocean Hill-Brownsville Demonstration School District. Even though the City Board authorized the District, nevertheless due to pressures it did not initiate the required decentralization. As McCoy tells it:

> "I'd call and ask. They'd say we'll see. And then I'd get a shipment of new directives. Now what in hell do I need a sheaf of directives for. I want teachers. I want books. I want equipment. Most of all, I want good teachers. Not the subs they kept sending by the carload. But I got directives." He concedes that most times, "I'd just tear them up 'cause my office is too damn small to hold them all, and when Nate Brown called to ask if I'd gotten them, I'd say, 'Directives? No, can't say that I have.' So I'd get another pack." But as for what he really needed says McCoy, "They weren't answering." (Feretti, p.27)

The tearing up of the directives indicates a move toward independence. Independence differs from decentralization. In decentralization connections between all parts of the system are enhanced but in independence they are severed. The dismissal of 1 principal, 5 assistant principals, and 13 teachers by the Governing Board was indeed a declaration of independence. This independence was to increase, since decentralization was furthered by a reconstituted City Board and by state legislation. If centralness of an educational organization is less than some value, then independentness of that organization increases, i.e.

$$\overline{CE} \Rightarrow I\uparrow$$

With increases of independence comes substandard stability. If independentness of an educational organization increases, then stableness of that organization is less than some value, i.e.

$$I\uparrow \Rightarrow \overline{SB}$$

New York's children being locked out of their schools three times by the strikers clearly attests to not enough stability in their schools.

Obviously there is more to explain with respect to the Ocean Hill-Brownsville School District Case. Let the above suffice, for it does fulfill the purpose of the paper. The SIGGS Theory does explain the Case by viewing educational organizations as social ones and by doing so within the context of general systems theory enhanced by set theory, information theory, and graph theory. Moreover, it now should be patent that if schools are to serve their people too much or too little centralization is not wanted.

REFERENCES

Buder, L., "Demands for Local Control Grow," *New York Times* (February 12, 1968).

Bundy, M., et al., "Letter to Lindsay," *New York Times* (November 9, 1967).

Farber, M.A., "Classes Are Held in Two Demonstration Districts," *New York Times* July 11, 1968).

Ferretti, F., "Who's to Blame in the School Strike?" *New York Times* 22-35 (November 18, 1968).

Hechinger, F.M., "Who Will Run Schools?" *New York Times* (June 2, 1968).

Kroeber, A.L., and Parsons, T., "The Concepts of Culture and of Social System," *American Sociological Review, 23* (1958).

Maccia, E.S., and Maccia, G.S., *Development of Educational Theory Derived from Three Educational Theory Models*, Final Report, Project No. 5-0638, Washington, D.C., U.S. Department of Health, Education and Welfare, Office of

Education (December 1966).

McCoy, R., *A Plan for an Experimental School District: Ocean Hill-Brownsville*, 3 pp. mimeographed.

Parsons, T., *Structure and Process in Modern Society*, Glencoe, Ill: The Free Press, Glencoe (1960).

"Text of Regents' Proposals for Decentralization of City School System," *New York Times* (March 30, 1968).

NOTES ON THE EDUCATION COMPLEX AS AN EMERGING MACRO-SYSTEM**

Michael Marien*

ABSTRACT

In a dynamic, knowledge-based era, new conceptual approaches are required to supplement the traditional focus on single organizations confined in spatial boundaries. Growing social complexity also necessitates transdisciplinary program overviews to aid public policy formulation. "The Education Complex" focuses on the interaction of four components (core educating systems, peripheral programs, selected suppliers, and organized beneficiaries), and places special emphasis on system relationships to informal education and international education. The initial task of entitation entails a consideration of functional, spatial, input and beneficiary boundaries. The resultant entity of the education complex is justified by exploring a fifth boundary - time - in which evidence is provided to support the view of an emerging macro-system. The suggested

---

* Research Associate at the Educational Policy Research Center, Syracuse University Research Corporation, 1206 Harrison Street, Syracuse, New York, 13210.

** This is a discussion paper, reproduced at the discretion of the author for private distribution and without benefit of formal review by the Educational Policy Research Center at Syracuse. The views contained herein are those of the author and are not necessarily endorsed by the Center, its staff, or its contracting agencies.

realities of emerging and overlapping macro-systems require certain adaptations for general systems theory: accepting a world of blurring boundaries and lagging concepts, and moving from a focus on isomorphic traits to one involving a systems spectrum. These adaptations will hopefully result in greater attention to understanding the systems most important to shaping the future human condition.

1. INTRODUCTION

The mission of general systems theory can be viewed with two contrasting concepts. The presently accepted statement of ultimate purpose is that of investigating the isomorphy of various concepts, laws, and models as a means of establishing a unity in science [1]. The contrasting view of a general systems "spectrum" suggests a focus on similarities *and* differences, and a continuum between the hard and soft sciences. The two views are not necessarily opposed, but may represent different stages in the growth of this interdisciplinary knowledge system. An isomorphic focus is necessary to establish the field by drawing together the parallel efforts of many scientific disciplines. Once achieved to some degree, the further sophistication of general systems inquiry may lead to a "spectrum" focus, and a knowledge system open to greater variation.

The purpose of this exploratory paper is to raise some fundamental and hopefully provocative questions. The basic argument is that there is indeed a spectrum between closed and open systems, and extremely open systems tend to be man-made, "unsystematic " [2],loose, only partially knowable, and subject to irreversible changes or high plasticity. To support this view, a tentative outline of "the education complex" is sketched to illustrate a system so open that serious attention must be devoted to various boundaries. This exercise in entitation [3] is supported with evidence that there are emerging properties of "systemness" and that the suggested array of components is not merely an arbitrary pattern-system but an action-system in formation [4].

2. AN OUTLINE OF THE EDUCATIONAL COMPLEX

The education complex includes all organizations and parts of organizations involved with the provision of formal instructional services that purportedly enhance the learning processes of students [5]. It includes four major components: the core educating institutions, the substitute and supplementary activities in "the periphery," the organizations and parts of organizations involved in supplying inputs to educating institutions, and the various interest groups especially benefiting from institutional outputs. In addition to including a peripheral component that is usually hidden from analysis, the process of entitation requires four boundary distinctions that carve out functional, spatial, input, and beneficiary lines. A fifth boundary (Part 3 of this paper), views an emerging system over the time period very roughly between 1950 and 1975. The system interacts with various aspects of its environment, noted in Chart 1 as family, culture, polity, economy, and community.

The Core Educating System

The leading sub-system of the education complex is the "core" of about 115,000 elementary and secondary schools, and 2300 colleges and universities. Since well before World War II, this grouping has been casually referred to as "the American educational system," although it has never been subjected to an analysis as such. The popular notion that learning takes place in, or only as a result of, schools and colleges (often reinforced by "educators" who seek an exclusiveness for their services), is, of course, highly misleading.

Discovering Hidden Components: The Periphery

There are considerable numbers of students in other programs entailing formal instruction. The programs are called adult education, continuing education, vocational training, management training, remedial training, re-

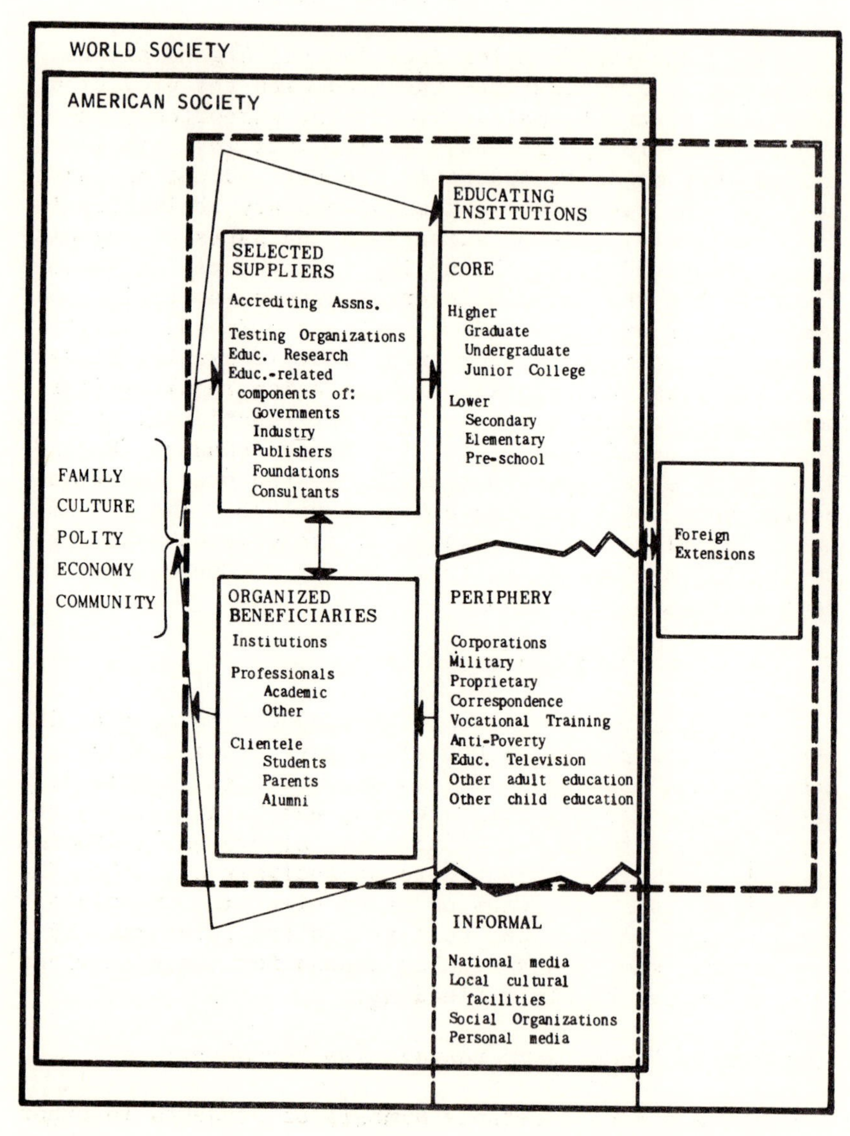

Chart 1. Overview of boundaries and components of the education complex

training, or youth activities [6].

In addition to a core enrollment of 53.6 million in 1965, there were 44.2 million students involved with formal instruction in the periphery. This included 14.5 million in organization training (business, government, and military; on-the-job training and formal classes sponsored by the organization itself or another institution); 7.8 million in proprietary or trade schools; 2.8 million in anti-poverty programs (including programs formerly in the Office of Economic Opportunity and programs under the Manpower Development Training Act); 5.0 million in correspondence schools; 5.0 million taking formal courses via television; and 9.1 million in other adult education (including programs conducted by core institutions, libraries, churches, Red Cross, Great Books, community centers, etc.) [7]. Because of obvious problems of duplication, estimates of other child education are not included in the above totals, although Scouts, Little Leagues, dancing classes, summer camps, etc. serve to supplement the educational services provided in core schools.

Beyond the Functional Boundary: Informal Education

The first boundary to be delineated between system and environment is the critical one of function. There are countless activities that are educational and that enhance learning processes. By the definition used here, formal education involves a teacher and students in a situation involving some reward beyond the inherent one of learning for its own sake, i.e.: diplomas, certificates, job access, promotion, licenses, merit badges, etc. Informal education is the residual category, involving national media (television and radio broadcasting, print, film, recordings), local cultural facilities (theaters, museums, zoos, planetariums, fairgrounds, historical sites, libraries), social institutions (family, church, the political process, clubs and associations, peer groups), and personal media (telephone, personal tape recordings, mail, etc.). Any full consideration of education must consider the content and forms of learning that takes place beyond classrooms.

Beyond the Spatial Boundary: International Education

Every system has a spatial boundary. In the present instance a national focus is involved, although the concept of an education complex could be applied on a world, regional, state or local level. The foreign programs of American universities, as well as foreign students and faculty in American institutions, are considered here as part of the American education complex. American students and faculty engaged formally in foreign institutions or informally through travel abroad are considered as outside of the complex. The knowledge inputs gained outside of national boundaries, however, are an important consideration in assessing American education.

Drawing an Input Boundary: Suppliers in the System

Although many groups are involved in supplying inputs to core institutions [8] (including core institutions, which train their own personnel), there are important groups that can be considered as part of the educational complex by virtue of full involvement in education-related matters, and easy interchange of personnel [9]. Through the provision of funds, policies, equipment, and/or knowledge, these groups exert a considerable influence on instructional services. There are a few organizations involved solely in education, such as the National Commission on Accreditation and the Educational Testing Service. But in many cases, education-related activities are only part of the activity mix of organizations such as governments, corporations, foundations, consulting firms, and publishers. This presents a dilemma in drawing system boundaries. Excluding all supplying organizations would present a narrow "world" of education, and draw the focus of attention away from some very important considerations. Admittedly, certain organizations such as government departments of education will always be included somewhere in any systems model. But other suppliers such as testing organizations are too often overlooked simply because system boundaries are not extended far enough to include them. On the other hand, to include all organizations that have anything whatsoever to do with supplying educating institutions, would ultimately involve the whole society. A balanced approach is taken here by drawing an arbitrary line through certain organizations. Thus, in the federal government, the U.S. Office of Education, House Education and Labor Committee, and military schools are considered as inside the system, while other components of the Department of Health, Education, and Welfare, the U.S. House of Representatives, and the Department of Defense are considered as outside of the system.

Drawing a Beneficiary Boundary: Organized Beneficiary Groups in the System

Suppliers are obvious beneficiaries of educating institutions, but there are many other groups that benefit in many ways from the outputs of educating institutions. Similar to the dilemma of locating suppliers within the system, the inclusion of all beneficiaries would again ultimately include the whole society. Nevertheless, there are certain groups that are explicitly organized to further the interests of educating institutions or to further group interests within educating institutions. Thus, there are groups of institutions (American Council on Education, National School Boards Association), professionals (National Education Association, American Association of University Professors, Modern Language Association, American Economic Association, Society for General Systems Research), clients [10] (National Student Association, National Congress of Parents and Teachers, various alumni clubs), recruitment groups (organization such as the Young Republicans and Future Farmers of America that serve to socialize the young into various realms of adult activity), and retention groups (organizations such as Hillel and various foreign student associations that serve to retain and promote certain values and identities).

The system components suggested here represent many necessary simplifications. Even with considerable

OUTSIDE EDUCATION COMPLEX (ENVIRONMENT)	INSIDE EDUCATION COMPLEX (SYSTEM)	CORE EDUCATING INSTITUTIONS — INPUTS: STUDENTS	FUNDS	GUIDANCE	PERSONNEL	PHYSICAL RESOURCES	KNOWLEDGE
CORE							
	LOWER (ELEMENTARY AND SECONDARY)	GRADUATES					
	HIGHER			ENTRY REQUIREMENTS	TEACHER TRAINING		CURRICULA TEACHER KNOWLEDGE
OTHER							
	PERIPHERY						STUDENT KNOWLEDGE
INFORMAL							STUDENT AND TEACHER KNOWLEDGE
INTERNATIONAL		FOREIGN STUDENTS RETURNING U.S. STUDENTS			FACULTY TRAINING		STUDENT AND TEACHER KNOWLEDGE PRINT
SELECTED SUPPLIERS							
GOVERNMENTS	EDUCATION DEPARTMENTS	STUDENT AID	APPROPRIATIONS	LAWS STANDARDS	STANDARDS		CURRICULA
	ACCREDITING ASSOCIATIONS			STANDARDS	STANDARDS		
INDUSTRY	EDUCATIONAL CORPORATIONS	SCHOLARSHIPS	GIFTS	IDEAS		EQUIPMENT	CURRICULA
PUBLISHERS	EDUCATIONAL PUBLISHERS						PRINT
INTEREST GROUPS		SCHOLARSHIPS	GIFTS	VALUES			PRINT AND BELIEFS
FOUNDATIONS	EDUCATIONAL FOUNDATIONS	SCHOLARSHIPS	GRANTS	PILOT PROGRAMS			
RESEARCH AND CONSULTATION	EDUCATIONAL RESEARCH			IDEAS			
CONSTRUCTION						BUILDINGS	
	TESTING ORGANIZATIONS	TESTING		SELECTION CRITERIA			
BENEFICIARY GROUPS							
	INSTITUTIONS			NEEDS			
	PROFESSIONALS			NEEDS	INFORMAL TRAINING; REWARDS		PRINT
	CLIENTELE		TUITION	NEEDS			
	ALUMNI		GIFTS	VALUES			
EMPLOYERS		TUITION AID		NEEDS			
ENVIRONMENT							
FAMILY		CHILDREN ACCESS MOBILITY	TUITION	VALUES	PARENT VOLUNTEERS		STUDENT KNOWLEDGE
CULTURE		YOUTH CULTURE		VALUES			BELIEFS
POLICY		VETERANS AND DRAFTEES TRAINEES	TAXES	GENERAL LAW			
ECONOMY		TRAINEES	WEALTH; INVESTMENT INCOME	LABOR FORCE REQUIREMENTS			
COMMUNITY		RESIDENTS	TAXES	VALUES	RESIDENTS	LAND; SUPPLIES	BELIEFS

Chart 2. Suppliers to the core and their inputs

refinement in "system mapping" it would be dangerous to utilize only one particular configuration. There are many speculative juxtapositions that can profitably describe the elements of the education complex. Education can also be viewed as part of other macro-systems, such as the communications complex, the science-government-education complex, the knowledge complex [11], the health complex, or the pervasive military-industrial complex.

A listing of components, however, does not constitute an action system (although any listing can be subjected to a systematic analysis). It is necessary to show how the components are interrelated, and to suggest how a particular configuration is more meaningful - or as meaningful - as any other grouping of entities. To facilitate this task, it is helpful to look at the components over a time period involving the recent past and the near future.

3. THE EDUCATION COMPLEX AS AN EMERGING MACRO-SYSTEM

More Systemness in "The System"

The elements of the core educating system are increasingly linking together in both vertical and horizontal dimensions. In the vertical or hierarchical sense, the growth of graduate school training has increasingly affected undergraduate work so that the primary purpose of "the university college" has become the preparation of students for graduate work of some kind [12]. At lower levels, the increased importance of high school graduates attending colleges has pushed the high school increasingly in the direction of college preparation activities. To insure an adequate and relevant preparation for college, liberal arts academicians have increasingly entered the field of preparing secondary school curricula, a task previously left to high school teachers and education professors. At still lower levels, the consolidation of school districts has enabled the coordination of elementary and secondary curricula.

On the horizontal level, there is more homogeneity of educating units, whether or not accompanied by formal linkages. One reason is that public institutions have enrolled an increasing proportion of all students. Thus, state university systems are rapidly growing, not only by building new campuses, but by absorbing private universities that have been unable to meet rising operating expenses. Even if colleges and universities remain private, the traditional distinctions between social class, religion, race, and sex are becoming considerably less sharp. At lower levels, enrollment in the Catholic schools (accounting for the major part of private enrollment) has declined in the past four years from 5.6 to 5.0 million, with the threat of further decline in the face of rising expenses [13].

At higher levels, formal and informal cooperative arrangements have been increasing rapidly in recent years. A 1965 study counted 1,017 consortiums among 1,509 responding institutions. One-third of these arrangements were multilateral [14]. At lower levels, a dramatic consolidation of school districts and school buildings continues to take place. In 1931, there were 128,000 public school districts controlling 238,000 elementary schools and 24,000 high schools. Today there are about 20,000 operating school districts controlling about 70,000 elementary schools and 27,000 high schools [15]. A few cities are already planning to consolidate local school system plants into one or several education parks.

Growing Significance of the Periphery

Despite the rise in core enrollment from 31.3 million in 1950 to 60 million in 1970, enrollment growth in the less visible realms of formal education has probably grown at an even greater rate. Preliminary data suggests that the ratio of periphery to core on a head count basis has grown from 0.71 in 1950, to 0.82 in 1965 and 1.02 in 1970. By 1975, it is estimated that there will be 1.32 learners in the periphery for every learner in the core [16]. Together this "learning force" (of core and periphery) will be substantially greater than the labor force, rising from a ratio of 1.02 in 1960 to a ratio of 1.59 by 1974 [17].

There is a high degree of overlap and consequent blurring as to what de-marks the core from the periphery. Equivalents of many vocational courses in high school are taught elsewhere (anti-poverty programs, proprietary or trade schools, industrial training) without legitimation of the high school diploma. Many vocational courses in universities and especially in junior colleges have little resemblance to the liberal self-development that one traditionally presumes of higher education. High schools and universities are increasingly involved in adult education, which may or may not carry degree credit. Universities carry certain programs of other organizations (most notably ROTC, which often carries college credit); and other organizations such as Bell Laboratories and the U.S. Army conduct programs that are as sophisticated as those offered by any graduate school.

Functional Blur in the Electric Society

As opportunities for non-structured learning continue to proliferate, there is not only a blur between core and periphery, but a growing blur between formal and informal education.

The most obvious instance of informal learning is television. Less than 10% of all households owned television in 1950, but by 1967, television was in 94% of all households, and 25% of all households owned two or more sets [18]. In the same period, the number of stations grew from 107 to 608, and this channel capacity will be vastly expanded by cable television (CATV). Culkin estimates that the average child spends some 3-4000 hours in front of a television set before entering the first grade: "By the time he graduates from high school he has clocked 15,000 hours of TV time and 10,800 of school time" [19]. The form of involvement is admittedly unsystematic and the content is largely superficial and innocuous. Yet, for better or worse, there has undoubtedly been a considerable amount of learning as a result of this time spent. Indeed, Gans argues that because the mass communications system is considerably more adaptive than the school system, it is probably a better teacher: "...the conception of the teacher as a professional who has a monopoly of knowledge about education, which emerged when the students came from immigrant, rural and frequently illiterate homes, is no longer applicable in an era when the mass media have informed both children and parents" [20]. At the university level, a similar point is made by Schrag: "When the university and its scholars lost their monopoly as disseminators of news and ideas - as purveyors of information - the halo began to tarnish. What printing and the Bible did to the church, mass media are doing to the university" [21].

Television has already made a major impact on society, but there are new forms of communication that portend even greater implications for the core educating system. For example, electronic video recordings (EVR) will be commercially available by 1971, enabling individuals to rent, borrow, or buy cassettes that will play through the home television set.

As a response to the increase of informal learning and the need for credentials, the Educational Testing Service has recently instituted the College Level Examination Program (CLEP), which, in cooperation with local universities, enables the informally educated to acquire college credit by passing an exam.

While opportunities for informal learning have greatly increased, direct educating activities within universities are accounting for a declining proportion of the sum of all university activity. In 1930, the ratio of expenditures for "Organized Research" to those for "Instruction and Departmental Research" was 0.08 for all institutions of higher education. In 1950, the ratio was 0.29 but by 1964 it was 0.70 and is doubtlessly climbing rapidly [22].

Whether or not one wishes to recognize it, there is

indeed a growing functional blur between formal and informal education, and the availability of mechanisms (such as CLEP) for legitimating informal learning can only serve to further this blur. It is therefore imperative that any consideration of the future of the education complex must also consider informal opportunities outside of the complex.

Spatial Blur in the World Society

Communications satellites, ocean-hopping jet airplanes, globe-spanning conglomerate enterprises, and the psychological implications of travel in the space beyond our global system have all aided the formation of a world society, or, from McLuhan's electronic perspective, a global village. Two well-known anthropologists predicted a decade ago that "It seems merely a matter of time before all of the cultural systems of the world will be different variations...of a single culture type." The emerging world culture is not necessarily stereotyped and malign, for it was pointed out that "increasing homogeneity of culture as the diversity of culture types is reduced (leads to) an increasing heterogeneity of the higher cultural type" [23].

Foreign students in U.S. institutions of higher education totaled 1.3% of enrollments in 1950, and 1.8% in 1968; however, at the graduate level, they totaled 6.0% of 1963 enrollments [24]. But the data do little justice in describing the dramatic internationalization of curricula, service activities, students, and faculty at the major universities [25]. According to one proponent of international education, the impact on American education - especially at the higher levels - will be pervasive by 1980: "No discipline will be unaffected, no student will be untouched" [26].

In 1968, only 21,500 American students were enrolled in foreign institutions (contrasted to 122,000 foreign students in American institutions), but informal education via travel has been increasing at a rapid rate, and lower carrier costs after the introduction of "the jumbo jet" should serve to further augment the flow of students. Even the high school students of our affluent society are also transcending national boundaries. In the summer of 1969, "there will probably be close to 30,000 students under eighteen years of age going 'abroad' to make a somewhat less grand tour, but one that nevertheless still proclaims education through exposure as a primary value" [27].

With a growing interpenetration of societies and educational systems, it is not surprising to find a similarity in the stresses faced by all national systems. In his highly competent international survey, Coombs found a "world educational crisis" resulting from disparities between system and environment in all nations, and increasingly aggravated by rising clientele aspirations, acute scarcity of resources, and the inertia of educational systems and their societal environments [28]. Despite the many failings of the American system, it may nevertheless be more open and subject to adaptation than any other national system. Already a leader in world education and especially in world science, American education may by default become even more so.

Emergence, Nationalization, and Agglomeration of Suppliers

Despite the wide diversity of supplying organizations, there is one consistent theme that all have in common: the recent and rapid emergence of national organizations which are involved with education yet far removed from the classroom. In many cases, such as the federal government and business corporations, there has been an increasing involvement of existing organizations.

In other instances, there has been a stimulation of new organizations by business, foundations, and government. For example, General Electric and Time, Inc. formed the General Learning Corporation in 1965, and the Ford Foundation created Educational Facilities Laboratories in 1958. The federal government, largely under the authority of the Elementary and Secondary Education Act of 1965, has created 20 regional educational laboratories, 12 research and development centers, 2 policy research centers, 46 research coordinating units for vocational education, 14 instructional materials centers for handicapped children, and 19 ERIC (Educational Resources Information Center) Clearinghouses. These organizations form the research, development, and dissemination component of the education complex, but would be neglected from consideration if one were to only look at core educating institutions.

National testing organizations first emerged after World War II. The Educational Testing Service was started in 1947, the National Merit Scholar Program in 1955, and the American College Testing Program in 1959. Today, all three enterprises are firmly established in their role of facilitating articulation between high school and college. Indeed, preparing for the College Board Admissions Tests administered by ETS may well be the most important goal of college-bound students. As pressures on graduate schools increase due to more potential students and more applications per student, a centralized admissions service may soon be established, parallelling the universal use at the present time of the Graduate Record Examination for admissions, and centralized language examinations for fulfilling Ph.D. requirements [29].

Professional and regional accrediting associations emerged after the turn of the century as a means of private governance. The National Commission on Accrediting was founded in 1949, and has steadily expanded its influence in coordinating the 6 regional and 25 professional accrediting agencies.

The recent and rapid trend toward agglomeration and conglomeration of American business enterprises (which, incidentally, has been largely if not completely ignored by economic, social, political and systems theory), has also had an effect on the provision of goods and services to the core educating system. Many hardware-developing firms, such as RCA and Xerox, have bought up publishing firms to provide a software component. The development of a "total systems capacity" has raised possibilities of sub-contracting the operation of entire school systems to commercial operators, as has already been the case for certain Job Corps centers. Expansion has also taken place from a publishing base; for example, Crowell Collier and MacMillan supplies all levels and functional areas of the education core, operates schools in the periphery, and supplies informal and international education.

A final mode of expansion is for isolated units to gradually form a national network. Libraries are increasingly involved in linking up to computerized data banks, and will very soon be facing the implications of facsimile text transmission, which, in turn, will be considerably enhanced by widespread utilization of the Picture-Phone (presently inhibited by high unit costs). "Regional, national, and special purpose libraries will become increasingly important" [30]. For example, the American Institute for Physics is forming a National Information System for Physics, while art museums are being linked up by the Museum Computer Network, which will provide a central archive of United States art resources. "In about ten years the network expects to be ready to store, retrieve and distribute information over a system of terminals placed throughout American museums, libraries and other educational institutions. The next step will be to dovetail data on works in European museums and in private U.S. collections into the network" [31].

The commercial components of knowledge dissemination have also been linking together. The National Association of College Stores was founded in 1923, but only had 125 member stores in 1945. By 1959 there were 923 member stores, and by 1968 membership totaled 1,739 - very nearly covering the appropriate universe. In 1967, NACSCORP, a wholly owned subsidiary of NACS, was formed as a paperback book distributing center, enabling member stores to order the outputs of 70 publishing houses from one single source [32].

Viewed together, these supplying organizations lend an appreciation to the concept of the education complex, and the value of extending system parameters beyond core institutions in order to understand what is happening in education.

The Spreading Web of Power

The non-hierarchical arrangement of components in the education complex is intentional. It does not imply that governments do not exercise certain powers over educating systems, for this is obvious; just as it is obvious that, as the United States becomes more of a national society, a greater proportion of government power is derived from the federal level. For those who look only at the lines of formal or legal power, the spectre of centralization inevitably will be found. But if the shifting configuration of all influences is fully examined, the notion of centralization appears as a gross simplification.

The web of power is simultaneously moving out and down from traditional centers. In moving "out," schools and colleges are more open to influence from their environments. Instead of power moving up a single hierarchical line, it is moving out along several lines involving various government levels and agencies, corporations, publishers, testing organizations, accrediters, foundations, and research units. Conceptually, it has become necessary to extend the parameters of the system to include certain salient organizations, and system boundaries may have to be further extended in the future to include new components.

In moving "down," power in educating institutions is dispersing, with faculty and students attaining a greater proportion of influence in the shaping of institutional directions. Black studies departments have not been initiated by law, budgeted funds, a book, a research report, or a faculty committee - but by highly visible and often violent pressures from students.

Thus, it is important to look at trends in the fourth component of the education complex: the organized beneficiary groups [33]. Of these groups, it is most important to look at trends in professional organizations and student power.

Modern professional associations are essentially guilds, forming "hidden hierarchies" of governance. Gilb states that "the twentieth century has seen a revolution in the dispensing of professional services in America comparable to the industrial revolution in the nineteenth century" [34]. The professions today include not only the traditional ones such as law and medicine, but every academic discipline. The trend is toward group practice and increasingly uniform national organizations with control over professional education through a variety of devices, systematic job placement mechanisms, and various services such as group insurance and travel. Uniformity in state regulations and the general trend toward secularization have furthered nationalizing tendencies among the professions.

In institutions of higher education, faculty power has been well established, especially in the major institutions [35]. Jencks and Riesman note that "large numbers of Ph.D.'s now regard themselves as independent professionals like doctors and lawyers, responsible primarily to themselves and their colleagues rather than to their employers" [36].

Unlike the many professional associations involved with higher education, teachers at lower levels join the National Education Association or, especially in urban areas, the more militant American Federation of Teachers. Faculty power at lower levels is considerably less than at higher levels, but is nevertheless beginning to grow, due to an increasing proportion of teachers joining the NEA or AFT, and an increasing number of unifying strikes called by both organizations. Between June 1967 and June 1968, there were 114 strikes, which together with the 34 strikes during 1966-67, constituted a total greater than that of the previous 26 years. Moreover, the latitude of conflict is widening from individual school and school district levels to statewide levels, with one statewide strike in 1964, one in 1966, and three in 1968. The magnitude of strikes in man-days lost is also growing [37].

In the past few years - and especially within the past year - student power on college and university campuses has increased significantly. Experimental colleges and curricula have been established (with or without academic credit), representation has been gained in academic senates and on boards of trustees, the various "in loco parentis" rules are being removed, courses and faculty have been formally evaluated, pass-fail grading and Ombudsman mechanisms have been established, and a greater number of black students have been admitted. In turn, black students have pressed demands for more relevant college experiences, and accommodation has been made with black studies programs and social facilities. At the time of this writing, it appears that 1969 will be noted for university uprisings, as 1967 was noted for ghetto rebellions. But unlike the more complex problems of black power in cities, the quest for student power on the campus may be more readily satisfied.

Students are also stirring at high school levels - and it has even been reported that a group of fifth and sixth graders presented "a list of 13 black demands" to the principal of a Berkeley elementary school [38]. A national network of perhaps 1,000 underground high school papers now exists, in addition to a proliferation of independent high school unions and high school chapters of Students for a Democratic Society [39].

Criticism at lower levels is against dehumanizing institutional conditions which inhibit curiosity, learning, and human dignity, whereas at higher levels the revolt is also against university linkages with the military-industrial complex. But whatever the criticism, the significance for the future of the education complex is that students are questioning classroom authority and asking serious questions about their education, whereas in the past there was only passive acceptance. This questioning by clientele may be a blessing (for, after all, students are supposed to ask questions), or it may be a Pandora's Box; in any event, now that it has started, it is hardly conceivable that this trend will reverse itself. Unless there is a widespread movement of extreme repression, students will increasingly participate in the web of power that determines the structure and activities of the education complex.

4. IMPLICATIONS FOR GENERAL SYSTEMS THEORY

This paper has briefly outlined the components of a macro-system and identified some salient trends resulting in emerging systemness. Because of formal linking and growing openness of lesser entities, it is suggested that these entities become increasingly unacceptable as units of productive analysis, and that some configuration approximating an "education complex" becomes increasingly desirable as a means of viewing American education as a system. Yet, the boundaries of this complex will always be imprecise because of functional and spatial blurring. As the system changes, the concept or model by which the system is described must also be changed if the investigator desires to accurately describe the present and to provide a realistic systems foundation for conjecturing about the future. Because of their fundamental importance, these problems are discussed below in somewhat greater detail.

Blurring Boundaries in an Age of Complexity

Although seldom analyzed in depth, one often hears that we are living with increased complexity, that systems are becoming more open and therefore characterized by greater interface or interpenetration [40]. It is uncomfortable to the cause of rigorous analysis to discover an erosion of meaning in well-established categories. Yet, the blurring continues at a rapid pace.

In the wider society, there is a blurring in sex

roles and (at least at the upper strata) in social class. Good and evil are no longer clear, as are distinctions between Creator and creator, and between Communism and Capitalism. With growing professionalism, distinctions between work and leisure become blurred. The growth of government sub-contracting, regulation, and funding blurs the distinction between public and private. In a dynamic era that thinks young, and where age may not lead to but inhibit wisdom, there is a growing blur between young and old (while in other respects the distinction is sharpening). The transplanting of body organs blur the boundaries of the individual. The trend from clear possession of material things to possession by credit, renting, or leasing, raises questions of ownership.

Within the core of educating institutions, the growing influence of graduate school on undergraduate education mixes realms which had once been isolated from each other. Boundaries between college and high school are blurred by granting college credit in high school, boundaries within the high school are blurred by flexible scheduling, and boundaries within grade schools are blurred by flexible classroom space and non-graded schools. The growing openness of universities to their surrounding communities leaves the closed image of the Ivory Tower as an anachronism. Indeed, one proposal for the urban university of the future is to create an institution in which there are no walls and no outsiders, and where the community itself is viewed as the campus [41]. The sharing of computer services and centralized libraries serves to blur local boundaries. The academic disciplines continue to expand and overlap with each other. And, as books and electronic media become more available to all, the distinction between formal and informal education becomes less clear.

This general trend to blurring between previously distinct categories raises serious problems for systems theorists, as well as for individuals who feel more comfortable with distinct and unchanging categories. Although science aims for precision, it is not scientific to impose precision where none exists. The argument that pure models can lead to an ultimate understanding of social complexity serves more to advance the interests of various academic disciplines at the expense of describing present realities and meaningfully dealing with pervasive social problems. Too often, the scientist is captured by the purity of his own models and inhibited from making the less precise modifications and the necessarily speculative leaps into the unknown. The problem is increasingly one of choosing between several appropriate entities for analysis, with no one entity being completely satisfactory - especially over time. A spectrum approach to systems theory would highlight this problem, whereas the less sophisticated isomorphic approach tends to ignore these intractable realities of man-made systems.

System Lag, System Change, Concept Lag

As presented here, the discussion of the education complex has been confined to the entitation and justification of the proposed entity as an emerging macro-system. There has been no mention of the activities, processes or functions of core educating institutions, or whether these activities enhance or inhibit the varieties of learning that are necessary to meet the changing needs of society.

A brief mention of the changing state of core system activity insofar as its congruency to environmental demands, can enable some tentative explanations as to why the boundaries of the American educational system are widening. Moreover, the peculiarities of the education complex raise several questions for certain systems concepts, further suggesting the desirability of employing a spectrum approach to systems study.

System Lag

In an age of rising human aspirations and technical needs, formal educating systems in every nation are in a serious state of lag [42]. In the United States, at least, it is not only education that is lagging, but police departments, courts, prisons, political parties, churches, labor unions, legislatures, hospitals, the postal system, public transportation systems, home-building, and urban renewal. The most adaptive social systems are commercial enterprises which can readily sense that survival in the face of competition demands radical changes in structure and activities. A simple indicator of the vast changes that have taken place in recent years is the hundreds of corporations that have changed their name and/or identifying symbol in order to attain a congruency with their environment and to reflect new organizational realities. On the other hand, the least adaptive social systems are various areas of government with functional monopolies. As Drucker argues, there is less motivation to innovate, to abandon programs, and to release personnel [43].

In instances such as education where there is not a complete functional monopoly, the function can be formally and/or informally performed - although perhaps not as well - by other social means. For example, dissatisfaction with political parties can lead to the formation of new parties; impatience with the city can lead to a movement to the suburbs; and dissatisfaction with law enforcement or the law can lead to taking the law into one's own hands; and hospital overload has led to a broadened concept of "Comprehensive health care systems." In education, the relatively faster growth of the number of students in the periphery may well be attributed in large part to various incapacities of the core. Corporations have set up various programs to provide their employees with necessary skills; while many of the war on poverty programs are aimed at drop-outs from the core system. The Department of Defense has taken on the function of training 100,000 "misfits," as well as providing civilian-related job training and career counseling services for servicemen returning to civilian life. Some core clientele, while still maintaining their formal ties to school or college, have set up "free universities" in order to facilitate a more relevant learning experience.

System Change

It would be inaccurate, however, to imply that core institutions are static or decaying. They have changed, and, relative to previous system states, it is suggested that conditions for learning are improved. At higher levels, there is a human input improvement in the quality of faculty and students, with a greater freedom and professional sense among the former, and a greater interest and ability for learning among the latter. It is a myth, according to Jencks and Riesman, that there ever was a "golden age in higher education" [44]. At lower levels, faculty are also increasingly professionalized. And there has been an improvement in curricula, methods, and facilities - especially in well-financed suburban school districts.

Assuming the validity of these premises, improvement in the midst of growing inadequacy might appear to be a paradoxical contrast. Yet it is merely a full view of a system adapting to a dynamic environment where demands are rising at a faster rate than the core system can respond. The outcome of this system stress is not merely changes in the core system, but a widening of formal education to fill functional gaps, a greater number of education-related organizations offering panaceas, and the increased concern and activity of beneficiary groups.

A complete and accurate assessment of the degree of inadequacy is made difficult by roles, values, and dispositions of the observer. Educating institutions, similar to any social system, protect themselves by highlighting positive changes. Timid observers avoid controversy by ignoring lags, while critics invariably find them. Political liberals are prone to looking for lags and advocating various reforms ranging from incremental to sweeping. In order to maintain their view that the system is hopelessly static, corrupt, and in need of violent overthrow, political radicals ignore on-going changes and possibilities for orderly change. In a contrasting light, political conservatives are highly sensitized to changes, viewing them as undesirable and advocating a return to some simpler state of affairs, such as an emphasis on "the three R's" and closed systems of higher learning. Finally, optimists are disposed to look at improvements, while pessimists worry over growing inadequacy.

Despite the thicket of controversy that faces the systems analyst, it is simply not scientific to avoid consideration of changing system states and the changing state of the system relative to its environment. Yet, despite the scientistic pretext of being "scientific," a great portion of systems theory omits both considerations. In the first instance, it is far easier to view any system at one point in time and to ignore the further complexities of system change. These static or stop-action views ignore the dynamic nature of modern social systems, and by the time that a systems study is preserved and hallowed in print, the reality of the system that it purports to describe may have changed dramatically. Consequently, there is a considerable danger in assessing the present - not to mention forecasting the future - with the "scientifically" established concepts of the past.

In the second instance, the consideration of system lag is often omitted by glibly asserting that systems adapt [45], or that they maintain a dynamic homeostasis [46]. The conveniently overlooped question is "when-if-ever-does the adaptation take place?" Even in a more sophisticated analysis, such as Young's, the assertion that "deficiencies in adaptation represent the necessary price of on-going viability" [47] tends to overlook the condition of growing lag and to justify the minor adaptations of the status quo. To talk of lag is controversial, and it entails some risk of being directed by values. But to ignore lag, however defined, is to support the values of the status quo - which may well be those of the observer. In that the degree of congruence to environment is a system property, the omission of this consideration is simply not in the interests of intellectual honesty, to say nothing of science. The "necessary" viability of negentropy of any sub-system of society may well be at the expense of the viability of the larger system. Conversely, the chaos experienced by any subsystem in adapting to environmental stress may be to the benefit of homeostasis in the larger system.

The isomorphic approach to systems theory does not underscore the high plasticity of social systems or the problems of system lag. To the contrary, the emphasis on unchanging properties and homeostasis tends to obscure what is considered here to be critical properties of modern social systems. A spectrum view of systems would highlight these problems and direct inquiry to the aspects of social systems that involve human problems.

Concept Lag

A dynamic, affluent, technical, and post-industrial society places severe strains on various social systems developed for an earlier age. Systems lag, but they also adapt to some degree. The perception of systems, however, is also capable of lagging. As noted by Vickers, "it is essential to any society that its appreciative system shall change sufficiently to interpret a changing world" [48]. Lagging conceptual changes [49] reflect the serious "intelligence gap" [50] in assessing the state of modern social systems. Social explanation, prediction, and prescription can only lead to disaster if these prossess are based on obsolete system descriptions.

This problem of "imperfect feedback" is still another facet of modern systems that is neglected by the isomorphic approach. Social systems adapt, and they also respond to feedback. But unlike natural systems, man-made adaptation may nevertheless be inadequate, and it may take unanticipated paths. Feedback, whether positive or negative, often involves considerable time periods. When and if feedback arrives to guidance components, it may be inadequate and inaccurate, followed by further lapses in corrective action. In education, there is very little sense of what is happening, and whether various activities enhance or inhibit the learning processes of students. Viewing the system as satisfying latent functions such as socialization and "baby-sitting" can serve to point out further complexities - but by failing to point out lags in performing necessary functions, such analysis tends to reinforce the status quo by implying that because a system does such and so, it has successfully adapted.

A spectrum approach to system theory emphasizes the lags, changes, and imperfect feedback that characterize contemporary social systems. The failure of the conceptual system of systems theory to adapt itself to a spectrum approach is itself an outstanding example of concept lag.

Are the Most Important Systems Studied Least?

Systems theory is a method of dealing with the complexities of human interaction and the interaction of man with the material and natural world. It is not only critical to scientific understanding but to informing policy-makers who shape human directions.

There are considerable lags in social information. The United States is an "Unprepared Society" [51] in an "Accidental Century" [52]. The Planning-Programming-Budgeting System (PPBS) has recently been instituted throughout the federal government and in several state and local governments. Its use has been advocated for university systems and for school systems [53], and it appears that some variation of this approach will inevitably be embraced by educational planners. PPBS, by asking questions of inputs, outputs, benefits, and goals, as related to various programs, is amenable to systems theory and represents the marriage of scientific theorists and decision-making practitioners. At least a foundation for acquiring necessary feedback has been laid.

Systems theory is also essential in forecasting the future. As noted by Helmer, "future developments cannot be forecast in isolation (and) a comprehensive and interdisciplinary integrated analysis is mandatory, even if the subject of immediate interest is relatively narrow. In other words, we have to seek methods of what may well be called 'systems forecasting' [54]. In turn, informed conjectures on alternative futures (both descriptive and prescriptive), are essential to the rational guidance of a dynamic society where the only certainty is that the future will be different from the present.

As a response to an age of complexity, these new methods are an indication of societal adaptation. Nevertheless, there are still serious lags. Systems theory may have informed certain decisions, but it is too early to assert, as one observer has that "the practical application... demonstrates that the approach 'works' and leads to both understanding and predictions" [55]. There is still a lag in systems theory [56] and in the development of PPBS [57]. The newly emerging field of future-study has yet to utilize a systems approach either as an empirical foundation for describing present conditions or as a means to integrate various forecasts. The development of the cross-impact matrix as a means of systematically assessing the effects of various interactions [58] is a movement in the right direction, but falls short of analyzing a social system.

These preliminary observations on the emerging education complex have been presented here not only as a stimulus to the "pure" study of new social forms, but also, in the "applied" sense, as an outline of an approach to a national goal program analysis. If, as asserted, there are a considerable number of emerging and overlapping macro-systems, and if it is necessary to develop concepts of these systems in order to deal with modern complexities, then the necessary and unfilled task of systems theory becomes quite evident.

Boulding has warned that "our research resources in particular are poorly allocated in the light of the importance for human welfare of the problems to which they are addressed... failures in physical and biological science would not be serious; failure in social science could well be fatal" [59]. The extremely open, loose, complex, and dynamic macro-systems, at the far end of the general systems spectrum, are perhaps the most important systems that the analyst might consider. But because of their ambiguity and the necessary imprecision in their study, the most important systems may well be ignored in favor of those more readily conceived and easily measured.

The fruition of a spectrum approach and the consequent study of macro-systems will require new scientific methods only faintly resembling the elaborate precision characterizing the "respectable" science of the present time. Far more attention will have to be given to pre-mathematical work such as entitation and the creative juxtaposition of system elements. Many overlapping and competing models will be required, and a willingness to adapt models to new realities must be paramount. This conceptual response, if it ever develops, will be the necessary adaptation to the problems of man in modern systems.

NOTES

[1] von Bertalanffy, L., *General System Theory: Foundations, Development, Applications*, New York, George Braziller (1968) p. 15, 38.

[2] Gross, M., *The State of the Nation: Social Systems Accounting*, London, Tavistock Publications (1966) p. 33; also in R.A. Bauer (ed.), *Social Indicators*, Cambridge, Mass., M.I.T. Press (1966) pp. 178-179.

[3] "The recognition or discovery of ... entities must precede their scientific examination ... This qualitative recognition of the important systems, which I find helpful to call 'entitation', is far more important than their measurement." R.W. Girard, Entitation, Animorgs, and Other Systems, in: M.D. Mesarovic (ed.), *Views on General Systems Theory: Proceedings of the Second Systems Symposium at Case Institute of Technology*, New York, John Wiley (1964) pp. 120-121.

[4] This paper is an extensive reformulation and condensation of a previous draft entitled: *The Education Complex: Systems Theory as a Heuristic Approach for the Study of the Future*, Educational Policy Research Center, Technical Memorandum No. 3, Dec. 20, 1968, 68 pages, mimeo. This also serves as an overview of a larger project, *The Education Complex: Emergence and Future of a Macro-System*, to be published by The Free Press in Spring 1970.

Considerable credit for the concept and the general approach to its delineation is due to Bertram M. Gross via both his writing and his academic leadership.

[5] A complex should be distinguished from clusters or constellations (as proposed by Bertram M. Gross, The Guidance of System Constellations, AAAS Annual Meeting, Dec. 1967), the latter two terms suggesting some functional homogeneity in the associated organizations. Gross, however, defines constellation as a complex is defined here. It should also be noted that a conglomerate is simply a large single organization with heterogenous functions. The term "macro-system" or "macro-organization" is proposed as a generic category for these three types of systems.

There are various precedents for using the noun form of "complex" in reference to large social systems. The "Military-Industrial Complex", coined by Malcolm Moos, has become firmly imbedded in the language of political discussion, in the years following President Eisenhower's 1961 farewell address. Robert Weaver wrote about *The Urban Complex* (Doubleday Anchor, 1964), and Micheal Harrington has recently warned of the dangers arising from "The Social-Industrial Complex" (*Harper's*, Nov. 1967). Despite the widespread use of the concept - even von Bertalanffy mentions the "industrial-military complex" in passing ([1], p. 4) - there has been little effort to explicate the profound meanings suggested by this concept. An initial exploration of the military-industrial complex has been started by W.D. Phelan, Jr. (The 'Complex' Society Marches On, *Ripon Forum*, *1* (Jan. 1969) pp. 9-21), but the purposes of this effort are more political than scientific. Unlike the invidious attributes which are popularly associated with a "complex," the use of the word here is intended to be neutral. Another difficulty with the word is that every system may be defined as a complex: von Bertalanffy defines a system as "a complex of interacting elements" ([1], p. 55). Still, despite these semantic hazards, "complex" is found preferable to the above mentioned alternatives.

[6] In their models of the educational system, P.H. Coombs

(*The World Educational Crisis*: *A Systems Analysis*, New York, Oxford (1968)) and M.B. Miles (ed., *Innovation in Education*, New York, Bureau of Publications, Teacher's College, Columbia University (1964)) refer to these activities as "nonformal." There is indeed a shading off into loosely structured activities in some instances, but for the most part, these activities are as highly organized as any to be found in the core. "Nonformal" is a misleading term, a more useful distinction between core and periphery being only whether the activities lead to a degree. A forthcoming volume by W. Cohen, B.M. Gross and S. Moses (*The Learning Force*, Basic Books (1970)) will explore this sorely neglected area in greater detail.

[7] National Planning Studies Program (Syracuse University) and Educational Policy Research Center at Syracuse, preliminary data, June 1968. An earlier tally by Wilbur J. Cohen (Education and Learning, in Bertram M. Gross [ed.]: Social Goals and Indicators for American Society, V. II, *The Annals of the American Academy of Political and Social Science*, *373*, Sept. 1967) estimates a lower total in the periphery. The data presented here are head-count figures; if reduced to full-time equivalent, the number of students in the periphery would be considerably smaller. There is also some double-counting here, but an estimate as to its extent cannot be provided at this time.

[8] The model should ideally look at suppliers to both the core and the periphery. However, in that it is a major task to simply locate the periphery, the additional task of locating its suppliers will not be undertaken. In many instances, core suppliers of items such as books and equipment are also suppliers to the periphery.

[9] The power of the "military-industrial complex" concept is to highlight the widespread movement of retiring military personnel to top positions in defense-related industries. In education, for example, "educators" staff the U.S. Office of Education, and a former head of USOE went on to the stewardship of the General Learning Corporation.

[10] Ideally, the most important beneficiaries are students. Although there are many student organizations, most are controlled in varying degrees by adults. It has only been in recent years that adult-free student organizations (*ad hoc* or continuing, violent or non-violent in methods) have sought to seriously influence institutional policy.

[11] See Fritz Machlup, *The Production and Distribution of Knowledge in the United States*, Princeton University Press (1962), for an economist's exploration of the "knowledge industries." Such a viewpoint could provide a fertile foundation for building a macro-system model.

[12] Jencks, C., and D. Riesman, *The Academic Revolution*, Garden City, New York, Doubleday & Co (1968) p. 24.

[13] *New York Times*, April 6, 1969, p. 1.

[14] Moore, R.S., *Consortiums in American Higher Education: 1965-66*. Washington, U.S. Office of Education, OE: 50055 (Sept. 1968).

[15] U.S. Department of Commerce, *Statistical Abstract of the United States*, (1968 ed.), p. 105. Current data are estimated.

[16] Based on unpulbished data, National Planning Studies Program and Educational Policy Research Center, Syracuse University, June 1968. An earlier tally by Cohen ([6], p. 84) estimates a lower total for the periphery. Coombs ([6], pp. 139-140) is also thinking along these lines: "It is entirely possible that in some countries the aggregate of economic resources and human energies already commited to these part-time programs approaches the total involved in full-time formal education."

[17] Cohen, [6], p. 85.

[18] U.S. Department of Commerce, *Statistical Abstract of the United States*, (1968 ed.) p. 505.

[19] Culkin, J.M., S.J., A Schoolman's Guide to Marshall

McLuhan, *Saturday Review*, March 18, 1967. J.I. Goodlad (*The Future of Learning and Teaching*, National Education Association (1968) pp. 14-15) argues the same point.

[20] Gans, H.J., The Mass Media as an Educational Institution, *The Urban Review, 2* (1) (Feb. 1967) p. 10.

[21] Schrag, P., The End of the Great Tradition, *Saturday Review*, (Feb. 15, 1969) p. 103.

[22] U.S. Office of Education, *Digest of Educational Statistics* (1967 ed.) p. 97.

[23] Sahlins, M.D., and E.R. Service, *Evolution and Culture*, Ann Arbor, University of Michigan Press (1960) p. 74; p. 92.

[24] U.S. Office of Education, *Digest of Educational Statistics* (1967 ed.) pp. 115-116.

[25] Education and World Affairs, *The University Looks Abroad: Approaches to World Affairs at Six American Universities*, New York, Walker and Company (1965) p. 268.

[26] Marvel, W.M., The University and the World, in Alvin C. Eurich (ed.), *Campus: 1980*, New York, Delacorte Press (1968) p. 81. A more modest and discriminating forecast might describe a spatial classification of higher education, with major universities strongly international in clientele and concerns, state colleges aimed at regional concerns, and junior colleges retaining their community orientation.

[27] Roberts, W., Thirty Thousand Innocents Abroad, *Saturday Review* (Feb. 15, 1969) p. 61.

[28] Coombs, [6], pp. 3-4.

[29] Cartter, A., Graduate Education and Research in the Decades Ahead, in Eurich, *op. cit.*, p. 261.

[30] Educational Facilities Laboratories, *The Impact of Technology on the Library Building*, New York, EFL (July

1967) p. 14.

[31] *Newsweek*, Dec. 25, 1967, p. 42.

[32] National Association of College Stores, Oberlin, Ohio. Unpublished data.

[33] Although organized beneficiary groups could be seen as the suppliers of information and policies (either directly or indirectly through influencing lawmakers), the members of these groups also constitute the human component of educating systems. Because of their interest in outputs and outcomes, and not merely inputs, groups of beneficiaries are worthy of a separate category.

[34] Gilb, C.G., *Hidden Hierarchies: The Professions and Government*, New York, Harper & Row (1966) p. 104.

[35] Clark, B.R., Faculty Authority, *Bulletin of the American Association of University Professors, 47* (4) (Winter 1961).

[36] Jencks, C., and D. Riesman, The Triumph of Academic Man, in Eurich, *op. cit.*

[37] *Teacher Strikes and Work Stoppages*, Washington: National Education Association Research Memo, 1968-15, Nov. 1968.

[38] Maynard, R.C., A Black Study Quandary, *The Washington Post*, March 16, 1969, p. B2.

[39] See for example, D.Divoky, The Way It's Going to Be: Revolt in the High Schools, *Saturday Review*, February 15, 1 1969. This report may already be outdated.

[40] An excellent overview is provided by S. Terreberry, The Evolution of Organizational Environments, *Administrative Science Quarterly, 12* (4) (March 1968).

[41] Educational Facilities Laboratories, *A College in the City: An Alternative*, New York, EFL (1969).

[42] Coombs, [6], p. 4. " ... the growing disparity between education and society will inevitably crack the frame of educational systems - and, in some cases, the frame of their respective societies." D.N. Michael (*The Unprepared Society*, Basic Books (1968) p. 112) has similar fears: "While schools have been turning out mostly the conventional suburban or slum products of education, the world, as we have seen, is getting much more complex. The upshot may well be that the majority of the new generation of adults and those new ones to come for several years yet will be more out of touch with their world than was, say, the correspondingly educated generation of an earlier day."

[43] Drucker, P.F., *The Age of Discontinuity: Guidelines to our Changing Society*, New York, Harper & Row (1969). See Chapter 10, The Sickness of Government; also appearing separately in *The Public Interest, 14* (Winter 1969).

[44] Jencks and Riesman, *The Academic Revolution*, p. 199.

[45] Buckley, W., Society as a Complex Adaptive System, in Buckley (ed.), *Modern Systems Research for the Behavioral Scientist*, Chicago, Aldine Publishing Co. (1968).

[46] Herriott, R.E., and B.J. Hodgkins, *Socio-cultural Context and the American School: An Open-Systems Analysis of Educational Opportunity*, Tallahassee, Fla., Florida State University, Institute for Social Research, Center for the Study of Education (Jan. 1969) p. 58.

[47] Young, O.R., The Impact of General Systems Theory on Political Science, *General Systems, IX* (1964) p. 252.

[48] Vickers, Sir G., *Value Systems and Social Process*, New York, Basic Books (1968) p. xviii.

[49] Gross B.M., The Coming General Systems Models of Social Systems, *Human Relations, 20* (4) (Nov. 1967) p. 359. Terreberry ([40], p. 613) makes a similar point about the lag in theorists' ability to comprehend an evolving world.

[50] Gross, B.M., and M. Springer, New Goals for Social Information, in Gross (ed.), *Social Goals and Indicators for American Society, Vol. II.* The Annals of The American Academy of Political and Social Science, *Vol. 373* (Sept. 1967) pp. 209-213.

[51] Michael, *op. cit.*

[52] Harrington M., *The Accidental Century*, New York, Macmillan (1966).

[53] Hartley, H.J., *Educational Planning-Programming-Budgeting: A Systems Approach*, Englewood Cliffs, N.J., Prentice-Hall (1968).

[54] Helmer, O., Methodology of Societal Studies, Santa Monica, Calif., The RAND Corporation, P-3611 (June 1967) p. 5.

[55] von Bertalanffy, [1], p. 196.

[56] For example, see E. Clark, Systems Analysts Are Baffled By Problems of Social Change, *The New York Times*, March 24, 1968, p. 28. At the meeting that reflected this concern, a questioner in the audience proposed a synthesis of a very diverse network of non-homogenous complexities - essentially the task of describing the education complex.

[57] Gross B.M., The New Systems Budgeting, *Public Administration Review, XXIX* (2), (March-April 1969) pp. 113-137.

[58] Gordon T.J., and H. Hayward, Initial Experiments with the Cross Impact Matrix Method of Forecasting, *Futures, 1* (2) (Dec. 1968) pp. 100-117.

[59] Boulding K.E., Knowledge and Technology, in E.L. Morphet and O. Ryan (eds.), *Prospective Changes in Society by 1980*, Vol. 1 in: Designing Education for the Future, series, New York, Citation Press, (1967) pp. 205-206.

TOWARDS BUILDING A GENERAL URBAN SYSTEMS THEORY

Peter House*

ABSTRACT

A framework is developed for considering the city as a system. While techniques of systems analysis are desirable, due to the nature of the city, it is necessary to also approach it from a wide interdisciplinary point of view. The major elements of the framework are two, first the environment and the relations between it and the people, and second, the people and their relations with one another. The elements are broken down into even more detail, and on the basis of the model a game is constructed with players representing the various decision-makers, with the relationships of the city system built into a computer program. In this way, various policies for city development can be planned and tried and actual policies selected for implementation on the basis of computer playouts.

1. INTRODUCTION

In January 1966 a group now called Envirometrics was

---

*President of Envirometrics , 1100 17th Street, N.W. Washington, D.C. 20036.

The author wishes to thank the Envirometrics staff, particularly Dr. Philip D. Patterson, Janice Cooper, William F. Arnold, and Ben Laime, for helpful criticism of this paper.

organized at the Washington Center for Metropolitan Studies to design, program, and operate a model of the metropolitan area. This model was to be different from existing models in that it was to integrate a number of gaming and modeling techniques.

The developed operational simulation model, *City I*, made use of mathematical, systemic, and role-playing modeling methodologies to varying degrees. We believe it is opening a new way for social scientists to research urban problems, the current model being the first step in the developmental process.

This paper is our first attempt to formulate a general theoretical framework using our experience in designing the model. The framework presented here is presumed to embody all of the components of an urban system at a high level of abstraction. We have two prime purposes. First, we wish to present a framework for discussing urban areas in general. The framework is not presented as a general systems theory, although it is offered as a structure from which such a theory can be built. Secondly, we want to relate this work to operational simulation and discuss a particular simulation model available at present.

The ideas we introduce here are not new to many social scientists. However, we believe that the way in which they are combined is unique, and this combination, using the methodology of operational simulation, shows great promise for developing a usable urban systems theory.

2. THE NEED FOR A FRAMEWORK FOR UNDERSTANDING THE URBAN SYSTEM

Newspapers, professional journals, and popular magazines remind us continually these days that social science is beginning to study the urban area as a total entity. In comprehending the complex urban-metropolitan area, social scientists seek a rational, quantifiable, multi-discipline framework in research, as well as in the fields of administration and education. For example, on a national level, the Nixon Administration has set up an overseeing urban affairs council headed by Daniel Moynihan to coordinate the efforts of the Departments of Agriculture; Health, Education and Welfare; Housing and Urban Development; and Transportation, relating to urban programs. In education, many universities have introduced interdisciplinary courses, while others are setting up interdisciplinary departments focused on the urban area.

The strong emphasis on multidisciplinary approaches is relatively new. Most of the early research in urban affairs was undertaken from a single point of view. The economist, the sociologist, the political scientist and the administrator studied one or more problems, suggested solutions, or attempted explanations, each from his own point of view. To do so, he had to say, "other things being equal." But, other things seldom are equal. In studying the urban area the researcher finds that his discipline alone is not enough. For example, the geographer, who used to study climate, land mass, and water bodies, now studies a field he calls human geography which relates man to his environment, although the focus is still more on the environment. His intellectual brother, the ecologist, who used to study the relationship of the parts of nature to each other, now focuses on the relationship of man to his environment. Documents put out by the two disciplines seem to suggest that there could be a natural wedding of their fields with very little real differences in general subject matter. The macrosociologist , for another, finds himself reading in social psychology, human ecology, political science, public administration, economics and sociology.

However, although social scientists are increasingly incorporating knowledge from other disciplines into their studies, many are "hanging back" from using a true multidisciplinary attitude in looking at the urban field. They still see a particular problem as primarily economic, political, or sociological in nature. The reasons for this are probably many. Three seem relevant here. First, tradition or conservatism hinders us all from changing our ways or freely embracing new ideas. Secondly, division of labor, where specialists profitably proffer those goods and services they produce best, works as well in the social sciences as it does in the marketplace. And third, social scientists today lack a general urban theory in which to incorporate multidisciplinary approaches. Despite today's trend toward generalized or holistic thinking (although one can name Talcott Parsons, Max Weber, Adam Smith, and even Plato, as multi- or predisciplinary holistic thinkers), a generally accepted urban theory is lacking. Among other things, such a theory would allow us to break clean from the proliferation of narrow research papers and speeches (of which this is another) which keep us knowing more and more about less and less. A comprehensive theory would let us embrace an approach large and usable enough to handle the urban problem.

In looking for an overall picture of urban problems, social scientists reacted to the publicity and the recognized success of the technological programs fostered by our space and defense industries. Daily we are inundated by such concepts as fuel systems, information systems, flight systems, trajectory systems, and so on. We began to wonder: If a computer-aided systems orientation can work in space, why can't it work in our cities?

Concurrently, Robert S. McNamara was installing in the Defense Department a sophisticated budgeting technique called Program Planning Budget System (PPBS). It seemed to work like the famous "new broom" in attacking the shibboleth of government waste and inefficiency. More popularly known as "systems analysis," the new methodology seemed to use a great deal of rational, self-evident philosophy to encompass and control large, complex problems.

Another recent approach is called operations research. This technique relates sophisticated mathematical constructs to decisions involving quantifiable inputs and outputs. It makes use of statistical techniques, such as linear programming, to handle problems of resource allocation through highly rational budgeting plans and suggests optimal paths for decision schemes. Operations research is a technique which offers promise toward helping solving problems in all areas of social science research.

Systems analysis techniques and operations research methods coupled with the strong need for a general urban systems theory should eventually, we think, permit the development of such a theory. We think that a holistic approach is promoted by using operational simulation. We believe that a general urban systems theory will come from and help develop such a simulation. From our experience so far with man-machine modeling at Envirometrics, we would like to offer some first thoughts and discoveries in building a general urban systems theory.

3. THE COMPONENTS OF A SYSTEM

Our first problem is to define a system. Many times the definition depends on the problem at hand. It becomes a functional definition. For some problems a stylized urban system may be defined best as neighborhood within a city; for others, a city itself within a metropolitan area; while some may be that metropolitan area within a region. Certain factors may help chart its boundaries, such as population, labor market, natural boundaries and the like.

Initially, we may decide that the components or factors of the particular system we may define will act more "wholly" or as a unit than they will with factors "outside" the system. That is, they tend to affect and be affected by other factors within the prescribed system more than they do from outside forces. As an example, the neighborhood druggist appears to be more a component of his sub-city area, economically, socially, and politically, than he does of a neighborhood across town, or in another town, or in a metropolitan area.

Assuming that we can define a particular urban system in a geographic and functional sense, we must then define the larger system in which it works. This task is no easier than the first. For a neighborhood, the larger system might be a city or metropolitan area; for the city, a metropolitan area, then regional area, super region, a country, and possibly the world.

For our purposes here, we will assume that the larger system is the totality of all the systems outside the specific system under consideration. The larger system provides several major functions for the specific urban system. First, it determines the political, legal, and social institutional limits within which the particular urban area will operate and develop. Legally, the local area is bound by national and state law. Local cultural and social norms are affected by the larger system.

Secondly, the larger system affects the technological framework in which the local system operates. The local system is both an innovator and a receptor of changes to and from the outside. All technical changes made on the larger system eventually filter down to each local system. Historically, the consequences of the development of DeWitt Clinton's steamboat in one subsystem eventually filtered to all local areas through organizations set up on the larger system. The larger system's role changed from that of initiator of inventions to that of receptor of inventions made at one local area. The larger system, in terms of innovation and change, can be thought of as a huge clearing house where innovations (as well as other goods and services) made in one local area are exchanged for innovations of another.

Third, the population growth of the local system depends in part upon factors at the larger system level. The larger system supplies people for local areas and in turn receives people who move out of the local area. An example is the migration of people in this country to the West Coast. The larger system's social customs and mores tend to become "nationalized" within the smaller systems, which in turn also supply the larger system with trends toward novelty and change.

Lastly, the existence of a larger area provides an economic rationale for the smaller system in that it allows the smaller system to specialize, hopefully in an optimum manner. The smaller system produces a particular type of goods and services at a comparative advantage and trades these items for those produced by other small systems. The larger system provides the organization for exchange. The local economic system is also affected by the business cycle, monetary policy, and fiscal policy of the larger system.

4. THE STRUCTURAL FRAMEWORK

The following is an attempt to disaggregate an urban system as a local system into its major components. We think this structural framework will provide a useful way to view an urban system.

The framework consists of two basic parts, the environment, whose components are the natural resources, man-made structures, internal connectors, and technology; and population, which includes people and the organizational linkages among people. How the population acts with and within the environment gives the urban system its personality, or persona. We introduce persona as an important factor to be considered when studying urban areas.

5. THE ENVIRONMENT

The Natural Resources

The landscape is nature's endowment to a particular urban area. It includes such things as terrain, slope, soil type and mineral content, weather and climatic conditions, amount of available water, and so on. It helps characterize the urban system.

The climate and the natural foliage affect the style and living patterns of the inhabitants. The presence of valleys, rivers, mountains or plains influences the use and type of buildings, the size of the city, the placement of streets, and even the number of people. And, the availability of water may also prove to be a limiting factor on present and future urban forms.

The economic base of an urban system is affected by the type of natural area in which it is situated. The presence of a deep water harbor allows particular industries which require bulk raw materials and bulk shipment of output. Natural resources, such as large nearby deposits of coal and iron, permitted the growth of a steel industry in Pittsburgh and allowed other related types of manufacturing to be located nearby. While it is not the only defining feature, the natural environment both limits and sets potentials for urban areas.

Concretions

The rest of man's environment he creates himself. These man-made structures include homes, stores, factories, plazas and the like. For simplicity's sake, we will call these creations of man's which change the natural environment "concretions".

Prior to the construction of any structures at all, the early settlers had broad freedom to design and shape their city. They were limited only by the building techniques available at the time, their imaginations and needs, funds and materials, and, most of all, by the natural landscape. However, as a city grows and ages, the importance of the natural landscape begins to diminish and the new urbanites become more limited by the concretions. Generally, new development in an urban area not only follows the natural terrain (since it is the path of least resistance, it is the easiest to deal with), but it also follows the paths of least resistance delineated by existing man-made structures for the same reasons. Telephone lines, water and sewer mains, streets, sidewalks, overhead lights and many other capital improvements have been invested in the present city form. These, not to mention stores, homes, and office buildings tend to limit the number of alternatives open to building or changing the face of a city.

In some ways, the larger and more developed a city, the more difficult it is to overhaul it. The more developed a city is, the more "set" it becomes, because each subsequent construction is built in relation to the ones before it. Each subsequent development adapts to its natural environment - the concretions - just as the first man-made development adapted to the real natural environment.

Thus, concretions become a shaping force on a city, just as natural environment does.

Internal Connectors

Transportation and communications are the two kinds of internal connectors. They tie the various components and sub-systems of a city together. Transportation links include roads, rivers, rapid transit, railroads, and other pathways for travel. They also include the vehicles themselves, such as cars, boats, planes, trucks. The communication links include all forms by which people exchange information with one another, such as telephone, telegraph, radio, television and newspapers.

In all cases of transportation and communications links, we are not talking about the goods or information which flows across these lines. We are interested in the physical structure which allows the two forms of links to function. The messages and goods sent by the people are a feature discussed later in this paper.

Technology

As the last part of environment, technology defines and delimits the natural landscape, man-made structures and the physical links between them. In order to describe the structures of the city itself, we must set this environment in a time period by studying the level of technology used in the city at a given time.

In most cities, the central core was developed many years earlier than the suburban fringes. Such things as layout and capacity of the streets, the age and materials of the buildings, and the height of buildings depend upon the technology at the time in which the city developed. The size and shape of concretions and linkages in any specific city depend greatly on the level of technology available when the city matured.

Due to changes in building technology and styles, many cities are layered, much like the growth rings on trees. The central core contains tightly-packed residential, industrial and commercial sectors, most often showing older building methods jury-rigged to conform with modern safety practices. The sprawling suburbs show newer life styles and means. In transportation, congested Boston and New York matured during the horse and buggy era, whereas Los Angeles is still maturing with the automobile (or maybe learning to live with it).

Cities are dependent on new technology and engineering methods not only to handle the consequences of the latest technology but also to support increasing populations in living patterns they prefer. New methods of treating sewage or importing water will permit more people to live in the urban system. For future technology a major problem is how to dispose of the avalanche of garbage which threatens to impede urban growth in some areas.

6. POPULATION

People make up a city and its environs, so much so that the city has been called man's greatest monument to himself. We are concerned here with population as a numerical index, similar to that defined by the Bureau of the Census. Population so defined includes such things as the numbers of people and their distribution and density, their rates of growth, their composition (including racial and class mix), and changes in these categories.

These basic measurements are necessary for a probing study into urban systems. For example, it seems that a community must have a certain number of people before a division of labor optimally occurs in its economic and other functions. Not only do people become specialists in performing specific duties, but land uses become specialized. Finding the optimal population size in any given urban system would be useful. It seems that the larger the city the more the tendency there is for it to grow in rings with separate central core and suburban jurisdictions.

Other measurements, such as sex, race, and age, help the researcher determine the basic work force, the potential work force, the welfare program, the types of service and entertainment industries, the extent, cost and quality of education. They may also give a clue as to the social and political forces active within the urban system. Furthermore, such measurements may help the researcher diagnose the city's present "health" and speculate on its future condition.

Like the rest of the structural framework, the population is subject to change. Today's populations are very mobile. Active in- and out-migration can change the composition of an urban community in a short period of time.

Organizational linkages

The measurements of an urban system discussed up to now for the most part involve mechanical and structural interrelationships. We have described the linkages between the natural landscape and man-made structures in terms of transportation and communication links. Similarly, we find that there are linkages within the population and between the population and the environmental structure. They exist and function similarly in every way and operate irrespective of particular individuals. These linkages evolve functionally and necessarily through the increasing complexity of the urban system. They create more or less permanent "slots" which individuals fill. This is the essence for example, of a bureaucracy, one of many organizational structures. For convenience, we will break these into two broad categories: organizational structures which relate people to people, and those which relate people to their environment.

People to People

The organizational channels which relate people to people can be either formal or informal. Formally, for example, portions of tort law are largely concerned with injuries to persons instead of damage to property. Informally, all of our manners and customs which prescribe what we wear, what we eat and drink, what we say to each other, are organizational linkages, too. They provide continuity and change relatively slowly over time. Our political structures and forms, at local and national levels, along with the style of politics which we prefer, are manifestations of our system's attempt to organize itself so that its population can function smoothly. The business community with its formal structure and methods of operation shows similar arrangements, including hiring and firing policies, number of working days, coffee breaks, and the like.

People to Environment

There are just as many organizational linkages set up between people and their environment as there are between people. Simple mathematical calculations like profit and loss, or supply and demand concepts, or even the business organizational structure itself, are artifacts set up by man to score and to organize himself in the business community. The business community, in turn, is organized so that it can operate in and within the larger urban system. It seems that even the concept of "public" and "private" is designed to relate man to the natural and man-made environments.

Other examples of the ways we agree to operate conveniently within the structure include the way we keep time, our weights and measures, local regulations concerning building heights and setbacks and national regulatory laws.

7. PERSONA

The final portion of the urban systems framework we

call persona, an endogenous defining factor which can be described as the ethos or general method of operation for each particular urban system. It is the social norms and culture of the decision-makers of the local system.

We think that although the concept of persona or personality for a city may seem obvious when mentioned, it is an important component of an urban system which may be easily overlooked. And, if overlooked, it would render any urban systems study and resulting theory invalid.

The outward signs of persona differ so much among cities that they can become attractions to that area. For example, consider the French tradition of New Orleans, the Mediterranean avant garde atmosphere of San Francisco, the mercantilism of New York, and the historical atmosphere of Boston. The "mood" of these cities influences the way that they, as urban systems, develop. We know that persona influences the structural framework by helping to define the architecture of the area and the overall design of the city. It affects the population itself by setting local culture, the character of its people (who help create it while also receiving part of their identity from it) and their "action".

Somewhat like persons, cities behave as organisms in that they respond and probe and react to internal and external forces depending upon where they are (environment), how they are (state of well-being), when they are (technological level), why they are (reason for being), and who they are (persona). Persona then has the ability to influence every decision made in the urban area and all the variables discussed earlier.

Persona cannot yet be measured nor can it be segregated from the rest of the population-environment components, but we believe that with our recently evolved operational simulation methodology, we can begin to separate it and to examine it in isolation and in relation with the other components.

8. THE URBAN SYSTEM

What we have been describing here is not meant to be a theory but a means of organizing relationships which are not now precisely known or quantified. These structured relationships, we think, can be studied, quantified, and extrapolated so they can be used to construct an urban systems theory.

For clarification, we submit the following diagram (Figure I) which is designed to show the interrelationships of the components in an urban system. First, the urban system is surrounded by the larger system of which it is a part. The larger system affects the urban system through its political-legal institutions, migrations of people, social norms and culture, technology, and economic institutions.

Within the urban system there is the environment* of the urban system, the main parts of which are the natural resources and the concretions held together and intertwined by transportation and communications links. It is through such physical linkages that goods and services flow and help define environment. Included also under the environment is the state of the local technological art which acts as the exogenous limiting and defining factor of the particular environment at any given point in time. New inventions or new adaptations added to the technological art have helped establish a living environment for man.

Also within the urban system are the organisms living within this structure - people. The interrelationships among the people are facilitated by organizational linkages which are economic, political, social, and legal in character. The relationships between the people and the environment are also facilitated by the organizational linkages. People provide the animate portion of an urban system and are added to or subtracted from the urban system by natural population change and migration. The persona provides the character, ethos, culture, and style for the urban system.

*as defined on page 225.

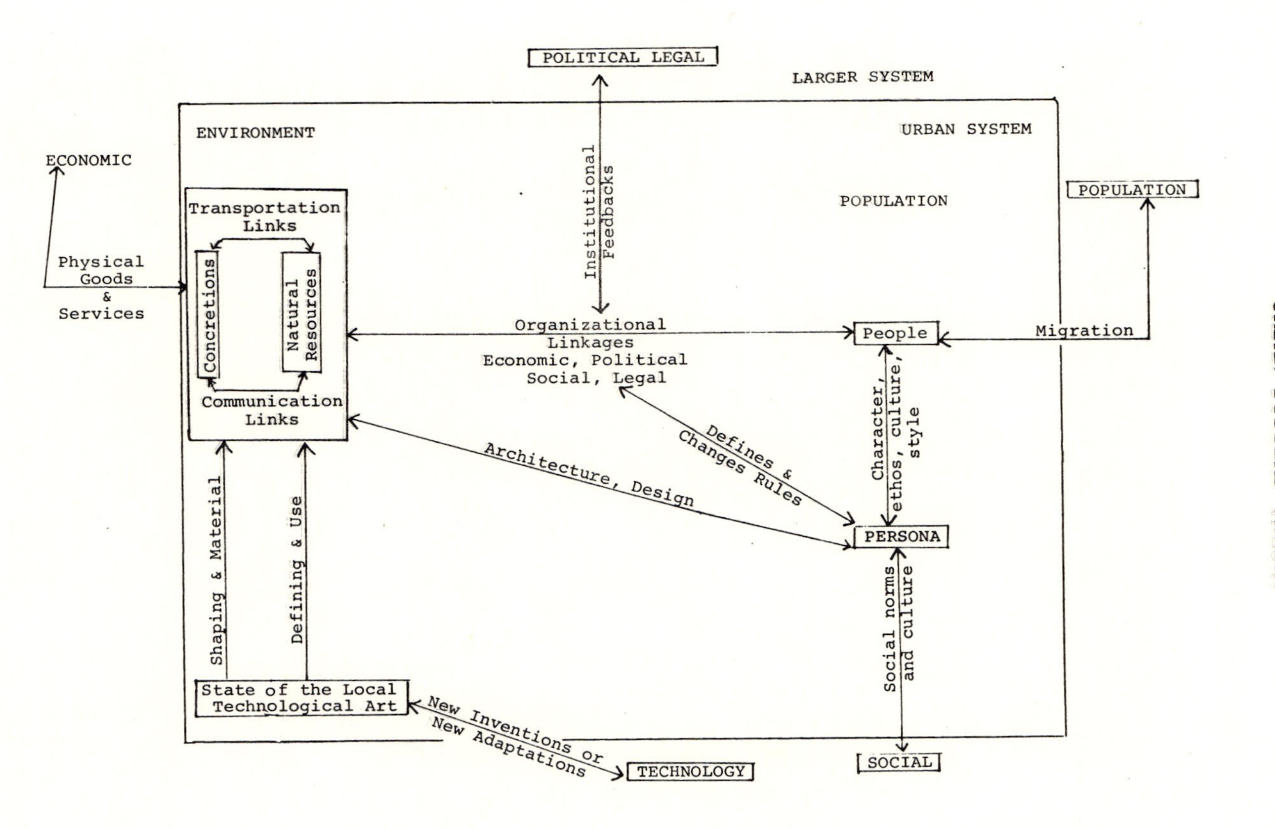

FIG. I. Interrelationships of the components of an urban system

9. UNIFIERS WITHIN THE THEORY, THE SYSTEM, AND THE MODEL

With the diagram we focused on the components and their interrelationships within the urban system structure. Here, we would like to clarify the concept of persona. We will use the concepts of status and role to help in the clarification.

Status may be defined generally as a somewhat fixed position, office or occupation in society. It is generally a steady or distinguishable rank which is knowable by all and is essentially part of an expected set of relationships. Role, on the other hand, is the participation of people acting in the status positions. The *office* of the President of the United States is a status position, whereas *how* an Eisenhower, a Kennedy, or a Johnson *act* within it is a role.

To translate this to our urban systems construct, the natural environment, concretions, internal connectors, technology, population and organizational linkages comprise the status part of the system.

We advocate that the "role" part of an urban system, which has not seen much research, can and should be studied. How persons act in their roles plays an obviously important part in urban dynamics. The "role" portion of real life is similar to "persona" in the theoretical framework. As a person acts within his status, so does a city within its environment. Furthermore, a city's persona may have an effect on its larger system in the same way that a people's persona has on a city.

An urban systems model, which would simulate an urban area and point toward a general urban system theory, should be analogous to an urban system as it is perceived and to the general theory as it develops. That is, it should contain the elements common to both if it is to be an effective tool. Moreover, it should unify research on a particular urban system with research toward a general urban systems theory. In a sense, it would be like, if not become, a learning experience, a research tool and a test vehicle for a theory wrapped up in one.

We are developing generations of gaming models, operational simulations, which show promise of leading toward such a unifying urban systems model. One completed early model, called *City I*, gives us a good start, although it is still very basic. It is a macro-model simulating a hypothetical county containing a central city (250,000 population) and three suburban jurisdictions (100,000 total population). The players act as decision-makers for the whole area. The interplay of conflicting and cooperative decisions made over time by the players, as they attempt to achieve a multiplicity of personal and social objectives, causes the simulated urban area to change in composition and size.

In *City I*, nine teams (three to five members per team) are the decision-makers in a partially urbanized county. The playing board is 25 × 25 squares (each square represents one square mile), most of which are unowned at the beginning of play. These land parcels may be purchased and developed by the teams during the course of the game. There are nine types of private land use which the teams can develop on a parcel of land: heavy industry, light industry, business goods, business services, personal goods, personal services, high-income residences, middle-income residences, and low-income residences.

Each of the nine teams is elected or appointed by elected officials to assume the duties of one of nine governmental roles, which are played simultaneously with the entrepreneurial functions common to all teams. The elected officials must satisfy the electorate (all teams acting as private decision-makers) in order to stay in office each round. The elected County Chairman team appoints other teams to function as the School, Public Works and Safety, Highway, Planning and Zoning, and Finance departments. The governmental departments build schools, provide utilities, build and upgrade roads and terminals, maintain roads, buy parkland, zone land, and estimate revenues.

Teams set their own objectives for both the public and private actions they undertake. Team decisions are recorded each round (approximately two hours in length) by a computer, which acts as an accountant and indicates the effects of the teams' decisions on one another and on the county itself. The interaction of public and private decisions and their influence over time is illustrated by periodic computer printouts. Even though conflicts may develop between urban and suburban interests, among businesses, and among governmental departments, teams often find that cooperation is equally as important as competition in fulfilling their objectives.

The portion of the model dealing with the city "status" is all of the givens faced by the players when they start to participate in a run of the model. These relationships are built into a computer program, represented on a board, as quantified on computer outputs which the participants have available to them. The role or "persona" portion of the model is provided by what the participants bring to the model themselves - their own personalities intermingled with those of the players in the other participating groups. It is these personalities which act and build upon the basic system and the basic city with which they start. The urban area represented in the *City I* model is lifeless until acted upon by people. Thus, every run of the *City I* model yields a different outcome, just as every actual city has grown somewhat differently depending on the major decision-makers in power. (For a more detailed description of the *City I* model and its relation to the theoretical framework, please see Appendix A.)

As Figure II indicates, the gaming concept exemplified here by *City I* contains the same basic components and relationships as the theoretical framework. The structure of the game parallels the theoretical framework that was shown more fully in Figure I. The players correspond to the people component, the game environment corresponds to the urban environment. How the players operate the game, "role," has much the same meaning as "persona" in a real urban system. The state of the gaming art would have the same effect on the game as the

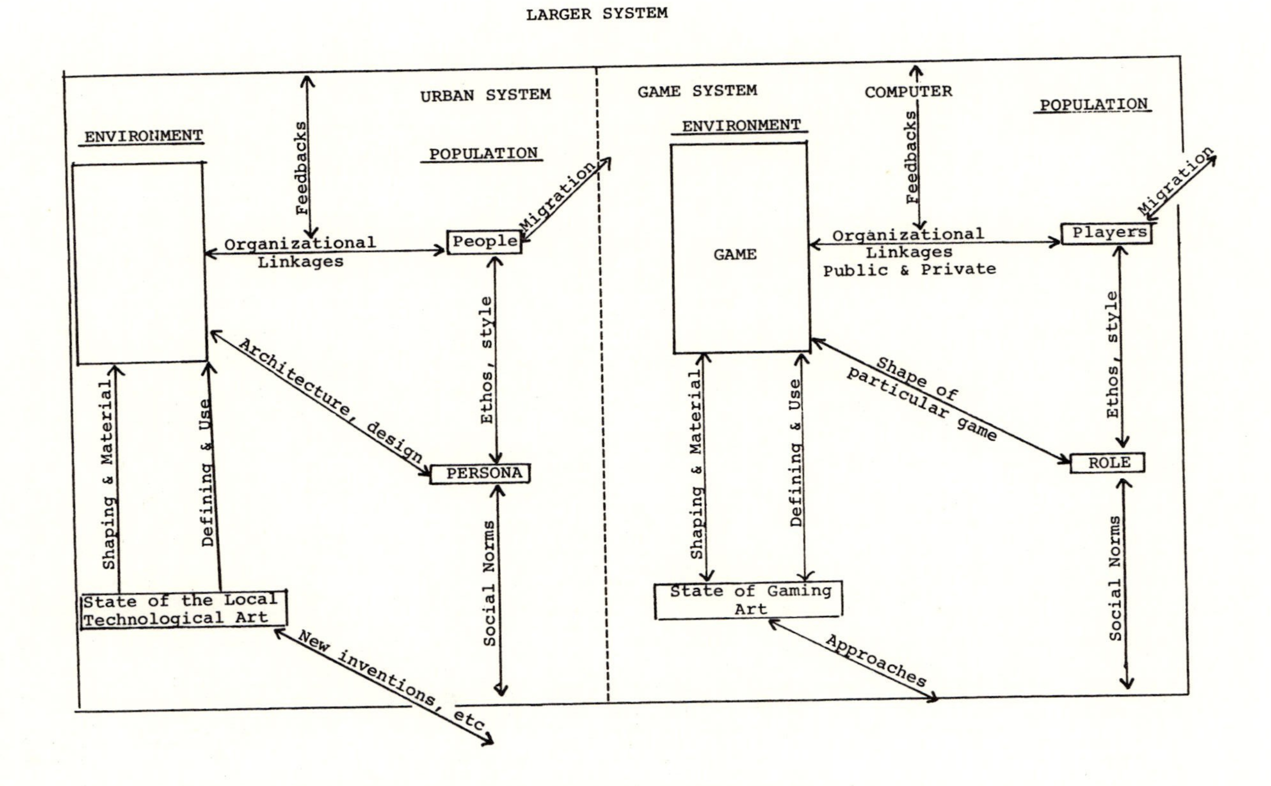

FIG. II. Symmetry of urban system components and game

state of the technological art would have on the environment. In the game, the computer takes the place of a larger system. If we social scientists can simulate the components and their relationships in a workable urban system game, we may be able to examine and gain a clearer understanding of those relationships.

The City I Model as an Example of the System Components

An example of the private development of a business goods establishment (BG1) in the *City I* model will illustrate the components of the system and the effect of persona on their inter-action in the urban system. (1) The BG1 is a *part of a larger economic system* than the local economy because its stock of raw materials and finished products (wholesale goods) is purchased from national markets and, in turn, sold to local industrial and commercial establishments. (2) The *natural resources* of the local community might influence the location of the BG1 within the local system. For example, the BG1 would not want to be located in a marsh land or on a steep slope. (3) The technology of the *City I* model requires that a BG1 cost a certain amount of money to construct and employ 1,000 workers from a high-income residence, 1,000 from a middle-income residence, and 1,000 from a low-income or slum residence. With this plant size and labor force, the BG1 is able to provide a given capacity of service. (4) Once the BG is developed, it becomes a *concretion*, that is, an actual building on the *City I* board, which is the physical representation of ownership and development. (5) Only if the owner of the BG1 sets a wage rate and employs a labor force from the population does the BG1 become more than a concretion and gain the capacity to provide business goods to the local system. (6) The wage rate and the price of business goods (also set by the BG1 owner) are part of the formal economic organization between the owner and the business community (potential employees and potential customers). *Transportation* of goods and workers and utility service for plant operation are also involved in the operation of the BG1 along the highways provided by the public decision-makers. (7) *Organization* is the means by which employees and customers are assigned to the BG1. The form that the organization takes in the *City I* model is the private market, that is, all customers choose to shop where they receive the best prices plus transportation charges to the BG, and all employees choose to work where they receive the best wage minus transportation charges to work. (8) The decision by the prospective developer concerning whether or not to build a BG1, the timing and location of the development, the method of financing the construction is the *persona* of individual decision-makers.

Another illustration of the system components and persona can be seen in the public sector of the model in the construction of a high-speed (Type I) road. The road itself is a *concretion* that is used for *communication* when vehicles operated by the *population* use it for travel. Where the road is constructed may be based on topographical or *natural resources* considerations, and the cost of the road and that it often requires land for rights-of-way is a function of the fixed *technology*. The road becomes *part of a larger system* when it is used and paid for in part by national taxpayers. Many of the decisions as to road location, timing of construction, and method of financing are made through the *persona* of the *City I* participants.

Values of the Theoretical Framework

We believe that modeling techniques will prove useful in developing an urban systems theory. With this paper we have attempted to show that one model contains and uses the same components and relationships that are found in an urban system. With development they should be able to realistically simulate those components and relationships. The theoretical framework has not been examined enough to see how far reaching it may be, but it is currently providing the Envirometrics staff with a basis for designing new models in order to gain a greater understanding of the urban system.

Models of this sort are designed and modified in a behavioral sense. That is, functional roles or the economic system or the social system are tested in the context of the rest of the model. Players from all walks of life are its critics. If a particular section of the model does not simulate reality to the satisfaction of the player and the designer, it can be changed or modified until, interacting with the rest of the model, it does simulate a real situation. Consequently, particular sections of the model can be continuously evolved and refined as experts and practitioners in particular fields try their hands at testing the reality of the model. The framework is, in effect, evolutionary.

Currently, we are finishing a refinement of *City I*, called *City II*, which should give us more experience with the theoretical framework. This model, as now planned, will add three new sets of variables: the use of three modes of transportation (car, bus and rapid rail) with the related consideration of cost in time and resources; a prejudice factor which will allow citizens to boycott stores, jobs, and residences; and a class-oriented leisure time-allocation factor. We are contemplating constructing still another refinement, which would add many more variables, such as migration, population characteristics, joint public and private land use and public institutions.

Downstream, we are planning a family of games, each one of which will cover a level of the country as a system (region, metropolitan area, city, and neighborhood). We are also working on a modular model concept. Under this concept the social, political and economic portions of an area would be broken down into categories by organizational type, activity, size, employment, and so on. If the players wanted to simulate Elmira, New York, for example, the operators would reach into the computer "bins" and plug in the right industrial mix, the right forms of government, the right population composition, using combinations of predetermined standard categories. Such a model would be truly "general" and would, for the first time, allow decision-makers, students and researchers in specific locales to simulate their particular area cheaply and efficiently. This methodology seems to show great promise for those who wish to try innovative planning concepts in a "model" city and test the results of alternative futures.

This, we feel, will be the ultimate value of urban systems simulation. This is the goal towards which Envirometrics is striving.

APPENDIX

Here we detail the breakdown of the components of an urban system as they are represented in *City I*.

Environment

I. Location within a larger system

A. The county area simulated in *City I* is located in the midwest area of the United States. The effect of the national system on the local system is explicit in the field of the economy:

1. The gross income earned by basic industries in the local system selling their products to national markets is dependent upon the state of the national economy as reflected in the national business cycle.

2. The interest rate on loans from national bankers depends upon the national business cycle.

3. The rate of return on national investments (speculative and conservative) is dependent upon the national business cycle.

4. The amount of federal-state aid received by the local government is a function of the local population.

B. The national system also has implicit effects on the local system:

1. The local system has a representative democratic government.

2. Capitalism is the economic organizing force for production and distribution of goods and services.

II. Topography and Natural Resources

The major topographical feature of the *City I* model is the location of the central city on a peninsula located on a lake. This feature restricts development from taking place to the west or north of the established city boundaries. Another constraint on development and, a political one, is the county boundaries to the east and south of the established city. Many topographical features, such as rivers, mountains, and marsh land, could be easily simulated in the model, but have not been up to this time.

III. Concretions

The fixed plant and equipment of the local system include large amounts of private and public development.

A. Private

1. Basic Industries
   a. Heavy Industry
   b. Light Industry

2. Commercial Establishments
   a. Business Goods
   b. Business Services
   c. Personal Goods
   d. Personal Services

3. Residences
   a. High-Income
   b. Middle-Income
   c. Low-Income
   d. Slum

B. Public

1. Transportation Routes (three types)

2. Terminals (three types)

3. Utility Plants

4. Utility Service

5. Municipal Service Units

6. School Units

These concretions provide the economic base, fixed investment, and social capital of the local system.

IV. Technology

Technological improvement is the means whereby given inputs produce greater or better outputs, or whereby new inputs are developed. The *City I* model assumes that private technology is constant, that is, the productivity of the workers and equipment does not increase, although if periodic renovation is not undertaken, productivity does *decrease*. Technology is also constant in the public sector: travel cannot be improved by introducing new modes of transportation. (This capacity for technological advancement is being included in a new model currently under development by Envirometrics.) Fixed technology is represented in the model by a number of constants, such as development costs, employment needs, and capacity limits.

V. Communications

There are several types of communication within the local system:

A. Transportation - The spatial arrangement of land uses within the local system is very much dependent upon the costs of transportation to necessary linking points (terminals, residences, and employment centers).

B. Mass Media - A team acts as the press, television, and radio services by using a blackboard to alert other participants to the newsworthy events of the local system.

C. Public Utilities - The public sector supplies utilities, representing water, sewerage, electricity, gas, and telephone systems, to the local system based upon supply and demand conditions.

D. Inter-team Communication - The teams use phones, forms, and verbal contact to accomplish deals, bargaining, etc.

E. Intra-team Communication - A team has a home-base desk where its members meet to set priorities and objectives, to decide on division of labor, etc.

Population

The residents of the local system are of four socio-economic classes: high-, middle-, low-income, and slum. The age, racial, and other characteristics of the population are based upon the national data summarized in the following table:

	High	Middle	Low	Slum
Population in a Residential Unit	3800	3400	2300	2300
School Age per Residential Population	760	850	690	690
Percent Non-White	5	10	20	40
Educational Attainment (Median School Years)	12	11	9	8

This population provides the labor force for production and distribution, the demand for local consumption of private and public goods and services, and the voting base for the local political-governmental system.

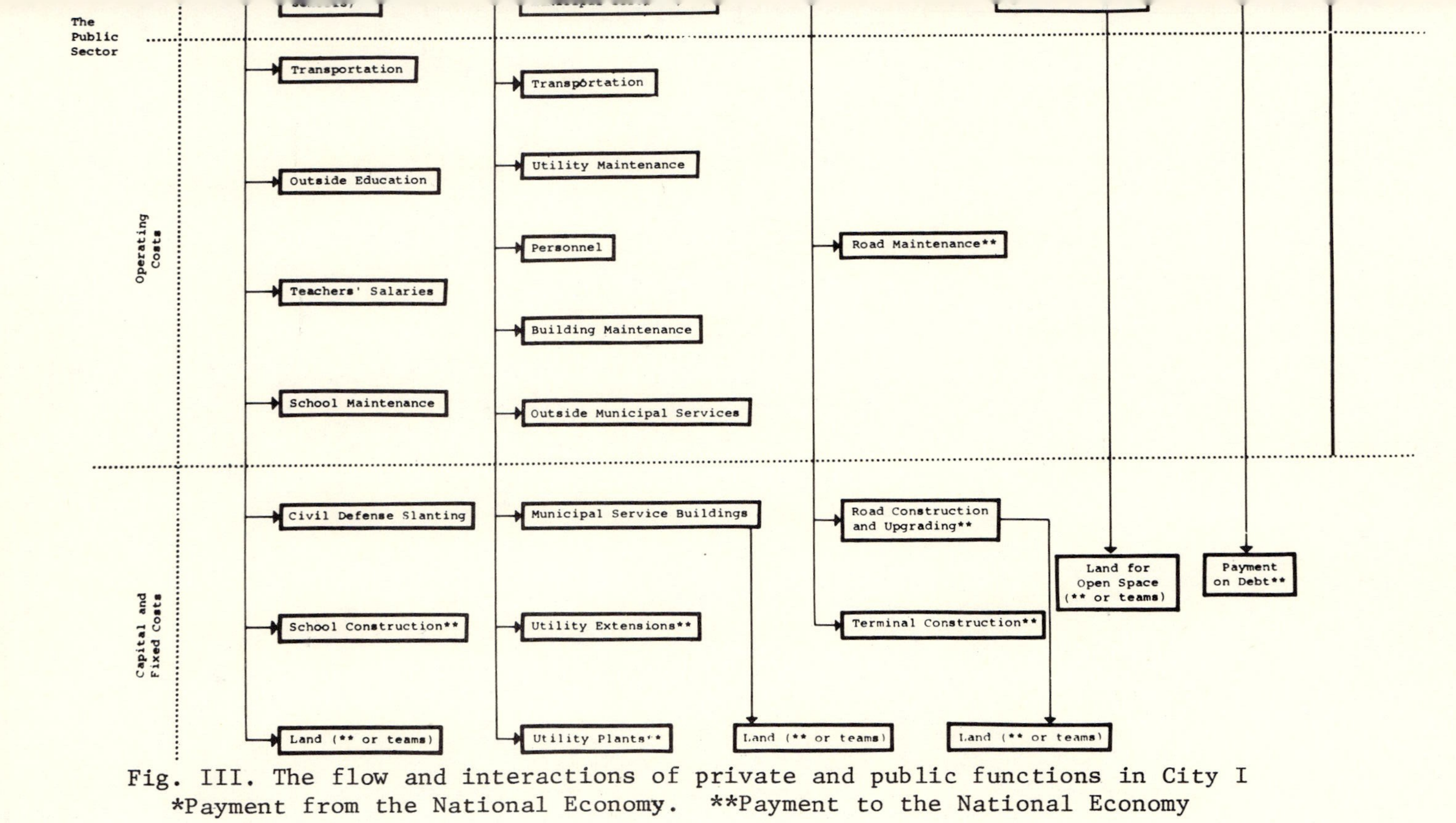

Fig. III. The flow and interactions of private and public functions in City I
*Payment from the National Economy. **Payment to the National Economy

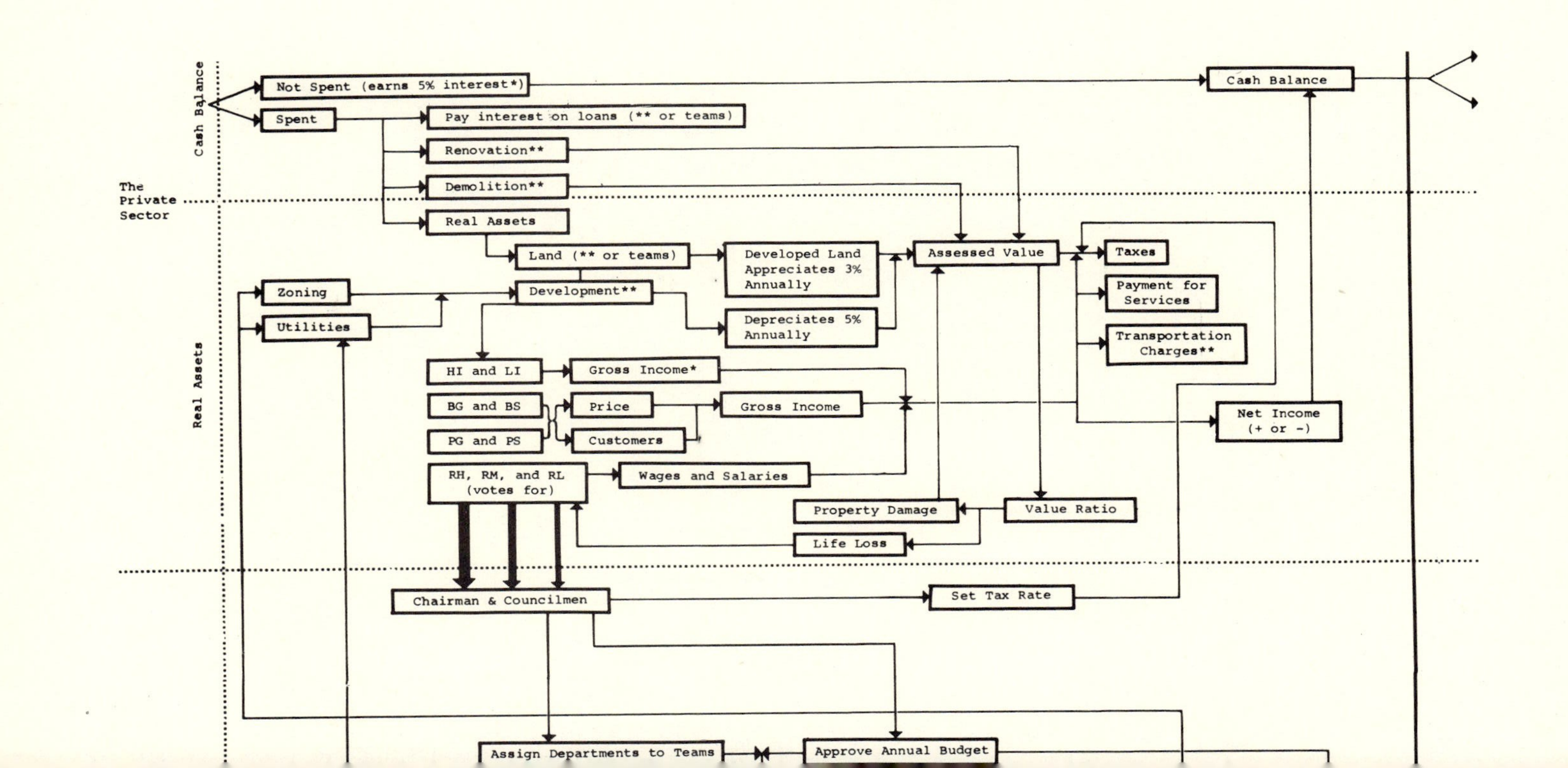

Cash Balance
The Private Sector
Real Assets
Not Spent (earns 5% interest*)
Spent
Pay interest on loans (** or teams)
Renovation**
Demolition**
Real Assets
Land (** or teams)
Development**
Zoning
Utilities
Developed Land Appreciates 3% Annually
Depreciates 5% Annually
Assessed Value
Taxes
Payment for Services
Transportation Charges**
Cash Balance
HI and LI
Gross Income*
BG and BS
Price
Gross Income
PG and PS
Customers
Net Income (+ or -)
RH, RM, and RL (votes for)
Wages and Salaries
Property Damage
Value Ratio
Life Loss
Chairman & Councilmen
Set Tax Rate
Assign Departments to Teams
Approve Annual Budget

Organization

Figure III represents the organization of the flow and interaction of both private and public functions present in the model. In many cases there are more detailed flows, but in no case are these pathways deterministic. The actual method of making operational any of these linkages depends on the players themselves.

SELECTED READINGS

General

Hauser, P.M. and Schnore, L.F., *The Study of Urbanization*, N.Y: John Wiley & Sons (1965).

Greer, S., *The Emerging City*, N.Y: The Free Press (1962).

Hirsch, W.Z., *Urban Life and Form*, N.Y: Henry Holt and Co. (1963).

Wirth, L., "Urbanism as a Way of Life," *American Journal of Sociology, XLIV* 1-24, (July, 1938).

Ullman, E.L., Presidential Address: "The Nature of Cities Reconsidered," *Papers and Proceedings of the Regional Science Association*, 9, 7-23, (1962).

Strauss, A.L., *Images of the American City*, New York: The Free Press of Glencoe (1961).

Lynch, K., *Image of the City*, Cambridge: M.I.T. Press (1960).

Banfield, E.C., "The Political Implications of Metropolitan Growth," *Daedalus*, XC, 61-78, (Winter 1961).

I. *The Search for a Unified Theory*

Klausner, S.Z., (ed.), *The Study of Total Societies*,

Garden City, N.J.: Doubleday (1967).

Duncan, O.D., "From Social Systems to Ecosystems," *Sociological Inquiry, XXXI* (2), 140-149, (Spring 1961).

Boulding, K.E., Toward a General Theory of Growth and Policy," in J.E. Spengler and O.D. Duncan (eds.), *Population Theory and Growth,* Glencoe, Ill., The Free Press of Glencoe, 109-124 (1956).

Szasz, T., *The Myth of Mental Illness,* New York: Harper & Row (1961).

Miller, J.G., "Living Systems: Structure and Process," *Behavioral Science, 10* (3 & 4), 337-379 (October 1965).

Boulding, K.E., *A Reconstruction of Economics,* New York: Science Editors, Inc., John Wiley & Sons (1962).

Smelser, N., *The Sociology of Economic Life,* Englewood Cliffs, New Jersey: Prentice Hall, Inc. (1963).

Von Bertalanffy, L., "General Systems Theory," *Yearbook of the Society for General Systems Research, 1* 1-10, (1956).

Buckley, W., *Sociology and Modern Systems Theory,* Englewood Cliffs, N.J.: Prentice Hall (1967).

Buckley, W., (ed.), *Modern Systems Research for the Behavioral Scientist,* Chicago: Aldine Publishing Co., (1968).

Bauer, R.A., (ed.), "The State of the Nation's Social Systems Accounting," B.M. Gross in *Social Indicators,* Cambridge: M.I.T. Press (1966).

II. *Topography*

Meggers, B.J., "Environmental Limitations on the Development of Culture," *The American Anthropologist,* Memoir 77 (1954).

Thomas, W.L. (ed.), *Man's Role in Changing the Face of the Earth,* Chicago: University of Chicago Press (1964).

Brown, H., *The Challenge of Man's Future,* New York: Viking Press (1954).

Martin, W.T., "Urbanization and Natural Power to Requisition Resources," *Political Science Review, V* 93-97,(1962).

Wolman, A., "The Metabolism of Cities," *Scientific American, XXCIII,* (3), 178-190, (September 1965).

Bollens, J.C. and Schmandt, H.J., "Nature and Dimensions of the Metropolitan Community," *The Metropolis; Its People, Politics, and Economic Life,* New York: Harper & Row, 32-57 (1965).

Fisher, J.L., "Natural Resources - Wise Use of the World's Inheritance," in W.R. Ewald, Jr. (ed.), *Environment and Policy, The Next Fifty Years,* Bloomington: Indiana University Press, 329-359 (1968).

Perloff, H. and Wingo, L. Jr., "Natural Resource Endowment and Regional Economic Growth," in J. Friedmann and W. Alonso (eds.), *Regional Development and Planning,* Cambridge: M.I.T. Press, 215-239 (1964).

Barnes, C.P. and Marschner, F.J., "Our Wealth of Land Resources," in *Land: The Yearbook of Agriculture, 1958,* Washington, D.C: Government Printing Office, 10-18, (1958).

Lynch, K., "The City as Environment," *Scientific American, CCXIII* , (3), 209-219, (September 1965).

Gibbs, J.P. and Martin, W.T., "Urbanization and Natural Resources: A Study in Organizational Ecology," *American Sociological Review, XXIII* 266-277, (June 1958).

III. *Technology*

Brown, H., *The Next Hundred Years: Man's Natural and*

Technological Resources, New York: Viking Press (1957).

Wolman, A., "Disposal of Man's Wastes," in W.L. Thomas, Jr. (ed.), *Man's Role in Changing the Face of the Earth*, Chicago: University of Chicago Press, 807-815, (1956).

White, L.A., "Energy and the Evolution of Culture," *The Science of Culture*, New York: Farrar, Straus and Cudahy, Inc. (1949).

Meier, R., "The Evolving Metropolis and New Technology," Chapter 2 in H. Perloff (ed.), *Planning and the Urban Community*, Pittsburgh: University of Pittsburgh Press (1961).

Cottrell, W.F., "Death by Dieselization: A Case Study in the Reaction to Technical Change," *American Sociological Review, XVI* , 358-365 (1951).

Ogburn, W.F., "Population, Private Ownership, Technology, and the Standard of Living," in J.J. Spengler and O.D. Duncan (eds.), *Population Theory and Policy*, Glencoe, Ill.: The Free Press of Glencoe, 152-158 (1956).

Kindleberger, C.P., *International Economics*, Homewood, Ill.: Richard D. Irwin, Chapters 5, 6, 7 (1968).

Meadows, P., "The City, Technology, and History," *Social Forces, XXXVI* , 141-147 (1957).

Moore, W.E., "Utilization of Human Resources Through Industrialization," in J.J. Spengler and O.D. Duncan (eds.) *Demographic Analysis*, Glencoe, Ill.: The Free Press of Glencoe, 518-531 (1957).

Meier, R.L., "Living with the Coming Urban Technology," in E. Geen, J.R. Lowe and K. Walker (eds.), *Man and the Modern City*, Pittsburgh: University of Pittsburgh Press, 59-70 (1963).

Fleisher, A., "The Influence of Technology on Urban Forms," in L. Rodwin (ed.), *The Future Metropolis*, New

York: George Braziller, 64-79 (1961).

Nelson, R.R., Peck, M.J. and Kalachek, E.D., *Technology, Economic Growth, and Public Policy*, Washington, D.C.: Brookings Institution (1967).

IV. *Location Within the Larger System*

Hawley, A., *Human Ecology*, New York: Ronald Press, 77-103 (1950).

Hatt, P.K. and Reiss, A.J., (eds.), *Cities and Society*, Glencoe, Ill.: The Free Press of Glencoe (1957).

Isard, W. and Kavish, R., "Economic Structural Interrelationships of Metropolitan Regions," *American Journal of Sociology, IX* , 152-162 (September, 1954).

Duncan, O.D., *Metropolis and Region*, Baltimore: Johns Hopkins Press for Resources for the Future, Chapter 3, (1960).

Bogue, D.J., *The Structure of the Metropolitan Community*, Ann Arbor: University of Michigan, Chapters 1 and 3, (1950).

Vining, R., "A Description of Certain Spatial Aspects of an Economic System," *Economic Development and Cultural Change, III* , 147-195 (1955).

Friedman, J. and Alonso, W., *Regional Development and Planning*, Cambridge: M.I.T. Press, Chapters 12, 13 and 15, (1964).

V. *Concretions*

Zarbaugh, H.W., "The Natural Areas of the City," in G.A. Theodorson (ed.) *Studies in Human Ecology*, New York: Harper & Row (1961).

Stanislawski, D., "The Origin and Spread of the Grid-

Pattern Town," in G.A. Theodorson (ed.), *Studies in Human Ecology*, New York: Harper & Row, 294-303 (1961).

Quinn, J.A., "The Burgess Zonal Hypothesis and Its Critics," *American Sociological Review, III*, 210-218 (1940).

Davie, M.R., "The Pattern of Urban Growth," in G.A. Theodorson (ed.), *Studies in Human Ecology*, New York: Harper & Row, 77-92 (1961).

Madden, C.H., "Some Spatial Aspects of Urban Growth in the United States," *Economic Trends and Cultural Change, IV*, 371-387 (July 1956).

Beaujeu-Garnier, J. and G. Chabot, *Urban Geography*, New York: John Wiley & Sons Inc., 205-276 (1967).

Blumenfeld, H., "On the Concentric Circle Theory of Urban Growth," *Land Economics, XXV*, 209-212 (May 1949).

Burgess, E.W., "The Determination of Gradients in the Growth of the City," *American Sociological Review, XXI*, 178-184 (1956).

Ross, E.A., "The Location of Industries," *Quarterly Journal of Economics, X* (3), 247-268 (April 1968).

Green, C.M., *The Rise of Urban America*, New York: Harper & Row (1965).

__________, *American Cities in the Growth of the Nation*, New York: John DeGraff (1957).

Lynch, K., "The Pattern of the Metropolis," in L. Rodwin (ed.), *The Future Metropolis*, New York: George Braziller, 103-128 (1961).

Weber, M.M., *Explorations Into Urban Structure*, Philadelphia: University of Pennsylvania Press (1963).

VI. *Transportation and Communication*

Colley, C.H., "The Theory of Transportation," *Sociological Theory and Social Research*, New York: Henry Holt & Co. (1930).

McKensie, R.D., "Spatial Distance and Community Organization," *American Journal of Sociology, XXXIII*, 28-42 (1927).

Meier, R.L., *A Communications Theory of Urban Growth*, Cambridge: M.I.T. Press (1962).

Bello, F., "The City and the Car," in Editors of Fortune Magazine (eds.), *The Exploding Metropolis*, Garden City, N.Y.: Doubleday, 32-61 (1957).

Wingo, L., Jr., *Transportation and Urban Land*, Washington, D.C.: Resources for the Future (1961).

Gilman, R.H., "Terminal Planning," in H.W. Eldredge (ed.), *Taming Megalopolis: Vol. I: What Is and What Could Be*, Garden City, N.Y.: Doubleday, 383, 399 (1967).

Gakenheimer, R.A., "Urban Transportation Planning: An Overview," in H.W. Eldredge (ed.), *Taming Megalopolis: Vol. I: What Is and What Could Be*, Garden City, N.Y.: Doubleday, 392-411 (1967).

Weber, M.M., "The Roles of Intelligence Systems in Urban Systems Planning," in H.W. Eldredge (ed.), *Taming Megalopolis: Vol. II: How to Manage an Urbanized World*, Garden City, N.Y.: Doubleday, 644-666 (1967).

Schrodinger, E., "Order, Disorder and Entropy," in W. Buckley (ed.), *Modern Systems Research for the Behavioral Scientist*, Chicago: Aldine Publishing Co. (1968).

Raymond, R.C., "Community, Entropy and Life," in W. Buckley (ed.), *Modern Systems Research for the Behavioral Scientist*, Chicago: Aldine Publishing Co. (1968).

Brillouin, L., "Thermodynamics and Information Theory," in W. Buckley (ed.), *Modern Systems Research for the Behavioral Scientist*, Chicago: Aldine Publishing Co. (1968).

Ackoff, R.L., "Towards a Behavioral Theory of Communication," in W. Buckley (ed.), *Modern Systems Research for the Behavioral Scientist*, Chicago: Aldine Publishing Co. (1968).

Owen, W., *Cities in the Motor Age*, New York: Viking Press (1959).

———, "Transportation," *Annals of the American Academy of Political and Social Science, CCXIV*, 30-38 (November 1957).

Deutsch, K., "On Social Communication and the Metropolis," in L. Rodwin (ed.), *The Future Metropolis*, New York: George Braziller (1961).

Lansing, J.B., *Transportation and Economic Policy*, New York: Macmillan, 167-400 (1966).

Dyckman, J., "Transportation in Cities," *Scientific American, CCXIII* (3), 162-177 (September 1965).

Feldman, M.L., "Transportation: An Equal Opportunity for Access," in W.R. Ewald (ed.), *Environment and Policy: The Next Fifty Years*, Bloomington: Indiana University Press, 167-208 (1968).

Sampson, R.J. and M. Farris, *Domestic Transportation, Practice, Theory, and Policy*, Boston: Houghton Mifflin, Parts 1, 2; Chapter 14, Part 4 (1966).

VII. *Population*

Hawley, A.H., *Human Ecology*, New York: Ronald Press, Chapters 7, 8 and 9 (1950).

Taeuber, C., "Taking an Inventory of 180 Million People: The U.S. Censuses," in R. Freedman (ed.), *Population: The Vital Revolution*, New York: Anchor, 84-99 (1965).

Ogburn, W.F., "On the Social Aspects of Population Changes," in J.J. Spengler and O.D. Duncan (eds.), *Population Theory and Policy*, Glencoe, Ill.: The Free Press of Glencoe, 435-440 (1956).

Cowgill, D.O., "The Theory of Population Growth Cycles," in J.J. Spengler and O.D. Duncan (eds.), *Population Theory and Policy*, Glencoe, Ill.: The Free Press of Glencoe, 125-134 (1956).

Hauser, P.M., "The Labor Force as a Field of Interest for the Sociologist," in J.J. Spengler and O.D. Duncan (eds.), *Demographic Analysis*, Glencoe, Ill.: The Free Press of Glencoe, 484-491 (1957).

Boulding, K., "The Malthusian Model as a General System," *Social and Economic Studies, IV*, 195-205 (September 1955).

Saunders, H.W., "Human Migration and Social Equilibrium," in J.J. Spengler and O.D. Duncan (eds.), *Population Theory and Policy*, Glencoe, Ill.: The Free Press of Glencoe, 219-229 (1956).

Macinko, G., "Saturation: A Problem Evaded in Planning Land Use," *Science, CXLIX*, 516-521 (July 30, 1965).

Carrothers, G.A.P., "A Historical Review of the Gravity and Potential Concepts of Human Interaction," *Journal of the American Institute of Planners, XX* (2), 94-102 (1956).

Boulding, K., "Toward a General Theory of Growth," in J.J. Spengler and O.D. Duncan (eds.), *Population Theory and Policy*, Glencoe, Ill.: The Free Press of Glencoe, 109-124 (1956).

Barclay, G.W., *Techniques of Population Analysis*, New York: John Wiley & Sons (1958).

Thompson, W.S., *Population Problems*, New York: McGraw-Hill (1930).

George, T.A., "What is a Population?" *Philosophy of Science, XXII*, 272-279 (1955).

VIII. *Organization*

Hawley, A.H., *Human Ecology*, New York: Ronald Press, Chapters 10-12 (1950).

Reiss, A.J., Jr., "The Sociological Study of Communities," *Bobbs-Merrill Reprint Series, S-233.*

Ramsoy, O., *Social Groups as Systems and Subsystems*, New York: The Free Press (1963).

Meier, R.L., "Explorations in the Realm of Organization Theory III: Decision Making, Planning, and the Steady State," *Behavioral Science*, 235-244 (July 1959).

Feiblemann, J. and J.W. Friend, "The Structure and Function of Organization," *Philosophical Review, LIV*, 19-44 (January 1945).

Terrien, F.W. and D.L. Mills, "The Effect of Changing Size Upon the Internal Structure of Organizations," *American Sociological Review, XX* (1), 11-13 (February 1955).

Duncan, O.D., "Optimum Size of Cities," in P.K. Hatt and A.J. Reiss (eds.), *Cities and Societies*, Glencoe, Ill.: The Free Press of Glencoe, 759-772 (1957).

Clark, C., "The Economic Functions of a City in Relation to Its Size," *Econometrics, XIII* (2), 97-113 (April 1945).

Schlettler, C., "Relation of City-Size to Economic Services," *American Sociological Review, VIII*, 60-62 (1943).

Vance, R.B. and S. Smith, "Metropolitan Dominance and Integration," in P.K. Hatt and A.J. Reiss (eds.), *Cities and Societies*, Glencoe, Ill.: The Free Press of Glencoe, 189-200 (1957).

Parsons, T., *Structure and Process in Modern Society*, New York: The Free Press (1960).

Thompson, W.R., *A Preface to Urban Economics: Toward a Conceptual Framework for Study and Research*, Baltimore: John Hopkins Press for Resources for the Future (1965).

Vernon, R., *The Changing Economic Function of the Central City*, New York: Committee on Economic Development (1959).

Andrews, R.B., "Mechanics of the Urban Economic Base: Historical Development of the Base Concept," *Land Economics, XXX*, 164-172 (May 1954).

Tiebout, C.M., "The Urban Economic Base Reconsidered," *Land Economics, XXXII* (1), 95-99 (February 1956).

Isard, W. and R. Kavish, "Economic Structural Interrelations of Metropolitan Regions," in H.M. Mayer and C.E. Kohn (eds.), *Readings in Urban Geography*, Chicago: University of Chicago Press (1959).

Banfield, E.C., *Political Influence*, New York: The Free Press (1961).

IX. *The Urban Social Problem*

Boskoff, A., *Sociology of Urban Regions*, New York: Appleton-Century Crofts (1962).

Parsons, T., "The Principle Structures of Community," *Structure and Process in Modern Society*, New York: The Free Press, 250-279 (1959).

Reiss, A.J., Jr., "Functional Specialization of Cities," in P.K. Hatt and A.J. Reiss (eds.), *Cities and Society*, New York: The Free Press, 555-575 (1956).

Berry, B., "City-Size Distribution and Economic

Development," *Economic Development and Cultural Change, IX*, 573-588 (July 1961).

Blau, P.M. and W.R. Scott, *Formal Organizations*, New Castle, N.H.: Chandler Publishing Co. (1963).

Easton, D., *A Systems Analysis of Political Life*, New York: John Wiley & Sons (1965).

Martindale, D., *Institutions, Organizations and Mass Society*, Boston: Houghton Mifflin Co. (1966).

X. *Persona*

Firey, W., "Sentiment and Symbolism as Ecological Variables," in G.A. Theodorson (ed.), *Studies in Human Ecology*, Evanston: Row, Peterson, and Co. (1961).

Myers, J.K., "Assimilation to the Ecological and Social Systems of a Community," in G.A. Theodorson (ed.), *Studies in Human Ecology*, Evanston: Row, Peterson, and Co. (1961).

Jonassen, C.T., "Cultural Variables in the Ecology of an Ethnic Group," in G.A. Theodorson (ed.), *Studies in Human Ecology*, Evanston: Row, Peterson, and Co. (1961).

Bennett, J.W., "The Interaction of Culture and Environment in the Smaller Societies," *American Anthropologist, XLVI* (1944).

Hughes, E.C., "The Cultural Aspect of Urban Research," in L.D. White (ed.), *State of the Social Sciences*, Chicago: University of Chicago Press, 255-268 (1956).

Chapin, F.S., "The Protestant Church in an Urban Environment," in P.K. Hatt and A.J. Reiss (eds.), *Cities and Society*, New York: The Free Press, 505-515 (1957).

Strauss, A.L., *The American City*, Chicago: Aldine Publishing Co. (1968).

SECTION IV

SYSTEMS RESEARCH IN ORGANIZATION AND MANAGEMENT

The utilization of general systems theory in the development of a comprehensive theory for the analysis, design, and administration of social organizations is currently the subject of extensive interest, study and research. The widespread interest and involvement in this subject area is reflected, in part, by the membership of the Society in the recent formation of the Organization and Management Studies Group and by the presentation of the initial symposium by this group at the fourteenth annual meeting. This section of the Proceedings includes selected papers from this symposium, which present the results of study and research concerned with the development of a comprehensive theory of and the application of general systems concepts to social organizations.

The principal approaches to organization research and the differences in these approaches is discussed by Stogdill, who points to the unifying role of general systems theory for the development of a general theory of organization. Each of these approaches, management theory, operations research, and behavioral science, differs in its tradition and methodology, giving rise to the tendency not to appreciate or utilize the research results of the other specialties. Although management theorists have made the most extensive efforts to incorporate concepts from the other specialties into their theories, general systems theory is suggested as the appropriate vehicle for achieving the desired theoretical integration.

Published studies that have combined management and behavioral variables as inputs are cited as examples of the trend toward the integration of two of the different approaches to organization research. Numerous additional variables are also identified, and each is described as being quantifiable and researchable. The development of techniques of measurement, multidimensional models, and a general theory of organization is a major challenge for general systems theorists, but there is also considerable promise for achieving significant success.

In a presentation that features a tutorial application of general systems theory, Duncan evaluates traditional approaches or mental models of organization and examines the suitability of the living systems model for organization analysis and design. Some additional aspects of living systems theory are also suggested for further research. The traditional approaches are viewed as successive amplifications of organization theory, but both individually and collectively these models are seen as being inadequate to provide a comprehensive theory of organization. General systems theory, however, is proposed as the foundation for a new mental model of organizations, for it seeks to develop a science of behavior.

Living systems concepts were utilized in a management training program at General Motors Institute as a framework for organization analysis, design, and problem solving. Student managers participating in a team project were divided into subteams that examined the production system, information needs of top management, information systems, and the structure of the management hierarchy, respectively. The integration of the results of the analyses provided student managers with the best insight into systemic interrelatedness.

The experience gained from the program indicates that more definitive mechanisms for modifying mental models are needed, cautions against the basic error of confusing the negentropic view of organizations with bureaucratic rigidity, but provides no conclusive evidence yet that organizational systems designed with a living systems orientation are more flexible than those designed traditionally, although it is hoped that such would be inherent in the design. To gain further insights into questions raised by the program experience, areas for future study and research including the effect of the living systems model upon organizational design, managerial performance, management education and development, and the theory of management are discussed.

An open-system model is described by Baker as the conceptual framework for the study of the organizational metamorphosis of a mental hospital to a community mental health center. In contrast to the closed-system approach of traditional organization theory, the open-system model permits the incorporation of inter- and extraorganizational relationships into a dynamic conceptual model of the adaptive process by which a changing mental hospital modifies existing goals and identifies and works toward the achievement of new goals. The open-system model is more comprehensive than the more traditional goal model, for it combines the final outcome analysis of the goal model approach with change process analysis which focuses on the intermediate outputs of the system.

A mental hospital system has multiple goals and simultaneous task performance that relate to patient admissions, therapy, research, education, and several other activities. These activities compete for resources, and goal priorities for resource allocation are influenced increasingly by belief systems, notably the belief system of mental health professionals that is termed the community mental health ideology.

The process flow of a mental hospital system consists of inputs of people, values, economic resources, facilities, and technology that are transformed or converted through subsystem variables, internal structures, attitudes, and activities into outputs of treated patients returned to the community, resident chronic patients that receive ongoing treatment, trained and educated personnel, and organizational morale. The functioning of the operating subsystems of input, conversion, and output find an appropriate conceptualization in equifinality, for identical final states can be achieved from alternative initial conditions and configurations of subsystems. The chief modifications in the managerial subsystem are expected to involve the decentralization of authority, coordination of subsystems, and interaction with the external environment. The strategies that are selected for relating to the community environment will affect significantly the nature of services and care given to the community.

The application of normative general systems theory to social systems, especially in the area of public policymaking, is advocated by Dror in the concluding selection. This view differs from that of many social scientists and policy practitioners, who feel that the diffuseness of characteristics, complexity of dimensions, and arbitrariness of events of social systems precludes formalization in terms of systems models and that such attempts are useless until sufficient hard data is obtained to permit rigorous analysis. In contrast, it is argued that the basic concepts of general systems theory can make significant contributions to the analysis of social macro-phenomena, but the philosophical assumptions, quantitative models, optimization techniques, and simulation methods are less useful in this type of normative treatment.

The public policymaking system is described from the general systems view as being dynamic, open, transient, and complex and is interwoven and overlapping with other social macro-systems. A need is expressed for a broad systems approach and several suggestions for more intensive rational-analytic approaches including explicit strategy decisions, learning feedback, policy analysis, education for policymaking, the development of policy science as a formalized professional field, and systematic meta-policymaking.

Dror suggests that with an intense effort to develop an interdisciplinary general systems approach to policy science to produce professional policymakers, modest changes in the quality of policymaking are to be reasonably expected. This limited degree of change, however, would be radical in terms of past experience. The political feasibility of accomplishing such changes is difficult to evaluate, but if one believes that the normative general systems approach can contribute to improved policymaking, then an effort should be made to help make such an approach feasible politically.

In conclusion, the papers in this section deal with social systems and the system models described are largely conceptual and are not concrete or physical. The discussion of mental models by Duncan is the most explicit in this respect, but the idea is more implicit in other presentations. Since the system models are largely conceptual, one might be led to ask along with Milton Rubin in his presidential address: "Where does the social system exist? Is it only a network of relationships that exists only in our minds? Is it then an abstraction if it exists only in our minds? Curiously then we are in the social system, yet the social system is in us. I am not talking about a theory of The Social System, which of course would be an abstraction and in us, but about The Social System, if there is such a thing." Perhaps a paraphrase of a statement by Stogdill is appropriate here in saying that the ability to verbalize the right questions offers the hope of finding the answers.

Lawrence L. Schkade

THEORY VERSUS MEASUREMENT IN RESEARCH ON ORGANIZATION

Ralph M. Stogdill*

ABSTRACT

Classical management theorists, operations researchers, and behavioral scientists have been concerned with different sets of variables and hold different attitudes toward measurement. They have been concerned respectively with structures, operations, and relationships. General systems theory would seek to incorporate structural, operational, and behavioral variables in the same system. The resulting system would be based on variables that are measurable and researchable.

No one has yet produced a complete *theory of organization*. The oldest body of knowledge, produced by *management theorists*, explains departmentation and the differentiation of positions with respect to status and function. That is, it accounts for the structure of authority and responsibility differentials. The *operations research approach* is concerned with the processes by means of which inputs are converted into outputs. *Behavioral theories* are concerned essentially with the relationships between managers, supervisors, and workers. If we wish to characterize each of these approaches with a single word, we might say that they are concerned respectively with structures, operations, and relationships.

* Professor of Management Science, College of Administrative Science, the Ohio State University, 11775 South College Road, Columbus, Ohio 43210.

Each of the three approaches has developed its own tradition of research methodology. Membership in none of the schools or traditions has been characterized by a high degree of respect for other approaches. The traditions differ particularly in their views concerning (1) the variables that are important for the study of organization, (2) the design of research, and (3) the use of measurement and mathematics in the analysis of organizational problems.

The operations researchers have been most successful in integrating *measurement* with mathematical models. They have probably been least successful in developing a body of explanatory theory. The management theorists have been most successful in developing theory, and least successful in applying measurement and mathematical models of their theories. The behavioralists occupy an intermediate position with respect to both theory and measurement. Their theories tend to be concerned with rather small sets of variables. The variables are usually capable of measurement. The analytical methods tend to be statistical in nature.

Many behavioralists are handicapped by their open contempt for the management theories, which they regard as outmoded and useless. They seem to overlook the fact that authority hierarchies, departmentation, and functional differentiation are realities of organization - realities which they accept readily enough in the university, where most of them are located. The work that people do and the operations of the organization are often regarded as unimportant. The engineering oriented researchers and model builders, on the other hand, tend to disregard the behavioral and structural aspects of organization on the ground that they are represented by judgmental variables which cannot be reliably measured. The classical theorists, caught in the cross fire, have been attempting to incorporate the behavioral and operational concepts in their theoretical systems. The end result of these differences should be the development of theories that are both more comprehensive and more adequate than those now available. General systems theory should be able to contribute toward the desired theoretical integration.

Parsons (1951), Homans (1950), and Stogdill (1959) have developed theories which combine the management and behavioral variables. Parsons, seeking to explain the societal system as a whole, and Homans, seeking to explain the human group, used *actions*, *interactions*, and *sentiments* as inputs. Stogdill used *performances*, *interactions*, and *expectations*, each carefully defined, as input variables for a theory of organization. It has been possible to conduct research which accounts for the structure of differentiated functions (Stogdill and Shartle, 1956), interaction structures and authority relationships (Stogdill, 1957), and the relations of management behavior to employee satisfaction and group performance (Stogdill, 1965). Classes of *variables* such as the following have been found to be quantifiable.

Level of position in the status hierarchy

Authority

Responsibility

Patterns of supervisory behavior

Patterns of work performance

Who works with whom

Member expectations

Member satisfactions

Group performance and outputs

When the above sets are broken down into their components, more than 100 variables are involved in an attempt to study superiors and subordinates on the same variables. A primary weakness in the above list is that it does not contain any variables that describe organizational operations. Research is being planned which will incorporate measures of physical inputs, operations or processing, and physical outputs. This is a particularly difficult proposal, since processing and units of output must be attributed to individual workers rather than to the plant as a whole. It will probably be necessary for the research staff to maintain its own accounting system in order to obtain valid indices of physical inputs, processing operations, and outputs.

Since organizations involve several diverse sets of variables, it has not as yet been possible to develop a single mathematical model that includes all of them. However, it has been demonstrated that all the variables are quantifiable and researchable. The integration of the various sets of variables into a single system stands as a challenge to the general systems theorist.

The student of organization is confronted by several difficulties. The first is that involved in developing theories that account for all the relevant dimensions of organization. Such theory is yet to be developed. The second major problem is that concerned with the development of measures for the different variables involved. A third, and by no means minor problem, is that involved in bringing diverse sets of variables together in the same analytical design or model. The very fact that we are able to verbalize these problems offers hope for their solution in the not too distant future.

REFERENCES

Homans, G.C. *The Human Group*, New York, Harcourt, Brace (1950).

Parsons, T., *The Social System*, Glencoe, Ill, The Free Press of Glencoe (1951).

Stogdill, R.M., and Shartle, C.L., *Patterns of Administrative Performance*, Columbus, Ohio State University, Bureau of Business Research Monograph No. 81 (1956).

Stogdill, R.M., Responsibility and Authority Relationships, *Leadership and Structures of Personal Interaction*, Columbus, Ohio State University, Bureau of Business Research Monograph No. 84 (1957).

_____________, *Individual Behavior and Group Achievement*, New York, Oxford University Press (1959).

_____________, *Managers, Employees, Organizations*, Columbus, Ohio State University, Bureau of Business Research Monograph No. 125 (1965).

TRAINING BUSINESS MANAGERS IN GENERAL SYSTEMS CONCEPTS

Daniel N. Duncan*

ABSTRACT

In their conceptualizations of the organization, business managers have used five traditional mental models. Research at the General Motors Institute indicates that general systems theory provides the foundation for a new mental model of the organization; a mental model which may overcome some of the limitations of the more traditional views or frameworks. This paper outlines the four assumptions behind and the central content of a training program for practicing management personnel which advocates a general systems mental model of the organization. The hypotheses which indicate further research conclude the discussion.

The complexity and dynamic nature of contemporary business organizations strains the limits of scholarly understanding and conceptual tools used by practitioners. The search for more powerful instruments of thought resulted in consideration of the theory and science of *system*. Four assumptions underlying the utilization of system theory in the understanding of organization and management are reported below.

---

*Research Specialist, Research and Development Staff, Management and Organization Development Department, General Motors Institute, 1700 W. 3rd Avenue, Flint, Michigan. 48502.

1. ASSUMPTIONS

First it is assumed that:

Managers of and scholars of the business organization hold, are influenced by, and use mental models.

What is a mental model and what is its source? Boulding (1965a, p. 5-6) discussed the subjective knowledge we have of the world and called this accumulation of perceptions the *Image*. Forrester (1961) used the phrase "mental model" to describe the image one holds of an organization:

"A mental image or verbal description in English can form a model of corporate organization and its processes. The manager deals continuously with these mental and verbal models of the corporation. They are not the real corporation. They are not necessarily correct. They are models to substitute in our thinking for the real system that is being represented. "[p. 49-50] Zalkind & Costello (1962, p. 222) reported that the term "stereotyping" was first used by Walter Lippmann in 1922 to describe "pictures in people's heads." Berelson & Steiner (1964, p. 99) reported that the facts of raw sensory data are themselves insufficient to produce or explain the coherent picture of the world as the normal adult experiences it. They (Berelson & Steiner, 1964, p. 99) commented that the perceived visual world, for example, contains many objects, not millions of discrete pinpoint impressions. Elaborating on Forrester's phrase, a mental model is a coherent picture or image, that is, it is a stereotype of the organization. The mental model is a perceptual image, or in Forrester's (1961) words, a "Mental picture of how the organization functions . . .[p. 57]."

Mental models are the after images of accumulated experience. They are images made from puzzle-like pieces of experience fastened together by bonds of learned associations. In Boulding's (1956a) words, "The image is built up as a result of all past experience of the possessor of the image [p. 6]." The constituent, puzzle-like parts of one's mental model are perceptions. A mental model is not a fixed or rigid mural but rather alters with experiences, or according to Boulding (1956a), "As each event occurs ... it alters my knowledge structure or my image [p. 6]." Some experiences may fail to alter one's mental model. Sensory inputs, according to Berelson & Steiner (1964), "Do not act on an empty organization; they interact with certain predispositions and states already there ... [p. 99-100]." It is contended that one's mental model may in part constitute the "predisposition and states already there" identified by Berelson & Steiner.

There is a discrepancy between concious experience and stimuli which can be accounted for in part by the perception processes of selection and distortion, involving perceptual organization and closure. In their discussion of selection, Berelson & Steiner (1964) reported,

"Which stimuli get selected depend upon three major factors: the nature of the stimuli involved; previous experience or learning as it affects the observer's expectations (what he is prepared or "set" to see); and the motives in play at the time, by which we mean his needs, desires, wishes, interests, and so on - in short, what the observer wants or needs to see and not to see." [p. 99-100].

If it can be assumed that one's mental model represents one's "experience or learning" or influences one's "set" then it can be concluded that the mental model has sizable influence on the selection of information that precedes perception formation.

To summarize, the mental model is a collage of perceptions, an image of the real world. The mental model is likely therefore to be a distorted image. Further, it is held that the aberrations in one's mental model are a systematic form of distortion, since it is the mental model itself against which real life events are reflected and through which information inputs are filtered. The mental model serves as a mechanism of focus, calling some input information to attention; and, as a filter, causing other input information to go unattended. The loop of systematic distortion is closed when the model - filtered input information contributes to the embellishment and elaboration of one's mental model. Therefore, the mental model tends to maintain itself and insure its own survival.

Decisions, and acts of behavior ensuing from decisions, are influenced by one's mental model. The mental model filters the inputs which signal the need for a decision, and filters the inputs acquired in the data gathering phase. Given a range of action courses, the mental model is then used as a test field and the various alternatives are mentally simulated to determine their effects. Reinforced by an anticipated positive outcome (based on the mental simulation) the subject makes his decision and initiates his behavior, fully anticipating a perfect correlation between the results of the mental model simulation and the results to be obtained in the real world. Feedback mechanisms which might tend to realign the decision and behavior are hampered by information filtering processes. The concept of mental model seems general enough to apply in numerous areas; however, this discourse is concerned with the mental models of business organizations held by managers of and scholars of such organizations.

The second assumption:

Several different mental models of the business organizations are held by, advocated by, and used by scholars and managers.

Scholars call their mental models *frameworks*, *theories*, or *concepts* of the organization. Gibson (1966, p. 234) identified three eras of organizational theory; the mechanistic tradition, the humanistic challenge, and realistic synthesis. Bennis (1959) simplified the argument and classified traditional organizational thinking into two polemics; "Organizations without people [p. 263]" and "People without organizations [p. 266.]" An identifiable mental model of the organization accompanies each of these categories of organizational theories.

George (1968 p. 136) identified four concepts of management and their major authors: (a) The traditional school: Scientific Management (Taylor, Gilbreth); (b) The behavioral school (Munsterberg, Gantt, Mayo, Follett, Sheldon, Barnard); (c) The management process school (Fayol, Mooney); and (d) The quantitative school. Management - whatever it is - is something done to an organization and at the same time something done within the organization. Any concept of management is intimately related to the context within which management occurs and the object over which it prevails. Implied by each concept of management is a model of the organization.

Five traditional concepts or mental models of the organization can be identified.

Initially the organization was seen as an instrument to extend the personal power of the leader or entrepreneur. This view of the organization can be traced to early tribal living when the chief was the most powerful man in the village and the tribe bent to his will. Power of leadership (sometimes reinforced by divine rights or mysticism) was frequently limited to blood line transfer. In commercial operations the *power view* emerged when opportunities for positional or economic wealth exceeded one man's capacity for control or productivity. The alternatives were to remain at a low level of status or wealth, or to have subordinate others perform the more demeaning tasks. By giving others the right to join him in the enterprise (so long as they submitted to his control) and through division of labor (based on his criteria) the leader was able to extend his geographical or commercial reach. The organization to a power-oriented manager is an extension of his arm or will. The power view of the organization resulted largely from the intact transference of military concepts to enterprise organizations and may have inspired Cardullo's (1914) observation that many industrial administration practices had their origin in "Greek slavery, Roman militarism, Saxon serfdom, the medieval guilds, and various other historical oddities . . . [p. 50]." As a pure model of the organization, the power era is largely over. The day of the bull-of-the-woods supervisor, though past, is not long past, and, most organizations continue to display their "organization charts" which dramatically identify those with power, those without, and the gradations in between. Present day thinking, both lay and scholarly is strongly influenced by the mental model which holds that the organization is mainly a vehicle for manifesting upper echelon will.

Subsequent to the power view, economic considerations engendered a concept of the organization that to date is the most detailed. Between the balance sheet and the operating statement a broad range of organizational facets are encompassed. The application of financial analysis resulted in what here shall be termed the *economic view* of the organization. The economic view gained a major impetus as slave labor declined. When leaders had to share income with subordinates, then methods of account were required so that shares of income were appropriately distributed. Note that the appropriateness of distribution was determined by the leader. This reflects a continuation of the power view; the power view was not replaced by the economic view, rather managers tried to use both. Taking this view the manager sees the firm as an object designed to amplify monetary inputs into monetary outputs of greater amount. Sloan (1963) noted that:

"The first purpose in making a capital investment is the establishment of a business that will both pay satisfactory dividends and preserve and increase its capital value. The primary object . . . was to make money, not just to make motor cars . . . . The problem was to design a product line that would make money." [p. 64].

Managers whose general orientation stems from the economic mental model typically detect problem situations by economic imbalance, evaluate alternatives by economic criteria, and advance solutions which address the economic aspects of the problem. This reiterates the earlier point that one's mental model becomes manifest in the problem solving and decision making process.

Practitioners of administration and management often work from a point of view articulated by Alfred P. Sloan

(1962) as he described management strategy after the boom years of 1928 and 1929, "Management should . . . direct its energies toward INCREASING EARNING POWER through IMPROVED EFFECTIVENESS AND REDUCED EXPENSE [p. 172]." It is held that this working point of view is an amalgamation of two mental models. First there is an indication of the economic mental model, and second, the regard for improved effectiveness turns us toward the mechanistic mental model, which according to Waldo (1962): "Taking efficiency as the objective, views administration as a technical problem concerned basically with the division of labor and specialization of function [p. 225]." Taylor will be given credit for siring the *mechanistic view* of the organization. This viewpoint assumes that a department of ten men would be an efficient department if each man was working with maximum efficiency. Little concern was given to the fact that the ten efficient men might be configured in a very inefficient pattern and thus achieve a minimal overall departmental efficiency. Among contemporary managers the goals or purposes of enterprise organizations are heavily influenced by the economic view of the organization. Managers, however, do not apply the economic view when they set about structuring their firms. Rather they work from the mechanistic functional view advocated by Taylor and his followers. Broadly abstracted the mechanistic view portrays the organization as a machine that contains functional elements which when logically united can be made more efficient by the application of time and motion study.

Upon the assimilation of Taylor's concepts into management thinking, the manager could view his organization as an instrument of *personal power*, as an *economic amplifier*, or as a *machine* that has some degree of tune. As each new view emerged, the newest edition did not replace the earlier perspectiveness. Rather, addition occured; that is, the organization was seen at the same time as an instrument of *personal power*, as an economic amplifier, and as a mechanistic thing.

After the articulation of the mechanistic view, another view of the organization began to emerge. Commencing with the work of Mayo, Rothlisberger, et al, at the Hawthorne plant of Western Electric and abetted by the subsequent work of Kurt Lewin, *a social-psychological view* of the firm emerged. In this view the in-human aspects of finance and time-oriented efficiency were de-emphasized in favor of social phenomena. The feelings of individuals, the norms and goals of small groups, and the manager's style of relating with people, took precedence over profit, or standard time, or volume.

Large-scale training in organizations has been a phenomenon of the second and third quarter of the twentieth century. Early in this period the "human relations" view of the organization emerged, and a great number of trainers adopted this orientation. To them, the firm is a social-psychological phenomenon and the trainers' main task is to improve the relationships between people and thereby achieve a better socio-cultural atmosphere in the firm. A sizable preponderance of training hours are devoted to "human relations" or closely allied areas. This stems from the human relations mental model held by most trainers and reflects their propensity to sense and diagnose human relations problems to the neglect of many others.

Presently a fifth view of the business organization may be forming which George (1968, p. 177) has identified as the *social responsibility view*. Poverty, balance of payments, consumer welfare, and other problems have caused administrators and scholars to consider the organization as a neighbor within the society's framework. Today, managers are being called upon, and sometimes forced, to consider the role of their organizations in the larger social context. The articulation of this view is not particularly new but its manifestation in organizational policies and practices seems to have high current emphasis. Consequently as firms were evaluated by their use of human relations training 20 years ago, today's firms are evaluated by their relationships with the public sector.

Thus a manager or scholar can view the business organization from five different vistas. The firm can be seen as an instrument of *personal power*, an *economic amplifier*, a *mechanistic object*, a *social-psychological phenomena*, or as a *neighbor in the societal context*. By combining and modifying the five basic views, a wide variety of branches and stems are possible so that today we find few examples of consistent pure view manifestation. Instead, we find most managers in fundamental agreement or compliance with the dictates of one view but constantly adjusting their basic model by adopting considerations central to other mental models when a situation demands.

As each of the five mental models emerged, it did not replace its predecessors but was merely added to the previous frameworks. Managers raised in the economic or mechanistic tradition of the early 20th century rarely abolished their frameworks and overnight adopted the humanistic or social responsibility views. Rather, those who saw value in the new views adhered to their previous viewpoints and gave the newer mental models some due by pointing out that the considerations engendered by the newer views were merely extensions of the older models which they practiced. Discussing the reaction of authoritarian type managers to newer frameworks, Beer (1966) observed, "It is more advantageous still for authority to allege that it has encompassed these ideas all along [p. 4]." Frederick Taylor held that application of his concepts would lead to improved supervisor - subordinate relations. Other managers, unable or unwilling to recognize the insights illuminated by the new mental models, ignored them as superficial fads unworthy of the time of solid businessmen. Still a third breed, a younger generation educated at the height of the humanistic mental model and at the onset of the social responsibility view, found both types of predecessor archaic and include power, economic and mechanistic considerations begrudgingly and often too sparingly to please their elders. Today managers represent the mental model to which they adhere by advocating that its considerations are the most important considerations in most problem/decision situations. They concede that other considerations (central to other mental models) are important and in some situations may even be dominant.

The five traditional views discussed above are mental models which have been used in perceiving the nature of business organizations. The general concept of mental model is held to be applicable to other forms and types of organizations. The task of identifying the mental models used by non-business administrators will be left to other scholars.

The elaboration of the five traditional views provides the basis for the third assumption:

Individually and collectively the traditional mental models are inadequate.

The organizational or management literature provides no general scheme for ordering the growing body of objective and subjective organizational and managerial information. (See Carzo & Yanouzas, 1967, p. vii; Frederick, 1963, p. 212; Meier, 1959, p. 185.) These remarks are not intended to condemn the many sincere efforts that have been made. The theories of serious students and insightful practitioners have been both useful and helpful in studying and building the complex organizations in today's society. As noted by Shull (1962), "Guided largely by instinct and limited experiences, the practitioner has made tremendous progress [p. 124]." In a later section the traditional theories of the organization and management will be shown to be more incomplete than inaccurate. To the degree of their span of applicability they are accurate and helpful.

The absence of a general theory of organization results in the continued use of the five traditional mental models, which in turn generates numerous problems.

First. It is held that an adequate concept of administration or management is dependent upon a framework for describing that which is administered or managed. At the same time that management is something done to an organization, it is also something done within an organization. Before we can articulate a theory of management we must understand the context within which management occurs and the object over which it prevails. Today there is a tangle of approaches to management (see Koontz, 1961) since it is possible to stem one or more theories of management from each of the five views of the organization. Until we know what we are managing - that is, until we answer the question, "what is the organization?" then we cannot answer the question "what is management?" Some readers might contend that the reverse tendency is more accurate. That is, that the manager determines the organization and that a theory of organization will emerge subsequent to a theory of management. This contention is discussed in the later section entitled Objections, Questions and Defense.

Second. Those who practice management are hampered by the lack of a pervasive or an encompassing mental model and an inclusive system of categories that would accompany that model. If a manager attempts to be a purist in the application of a traditional view he will most certainly omit numerous considerations in his decision making. On the other hand, a manager would be hard pressed to use all of the traditional views simultaneously since the foundations of the five traditional views are in many cases antithetical. That is, the management role or posture demanded by one mental model is denied by another. Some managers attempt to be eclectic and to oscillate from model to model according to the demands of specific situations. By its very nature, this approach is error-prone. First there is the task of identifying the problem situation, then selecting the appropriate model, and finally, implementing the model in a manner consistant with previous behavior. Asking a manager to oscillate from model to model is akin to asking a psychiatrist to oscillate from Freudian analysis to Rogerian therapy. In a similar vein, one can imagine the difficulty of building a science of astrophysics which would account for the views held by some that the earth is flat and the more widely held view that the earth is round. Such approaches are illogical and preclude consistency.

Third. Communications between managers (and between educators and managers, and between public officials and managers) are constantly hampered by the profusion of views of that which is being discussed. To date, managers have no common focus for their deliberations. Accountants can communicate effectively with other accountants, and engineers can communicate with other engineers. Among each of these groups there exists a common concept of the organization which facilitates communication. Difference in mental model of the organization has been the source of frequent difficulty between groups however, and considerable discussion time is often spent attempting to settle upon an orientation or view of a situation before an approach toward solution or decision is made. Managers often relieve such communication congestion by surrounding themselves with subordinates who see things "our way."

Scholarly communication is hampered as well. Bakke (1959) identified the effect of multiple views upon organizational and management literature:

"There is no reason, of course, why every exploration of organizational behavior should be premised by a statement of assumptions relative to the nature and structure of an organization any more than that every treatise on the relation of thyroid deficiency to musculature fatigue should be premised by a summary of anatomy. But those who explore the relationships between such biological variables have a common concept of the "organization" whose affairs they are talking about, the human body." [p. 95]

The adoption of a common mental model will not eliminate differences of opinion. Medical practitioners vary in their diagnosis and prescriptions even though they have a common view of anatomy and physiology. Choices between alternative treatments must still be made. A common mental model would merely shift the base of controversy from the fundamental nature of the organization to the more productive plain of diagnosis and treatment.

Fourth. Without a generalized and generally accepted concept of the organization and management, graduate and undergraduate education will continue to evolve by response to omission rather than by plan. That is, a curriculum is used (and defended) until an apparent omission appears at which point a response, consisting of a new course or program, is applied to obtain a fix. Furthermore, the curriculum itself is fractionated by the existence of camps which hold that various amounts of some traditional discipline or one of the new interdisciplinaries should play a greater role in managerial education. To a considerable extent this fractionization results from the absence of a central concept of organization and management.

Fifth. Without an adequate concept of the organization, in-house and continuing management development will function without guidelines and worse will be limited by the trainer's mental model. To the degree that the trainer is a purist in his orientation, then the considerations of other views will be omitted from his overall training plan. Training strategy based on a limited appraisal of the organizational situation can at best be limited strategy. Even a team of trainers whose membership represents each of the traditional views would fail to provide full appraisal since none of their individual traditional tools dictate that consideration be given to the relationships between the various disciplines. Yet the economic and human relations facets of an organization are fully interrelated.

Sixth. Most of the traditional theories of organization and management have had the effect of causing considerable ignorance of the larger system within which the firm exists. They have fostered an inward view to the exclusion of environmental factors. Katz and Kahn (1966) observed: "The dominant tradition is psychology has included the implicit assumption that individuals exist in a social vacuum [p. 1]." This same orientation has until very recently contaminated organizational thinking. Only the last of the traditional mental models - the social responsibility view - has generated an external concern.

The problems resulting from the numerous concepts of the organization may have their solution in an approach that at once rejects all previous views while also including the central issues of each of them. Concepts from systems science can be used to articulate a new view of the organization, that is, a new conceptual framework for understanding organizational phenomena. Systems science rests upon philosophical grounds apart from the Newtonian base underlying the five traditional views.

Thus the fourth assumption:

The concepts of general systems theory provide a foundation for the development of a new mental model of the organization.

Traditional or classical theories of the organization were largely modeled after and built upon the precepts of classical science. Classical science was described by von Bertalaffy (1962) as "essentially concerned with two variable problems,linear causal trains, one cause one effect, or with few variables at the most [p. 2]." But the precepts of traditional or classical science, epitomized by Newtonian mechanics and requiring the assumptions of linearity, single causality, and teleological explanations to account for vital phenomena, "could not deal adequately with problems of organization and structure [p. 2]." (Katz and Kahn, 1966). The general systems work of von Bertalanffy (1950, 1962), Rapoport (1966), Boulding (1956b) and others is seen as a starting point for a general theory of the organization. General systems behavior theory aims to develop a science of behavior. Living systems theory (see Miller, 1965a, b, c) is a most complete statement of general systems behavior theory. Living systems theory "is concerned with a special subset of all systems, the living ones [p. 193]." (Miller, 1965a). According to von Bertalanffy (1962), "living systems can be defined as hierarchically organized open systems maintaining themselves or developing toward a steady state [p. 7]."

It is the position of this paper that organizations of all types are living systems, and that living systems theory - briefly outlined below - provides a pervasive mental model applicable to all kinds of organizations. In line with the focus of this discussion the consideration of the living systems framework will be limited to its applicability to the business organization.

Miller (1965a) indicated that those systems which

are living systems meet the following criteria:

(a) They are open systems.

(b) They maintain a steady state of negentropy even though entropic changes occur in them as they do everywhere else. This they do by taking in inputs of matter-energy higher in complexity of organization or in negative entropy, than their outputs. Thus they restore their own energy and repair breakdowns in their own organization.

(c) They have more than a certain minimum degree of complexity.

(d) They contain genetic material composed of deoxyribonucleic acid (DNA), presumably descended from some primordial DNA common to all life, or have a charter, or both.

(e) They are largely composed of protoplasm . . . . and its derivatives.

(f) They contain a decider, the essential critical subsystem which controls the entire system, causing its subsystems and components to coact, without which there is no system.

(g) They also contain certain other specific critical subsystems or they have symbiotic or parasitic relationships with other living or non-living systems which carry out the processes of any such subsystem they lack.

(h) These subsystems are integrated together to form actively self-regulating, developing, reproducing unitary systems, with purposes and goals [p. 203-204].

von Bertalanffy (1950, p. 23) observed that a closed system *must*, according to the second law of thermodynamics, eventually attain a time independent equilibrium state, with maximum entropy and minimum free energy, where the ratio between its phases remains constant. Recall, however, the first criteria of a living system is that it is an open system. von Bertalanffy (1962, p. 7) observed that open systems show characteristics which are contradictory to the second principle and that such systems can maintain themselves at a high level, and even evolve toward an increase of order and complexity - as is indeed one of the most important characteristics of life processes. Thus living systems counteract entropy, or as Katz and Kahn (1966) noted, their "structures tend to become more elaborated rather than less differentiated [p. 19]." How does a system oppose entropic tendencies? Rapoport (1966) said that:

> Information can be used to reduce entropy. A simple analogy will serve to illustrate this principle. Consider a deck of playing cards as it comes from the factory, i.e., arranged in perfect order. If we know the order, we can name with certainty the card which follows any given card. Putting it another way, knowledge of what card has been picked gives us much information about what card follows.
>
> Let now the deck be shuffled by repeated cuttings of the deck. After only a few cuttings we can still frequently guess what card follows a given card (if the two happened not to have been separated by a cutting). However, as the number of cuttings increases, we shall make more and more errors in our guesses. Eventually the cuttings will completely "randomize" the deck and so we shall guess the card following a given card with no more than chance frequency (one in fifty-one times). That is, all the information provided by a given card about the next card has been destroyed by the shuffling. This process is analogous to the operation of the second law of thermodynamics. The deck goes from an ordered (improbable) state to a chaotic (probable) state. We cannot reverse this process by continued shuffling: the original order will almost certainly not be restored.
>
> We can, however, restore the original order by "feeding information" into the deck. This can be done in the following manner. Imagine a position assigned to each card according to the original order, i.e., 1-52. Look at each successive card of the shuffled deck; and

if it has moved forward from its original position, move it one position backward, and vice versa. In doing so we are "injecting" information (in the form of either-or decisions) into the deck. Eventually the original order of the cards will be restored. In other words, a process analogous to a reversal of the second law may well occur if interventions in the form of decisions are allowed [p. 6-7].

von Bertalanffy (1950, p. 26) stated that living systems, maintaining themselves in a steady state by the importation of materials rich in free energy, can avoid the increase of entropy which cannot be averted in closed systems. But importation of energy and material is not enough. Rapoport's example tells us that by injecting information into a system, we are able to reduce the entropy in that system. Living systems maintain their organized states against the everpresent onset of the entropic death of the second law of thermodynamics by importing energy and matter and then *organizing* those inputs by the use of information. Rapoport (1966, p. 7) notes that the most fundamental property of a living organism is its ability to maintain its "organized" state against the constant tendency toward disorganization implied by the operations of the second law of thermodynamics.

It is possible to identify living systems at seven levels: cells, organs, organisms, groups, organizations, societies, and supranational systems. The major concern here is the description of living systems at the organization level. Miller (1965b, p. 337) stated that all levels of living systems have the same salient characteristics of the subsystem and system-wide structures and processes. These characteristics are:

"1. *Structure*. The structure of a system is the arrangement of its subsystems and components in three dimensional space at a given moment of time [p. 209]." (Miller, 1965a)

"2. *Process*. All change over time of matter-energy or information in a system is process [p. 209]." (Miller,

1965a) Action, the movement of matter-energy over time or space, is one form of process. The communication process is the movement of information bearing markers over time or space.

"3. *Subsystems and components*. The totality of all the structures in a system which carry out a particular process is a subsystem [p. 219]." (Miller, 1965a) That is, "each subsystem carries out a particular process for its system and keeps one or more specific variables in steady state [p. 338]." (Miller, 1965b) Of the 19 subsystems (processes) identified by Miller, eight metabolize matter-energy. They are the Ingestor, Distributor, Decomposer, Producer, Matter-energy storage, Extruder, Motor, and Supporter. Nine subsystems metabolize information: Input transducer, Internal transducer, Channel and net, Decoder, Associator, Memory, Decider, Encoder, and Output transducer. The remaining two subsystems process both matter-energy and information: Reproducer and Boundary. A specific structure and process for each of the nineteen critical processing subsystems can be identified. Components are the "specific, local, distinguishable structural units [p. 218]." (Miller, 1965a)

"4. *Relationships among subsystems and components*." Guest (1962, p. 83) noted that the survival of any living organism depends upon the ongoing interactions of its component parts. Relationships among subsystems and components are either structural relationships or process relationships. According to Miller (1965b) structural relationships are "all spatial in character [p. 361]" while process relationships are either "purely temporal or they may involve a spatial change over time [p. 361]."

"5. *System processes*." Miller (1965b) identified six system processes which "concern multiple-subsystem units or total systems wherein inputs and outputs of the entire system or major parts of it are analyzed [p. 362]." These six processes are:

5.1 "Process relationships between inputs and outputs [p. 362]."

5.2 "Adjustment processes among subsystems or components used in maintining variables in steady states [p. 364]." These adjustment processes are" (a) matter-energy and information *input* processes, (b) matter-energy and information *internal* processes, (c) matter-energy and information *output* processes, and (d) feedbacks.

5.3 "Evolution processes [p. 371]."

5.4 "Growth, cohesiveness and integration [p. 375]."

5.5 "Pathology [p. 376]."

5.6 "Decay and termination [p. 378]."

Using the living systems framework a general definition can be built which is held applicable to any organization. As a living system any organization will have a structure. It will perform some process for its suprasystem. Nineteen critical subsystems can be identified and these are coupled by relationships between them. The relationships can be described in structural and process terms. At a larger scale than subsystem processes are system processes which involve: input/output process relationships; adjustment processes; evolution; growth, cohesiveness and integration; pathology; and finally, decay and termination. This then is the vehicle or theoretical framework for the description and/or analysis of *any* organization. Hereinafter living systems theory will be used as the theory of the organization. That is, the question, "What is an organization?" is answered with the statement, "The organization is a living system." To maintain the focus of this discussion, the previous statement can be read, "The business organization is a living system."

2. APPLICATION

The involvement of the Management Training Department of General Motors Institute in systems research began in 1965. Research commenced after several of the Institute personnel were given a short course in Industrial Dynamics (see Forrester, 1961). Retrospectively, the participants in the Industrial Dynamics seminar felt: (a) practicing managers are not going to carry out the mathematical model building required by Industrial Dynamics; but, (b) an order of thinking underlies Industrial Dynamics that is foreign to general management philosophy and (c) this "new" order of thinking seems to engender beneficial insights into the behavior of the organization. It was decided to investigate the concept of "system" and a research project was so chartered. Shortly thereafter, the research was realigned to extract from the topics of general systems theory, cybernetics, information theory, organization theory, and simulation those concepts which would help a manager view the organization as a system.

After considerable search of the literature and the formulation and subsequent negation of numerous approaches it was decided that a reasonable strategy was to try to implant the systemic mental model by providing student managers exposure to the organizational systems analysis tools of operations research and Industrial Dynamics, and the living systems framework advanced by James G. Miller, (1965a, b, c). The tutorial strategy was not to create systems scientists or research specialists but to expose the program participants to the degree that they would see the systemic nature of a business organization. Eighteen men from a General Motors automotive division near the General Motors Institute were chosen for pilot program participation. Most of the program participants were college graduates, held positions from the General Foreman to General Superintendent level of managerial responsibility, and managed a wide range of engineering and manufacturing operations.

Three texts were used during the pilot program Guest (1962), Carzo and Yanouzas (1967), and Sloan (1963). The standard classroom fare was complemented with case studies and textbook problems which necessitated the use of systemic thinking. During the preparation of materials for the training program considerable "translation" was made of the systems concepts to make the course perspective more easily understood. Once the program had begun, and after the bulk of the new concepts had been exposed to the pilot group, it was found that they were having considerable difficulty relating the systems concepts to the work-a-day world. Apparently the instruction had made them aware of new insights regarding the organization but the student managers, still using their traditional mental models, were having difficulty relating the new insights to the "real world" as they saw it. It was decided to assign the pilot group to a major organizational problem with the understanding that under the instructor's guidance the pilot group was to view the problem in a systemic framework and generate a proposal for top management reviews. The problem assigned to the pilot group was to design the administrative and informational system for a new manufacturing facility being built adjacent to the division's present quarters.

As the project work commenced, a change was noted in the general atmosphere of the program. No longer was it a teacher working with a group of students, but rather the atmosphere became one of a team of individuals attempting to apply new tools to a major organizational problem. At the conclusion of their analysis and design the program participants gave an hour and one-half presentation to local top management. This presentation consisted of a detailed description of the proposed management and information system. Most of the presentation consisted of the explanation of the ten-foot flow chart model (and overlays) which was required to describe the system.

The management and information system presented by the program participants was seen by the local top management as an innovative step forward in organizational design. Presently effort is being applied to obtain the full implementation of the plan. During the start-up phases of the plant's production it is impractical to implement the full design since it is a design established for full production and full manpower. As manpower is moved into the plant however, the proposed system is used as a guide for orderly growth and will prohibit a topsy-like explosion of departments, staffs and functions. The plant manager of the new plant, remarked that the system design provides a model or blueprint to work to. As time goes by, circumstances may cause alteration of specific details but on the whole the design will guide the future growth of the operation.

Following this pilot experience, additional trial groups were exposed to the management development program. With each trial program the educational strategy and technology were revised. In its final format (to date) the training program contains four discernable characteristics that might be considered novel: objectives, framework, projects, and mechanics. Each of these four characteristics are discussed below.

Objectives. The training activities are designed to obtain three objectives:

1. *To help managers view the organization as an integrated system*. A rewording of this objective would be: "To implant the translated living systems framework as a mental model of the organization." Thus it is one of the stated objectives of the training program to change the mental models used by managers.

2. *To provide managers with skill in analyzing and designing organizations from the systems framework*. One can draw an analogy between objective two and the objectives of a curriculum for preparing professional design engineers. When a professional engineer is given a design assignment he is prepared to approach the task from a scientific base. He must determine the function to be performed by the mechanism and the resources available (operating space, power, environmental conditions, etc.). Given these parameters he then designs a mechanism, remaining mindful of the scientific principles which pertain to mechanical things. Having obtained his design the engineer performs mental, computer, or test model simulations to obtain a test of the mechanisms behavior. Alterations may be required after simulated performance testing. The professional design engineer concludes by presenting his plans to management for their approval and subsequent adoption.

Graduates of the systems program should be prepared to professionally design an organization. Their training should help them determine the function of an organizational system and to identify the environment in which the system operates. Having determined these parameters the program graduates can design an organizational system bringing to bear the principles of systems science. Indeed, the science of system behavior is not as mature as mechanics or thermodynamics, but, nonetheless, managers are shirking their responsibilities if they fail to use that knowledge which is available. After design, the behavior of the organizational system can be evaluated by intuitive judgment or the system can be subjected to computer simulation for more extensive behavioral analysis. Program graduates are not taught how to program for simulation. It was held that if they could design the system that would be sufficient knowledge for managers and that if simulation were deemed necessary, staff specialists could be assigned the task. The design process is concluded with the steps of alteration and final presentation.

The third and final program objective:

3. *To apply the systems viewpoint and solve organizational problems*. In subsequent sections the topic of projects is discussed, and those remarks pertain to this objective. It should be noted however, that the third objective is a departure from traditional thinking and practice. In most training circumstances the classroom or conference time is seen as a productivity deficit. That is, the student is not productive during training but hopefully when he returns to the job he will be more capable of carrying out his duties. Objective three implies that a real-time and live organizational problem will be worked upon during the training activity. In this sense training time becomes "productive time." Further, the third objective provides for organizational change as well as individual change. That is the training results in changes both at the individual level and the organizational level.

Framework. Living systems theory is used as the

central conceptual framework of the program, although the original notions of Miller (1965a, b, c) are extensively translated for presentation to practitioners. In the final format of the program, the participant is taught that the systemic mental model can be obtained by gaining a depth understanding of the following six elements:

1. *Environment*. Each business organization has an environment and the environment provides, makes available, or denies certain inputs; and, demands, accepts, needs, or denies certain outputs from the system. This step in understanding the systemic nature of an organization reflects the thinking of Katz and Kahn (1966):

> "The first step should always be to go to the next higher level of system organization, to study the dependence of the system in question upon the suprasystem of which it is a part, for the supersystem sets the limits of variance of behavior of the dependent system." [p. 58]

In the training program, environment is defined as everything outside of the system and not controlled by the system's decision makers.

2. *Mission*. The mission of a system is a broad although precise statement which describes the system's process (see Miller, 1965b, p. 337), its reason for inception and continued existence, and its purposes and goals. The original charter of most organizations is of minimal help in determining system mission.

3. *Subsystems*. Several process performing subsystems can be identified in a system (organization). A subsystem is defined by the process it performs. Rather than identify the units in an organization on the basis of their place on the organization chart, or on the basis of their product contribution (direct vs. indirect labor), the living systems view identifies 19 critical processes that must be performed (see Miller, 1965b, p. 338). That is, every system (organization) requires the performance of certain processes which if omitted cause system breakdown.

4. *Geographical Units.* The processes previously identified occupy physical space in a non-random manner. Organizational systems contain plants, offices, departments, groups, people, and other easily identifiable units. These geographical units are arranged in three-dimensional space at any point in time. This element of the systems framework is a combination of Miller's concepts of structure (1965b, p. 337) and component (1965a, p. 218).

5. *Relationships.* The units of a system (geographical units and subsystems) are connected or coupled to obtain and maintain the integrated wholeness of the system. Such connections or couplings are relationships. The relationships within a system manifest themself most clearly in the sixth and final element of the systems framework.

6. *System-wide Processes.* The last major facet involves system-wide processes. The individual units of a system, and the somewhat single relationships between them have been dealt with. In considering system-wide processes, groups of subsystems and the multiple relationships between them are treated. A broad range of processes within a system cannot be attributed to single subsystems and upon investigation we find multiple subsystem arrays functioning. These can be classified into input/output processes, adjustment processes, growth, evolution, sickness and death. Here again Miller's work (1965b) provided the foundation from which translation was made.

Considerable elaboration occurs as each of the six elements is introduced into the management development program. For example, the teaching of the sixth element, system-wide processes, requires approximately 25 classroom contact hours out of a program total of 80 hours.

Projects. One major carry-over from the first pilot group is the continuance of project work as part and parcel of the training diet. Prior to program participation the students are divided into teams and assigned to analysis and design projects. After the formal presentation of each element of the systems framework, the teams turn to their projects and apply their newly gained insights. To supplement the formal concepts and to abet their implementation, a form of graphical flow chart modeling is taught. This flow charting is a synthesis of block diagramming for computer programming, Industrial Dynamics graphical modeling (see Forrester, 1961) and other graphical representations that were found needed. The graphical flow charting seems to enable the students to portray their system, maintain a mass of information available at a glance, proceed with logical order, mentally simulate the behavior of the system and finally communicate the system to others, including those not trained in the system's framework. The project work and modeling follows the sequence of presentation of the modified living systems framework. For example, geographical units are identified (unless they are a "given" in the problem specification) after subsystems have been identified. At the end of the project work the organizational system is represented by processing subsystems within geographical units. These in turn are integrated into a network of information, matter, and energy flows. Feedbacks, rates of flow, delays, decision processes, adjustment processes and other facets of the systems view are displayed as well.

The graphical model which results depicts an organization level system in line with Miller's (1965a, p. 212) concept of seven levels of living systems. Thus group level or organism (individual human) level phenomena are not represented in the model. This is not to say however that considerations are not given to systems below the organizational level. To the contrary, it is not unusual to hear a student ask his project co-workers, "How is the man (or group) performing that process going to feel if the feedback channel contains so much delay?" or "How are the people going to feel about trying to perform all of those processes in so little space?", or "If we don't channel the information through this point first, we're going to have trouble on our hands." It seems safe to say that in the project work, the student managers create organizational level systems which (a) are designed to accomplish the system's mission, (b) are cognizant of the constraints of the system's environment, and (c) to the best of the project groups ability, are void of configuration or demands which might generate interpersonal conflict or individual stress.

Program Mechanics. Supplementing the formal presentation of concepts are numerous case studies, simulations and other exercises. The simulations are treated somewhat uniquely. Training simulations have been criticized since they are frequently allowed to degenerate into competitive games among teams or between the teams and the computer. To avoid such a breakdown, graphical modeling of the simulated system occurs before, during, and after the computer runs. In this way, an effort is made to obtain maximum insightfulness or learning from the experience. Further, this provides the student an opportunity to practice modeling systemic behavior so that he can subsequently represent similar behavior in his project system.

3. OBJECTIONS, QUESTIONS, AND DEFENSE

In the following remarks, questions unanswered to date and anticipated objections are treated. The previously stated assumptions and the training application are reconsidered in light of the limitations visible to the author.

In the first assumption it was held that: managers and scholars of the business organization hold, are influenced by, and use mental models. The concept of mental model was equated to: Boulding's (1956a) Image [p. 5], and knowledge structure [p. 6]; Lippmann's stereotype (see Zalkind & Costello, 1962, p. 222); and Berelson & Steiner's (1964) coherent picture of the world [p. 99], predispositions and states already there [p. 99-100], and previous experience or learning or set [p. 100]. Equating the mental model to such a range of concepts is either profound synthesis or a pedestrian generalization. Unless a general synthesis has occurred, then the definitive discriminations of the aforementioned scholars have been compromised and a step backward has been made. A thorough analysis of the original research which lead to the concepts that have been equated to mental model would begin to illuminate the validity of the equivalency.

Zalkind and Costello (1962) observed, "Research thus suggests that our perceptions may characteristically be distorted by emotions we are experiencing or traits that we possess [p. 226]." Katz and Kahn (1966) related this same observation in the more limited dimension of perceiving role demands, "The ability to perceive accurately demands of a role was related to a measure of neuroticism based on self description [p. 194]." Both of these observations identify the influence of personality variables in the perception - decision making - behavior complex. The relationship between such variables and the concept of mental model needs further elaboration.

The second assumption identified five traditional mental models which managers have used to perceive the business organization. The concept of mental model is inferred from behavior, however this author is not aware of research which directly or indirectly suggests the existence of the mental models discussed. It is of course conceivable that the concept of mental model is but a convenient abstraction without potential for concrete measurement or validation.

The exposition of the five traditional mental models followed and implied a linear time sequence. It is possible that such a time sequence fails to represent the evolution of thought of practicing managers (see George, 1968) but rather indicates the evolutionary trend followed by the scholarly and lay writers in the field.

The third assumption contended that the five traditional mental models were inadequate as bases for managerial and scholarly conceptualizations. Managers and scholars alike have struggled with the five traditional views of the organization trying to prove one better than another or trying to obtain a blend or mix that eliminates the shortcomings of any particular one. The five traditional mental models are not inaccurate; rather, to the extent of their span of applicability, they are accurate and helpful. But, individually and collectively the five traditional views are incomplete. If a manager relies on any one of the traditional views, he will omit the considerations called for by one of the other views. Miller (1965a) stated that his purpose was "to produce a description of living structure and process in terms of input and output, flows through systems, steady states, and feedbacks, which will clarify and unify the facts of life [p. 234]." Living systems theory should help identify the strengths and weaknesses of traditional organizational concepts and theories. For example, theories of the organization, formal and lay, which center around the concept of profit can be seen as theories covering the flow of one form of information (money) through the organization. Profit theories fail to cover flows of other information and the flow of matter-energy.

The issue of translation and comprehensiveness requires consideration. That is, can the accumulated knowledge of the business organization and of management be translated into the living systems framework; and furthermore, is the systems framework sufficiently pervasive or comprehensive to encompass the present universe of knowledge? The discussion of interest payment and of participative management will serve to illustrate the issue of translation. Interest is a well-rooted "fact" of organizational life. From the living systems framework the borrowing of money is one form of adjustment process (increased importation of information). As with all adjustment processes there is a cost involved and interest is seen as the cost of an adjustment process. Considerable management literature has been devoted to participative management, which from the systems framework is an effort to perform the channel and net subsystem process within the organization for transmission of the decisions of lower "level" members. Furthermore, using the systems framework, participative management vs non-participative management, which from the systems framework is an effort cause it is impossible to keep lower echelon participants from involvement in some decisions. Managers can choose, however, the channels through which influence (information) will flow. The issue is not participation vs. non-participation, but rather internal (formal and informal) vs external channel routes. In sum, before the systems mental model can be held as a more pervasive or comprehensive framework it must indicate the capability of dealing with all of the insights generated by the five traditional views. Organizational power cannot be ignoted, nor will economics go away at the adoption of a new framework. These former insights must either exist alongside of or be encompassed by the systems mental model.

It was held that any concept of management stems from a concept of the organization. This contention might be questioned and the reverse offered - that is, that one's concept of organization stems from one's concept of management. In defense of the position of this paper it is suggested that in the typical case, the organization has more effect on the manager than vice-a-versa. One's mental model is not totally associated with or created by one organization or the organization within which an individual might presently be employed. The mental model is an accumulative collage and therefore the new entrepreneur brings to bear on the new firm a mental model formed in previous experiences.

The criticism of the traditional views was that they are individually and collectively incomplete. Their use causes even the most conscientious manager to omit or overlook important aspects of administrative situations. But, *what might the systems view bring into focus that was omitted by the traditional views?*

The systems mental model will engender consideration of these facets:

Wholeness. Initially and throughout the solution process, any problem or administrative situation is to be seen in its broadest light. Once focused on the broad mass of the problem, the manager must resist the temptation to boil the problem down to an over-simplified cause and effect. Considerations of wholeness alert the manager to the fact that the five "best" solutions to five apparently separate problems may congeal to produce unsatisfactory results. In a similar vein, wholeness dictates that the unstudied configuration of several efficient sections or departments need not result in an efficient plant or enterprise.

Steady State. Traditionally, managers have worked to achieve satisfactory balance between organizational ingredients. When a balance point was found, the manager attempted to lock all variables and thus hold the organization in that balanced state. With a machine (excluding frictional losses) such can be done. Organizations are not machines; they do not achieve a machine-like balance, but rather they move toward or away from an organic form of homeostasis or steady state.

Multiple Causality. Events in organizations rarely have single causes and the manager who continuously seeks them is deluding himself. Organizational behavior is usually generated by long chains of cause and effect building up in time or by multiple factors acting simultaneously. This is to say that the simplified mechanistic model of particle collision analyzed by vector forces fails to describe organizational phenomena.

Integration. The systems mental model of the organization portrays an integrated mass rather than a collection of independent units. From the traditional views, a decision concerning the engineering department was considered relevant only to that one department. From the systems view, the vast interdependence between organizational units comes into focus. When considering the consequences of a particular decision, the systems view urges cognizance beyond the immediate effects to the second and third level effects which arise as the decision permeates the entire organization.

Organized Complexity. Organizational behavior is more than the simple sum total of the behavior of the organization's units. When units interact in a complex manner, an emergent effect is observed. The behavior of the organization stems not only from the behavior of its constituent units, but stems from the relationships among the individual units, between the units and the whole, and finally between the whole and its environment.

Dynamic Relationships. Organizations have static relationships and dynamic relationships which involve the factor of time. Dynamic organizational relationships are often non-linear in that changes rerely generate consequences of equal or proportional scale. Dynamic cybernetic or feedback relationships occur throughout organizations. Some feedback relationships cause a dampening of system behavior, other feedbacks cause system behavior to deviate further from the desired range. Vicious circles are good examples of feedback-induced or feedback-propelled deviations. Dynamic cybernetic or feedback relationships are the mechanisms that enable a system to "home in" on a target or goal. In any feedback relationship there is always some delay in signaling the initiating mechanism. Organizations are exposed to detrimental delays in feedback, when, for instance, they depend on consumer usage to point up product shortcomings. By the time a field failure occurs, the pipeline from factory through warehouse to distributor is full of commodities that must undergo an expensive field alteration.

Subsystem Process Performance. Traditionally, managers have built their organizations using as few functions as possible. The strategy was to identify the purpose of the business (product or service) and then include only those functions needed to accomplish that purpose. This is a naive view of what it takes to make an organization function properly. College professors need muscles and field hands need brains. Steel mills need to have information processes performed within them and insurance companies need materials moved from place to place. From the systems view, the livelihood of the organization requires the performance of certain critical functions and processes. The size of the various functions will vary with the industry. Steel mills have more people devoted to material processing than do insurance companies who occupy most of their personnel in the processing of information. Fundamentally, however, every organization has similar life functions that must be carried out.

When a manager fails to include the performance of critical functions in planning his organization, then that organization will fail or the performance of the omitted functions will occur in a clandestine manner. Unions emerged after 19th century organizational planners failed to recognize that the formal chain of command would not provide adequate vertical information flow in large organizations with multiple layers of supervision, each of which filters the message it receives.

Adjustment Processes. From the systems view the organization's own adjustment processes come into focus. Managers who operate from the traditional views often attempt to solve problem situations by initiating some compensation program. The systems-oriented manager would often conserve valuable resources by recognizing and supplementing the organization's adjustment processes rather than confound a disturbed system by introducing some strange new procedure.

Termination. Systems die; empires are overthrown, nations fall, and organizations go bankrupt. Traditionally, organizational failure is ascribed to bad luck, poor economic control or bad market strategy. The manager who uses the systems mental model attends to those considerations which can cause a system to succumb. By knowing what causes a system to die, the manager can avert (or postpone) similar events in the organization under his command. Conversely, when the situation demands, the systems-oriented manager can be a strategic executioner and perform timely and effective system terminations where appropriate.

Flows and Decisions. From the systems view, the manager sees the organization as a collection of process performing subunits connected together by a vast network of information and material flows. The rate of flow in any pipeline is controlled by management decision.

Level. Although fundamental to systemic thinking, the concept of level and hierarchical ordering of levels is a new insight into organizations for most managers. Considering the organization as a discernible entity allows organization level problems to be identified and treated and should reduce the number of occasions wherein organizational level problems are attributed to lack of individual level skill or breakdown of group level interpersonal relationships. Intelligence quotient, amount of job knowledge or experience, "personality," and agreeableness have too frequently been the scapegoat for organizational level systemic breakdowns.

In summary, it is possible to identify eleven insights which are not made available by the traditional mental models. The eleven insights previously discussed can be integrated into a descriptive statement. Using the systemic framework the business organization is a discernible *level* consisting of an *integrated organized complexity*. *Holistic* characteristics may be observed apart from the characteristics of *subsystems* which are *statically* and *dynamically related*. *Adjustment processes* requiring *flows and decisions* operating under *multiple causation* obtain and maintain *steady state*. *Termination* is likely under certain circumstances.

The fourth and final assumption contended that the *concepts of general systems theory provide a foundation for the development of a new mental model of the organization*. Within the discussion of the fourth assumption, living systems theory was prescribed as the new mental model to be applied to business organizations. Other authors have articulated a systemic model of the organization. Bakke (1959) perceived the organization as an abstracted system and identified four features as the prime elements of his taxonomy: (*a*) The Organizational Charters, (*b*) The Basic Resources, (*c*) The Activity Processes, and (*d*) The Bonds of Organization. Forrester's proposals (1961) provide a second example. He advocated that the organization can be seen as an assortment of levels (of information, material, orders, money, personnel, and capital equipment) which are united by flows of these six commodities. The flows between levels are variable in rate and the rate is controlled by decisions. Delays in flow and in decision-making account for considerable system behavior. Forrester's concepts are especially valuable in that a system defined by these variables can be simulated through the use of a prepared software package Dynamo. Prior to accepting living systems theory as "the" mental model of business organizations, the contributions of scholars like Bakke and Forrester must be appraised.

From the viewpoint of this author, two important topics require treatment before the living systems framework is fully acceptable. Firstly living systems are purported to be negentropic. It would seem that one condition for the application of living systems theory to business organizations would be the acquisition of methods of measuring entropy in organizations. Secondly Miller's (1965b) treatment of relationships is concrete and therefore is presently restricted to those relationships which can be measured in terms of space and time. Until, as Miller (1965b, p. 362) indicates, more satisfactory methods are available for measuring those relationships which involve meaning, the researcher is forced to use human ratings or other techniques difficult to quantify.

A question can be raised at a more pragmatic or applied level. Von Bertalanffy (1962, p. 8) interprets one of Buck's (1956) criticisms of living systems theory as the "so what?" argument. In the context of this discussion the same question can be raised, i.e., "so what?": if a manager views the business organization as a living system, that doesn't change reality or reduce the hazards of the competitive environment. Indeed adoption of the systemic mental model will not directly change reality, but the manner in which scholars and practitioners relate with reality would change and such a change may be helpful. Von Bertalanffy (1962, p. 10) discussed the changes wrought by Copernicus, Einstein, and Darwin and noted that it was the changes in the general frame of reference that mattered. Hopefully a meaningful analogy can some day be drawn between changes of managerial mental model and changes in the reference frames used by scientists.

The fourth assumption advocates a general mental model, a systems mental model, for use with all organizations. No complete defense is available for supporting the advocacy of a single systemic mental model although three arguments will be advanced. First, the living systems framework is cross applicable and thus research findings from other levels and other disciplines can more readily be tested for their validity in business organizations. Second, the mobility of contemporary managers, between industries and within single firms, necessitates a general mental model which can be adjusted rather than specific mental models which are used and discarded. In the folklore it is held that a good manager would or could be a good manager anywhere. Such universally applicable skill would seem to necessitate a general mental model. Third, and finally, the adoption of a general systemic mental model does not preclude eclecticism or emphasis. Many organizational problems are specifically boundary problems or adjustment problems, or information breakdowns and in these circumstances the manager is justified and required to consider more microscopic details. The systems mental model should however preclude becoming blinded by details and other problems which such myopia generates.

Subsequent to the statement of assumptions, a description of application was made wherein the use of general systems concepts in management education was described. Four characteristics of the training program were discussed; objectives, framework, projects, and program mechanics. Objections to, questions regarding, and defense of these four program elements appear below.

Program Objectives. The first objective of the training program was to alter the student's mental model to the systems orientation, yet mechanisms for altering mental models are unfortunately based on trial-and-error, and more definitive theory and research is lacking.

Program Framework. The living systems framework (see Miller, 1965a, b, c) has been defined in the most explicit terms and as with any scientific work requires strict understanding of and adherence to definitions. In the process of translating the formal living systems theory into a meaningful message for practitioners, considerable hazard is encountered. Experience with program graduates will indicate the amount of translation loss and the effects of such loss.

Program Projects. Boguslaw (1965) criticized those who attempt to build new utopias based on systems analysis and computerization and who do so in ignorance of the mistakes of earlier utopian designers. The design of organizational systems within the context of the training program holds the potential for violating the admonitions of Boguslaw and other authors (see von Bertalanffy, 1968, p. 4-5). However, Boguslaw said, "Information necessary to take apart the clock of the contemporary world is simply not underscored in contemporary computer journals and works on system engineering, which remain devoted to the idols of physical efficiency." [p. 203]. Boguslaw's remark raises the possibility that some "systems" scientists are but retracing the steps of earlier authors, specifically Frederick Taylor. It is held that adherence to the living systems framework might enable system designers to avoid the mechanistic physical efficiency orientation of scientific management. Utilizing a general systems base, living systems theory hopefully provides a unique and far more comprehensive approach to system design.

4. FUTURE RESEARCH

In addition to answering the numerous questions raised in the foregoing section, research is planned to test the results of advocating the living systems mental model. Below, four general questions are asked and hypothesized answers to each of the four questions are given.

Question 1.

What are the effects on organizations when they are designed from the living systems mental model?

Hypothesis 1.1. Organizations designed from the living systems framework will indicate more comprehensive and more integrative initial planning and therefore will be subject to less subsequent overhaul and disruption.

Hypothesis 1.2. Organizations designed from the living systems framework will provide more satisfaction to employees resulting from reduced organizational conflict.

Hypothesis 1.3. Organizations designed from the living systems framework will show better organizational performance on traditional measures and as well will demonstrate faster organizational adjustment to internal and external stress.

There is little formal data available to support such far reaching hypotheses. However, several arguments can be advanced which indicate that these hypotheses are valuable to pursue.

Organizations evolve (see Miller, 1965b, p. 369) and in doing so undergo structural changes. Too frequently however, the organizational alterations necessitated by growth or a changing environment somewhat haphazardly occur in response to the demands of the day. Too frequently functional units have emerged to solve a particular problem or keep abreast of a fad and thereafter remained a part of the organization. Unfortunately, this topsy-like evolution often occurs without a general plan or framework to provide guidance, and in some cases with little consideration given to the effect some addition will have upon the initial organization. The evolution and growth of a small firm to maturity and size is only one of the phenomena which too frequently follows a random course. Today's huge enterprises have the capability of bringing about large complex organizations almost, instantaneously. When a large organization decides to build a new plant, it doesn't build a back alley job shop with one or two employees. To the contrary, it brings into being a full scale organization of the size not reached in the life time of many fledgling entrepreneurs. But how do the administrative systems of these new entities come into being? Without an administrative model or plan to guide the new system, the personnel rely on past memory and previously formed mental models. The new entity soon takes on a more than coincidental similarity to its predecessors which is a long lineage traceable back to the

Roman Army and earlier. Tools for organizational design are needed.

In the past some organizations have been designed or laid out before they were allowed to materialize. Well intentioned managers have had and often used several tools for laying the best of plans. But each of these tools stemmed from the traditional mental models and are open to the criticisms discussed earlier. In planning a new organization, for example, organization charts are often pre-determined (the power view); budgetary procedures and projections are prepared (the economic view); a detailed plant layout is prepared and production processes and methods are planned (the mechanistic view); employee benefits, facilities and styles of leadership are determined (the human relations view); and finally speeches are given by the top management to local service groups and the mayor is invited to the ground breaking (the social responsibility view). Well intentioned as they may be, users of the traditional mental models will omit numerous considerations in their decision making. Miller (1965a, p. 204) noted that all living systems have templates, blueprints, charters, or programs. It is the position of this author that traditional templates have been obscure, inadequate, open to wide interpretation, fragmented and usually sufficient only to survive the first day of business. Thereafter, the system was left to evolve with little semblance of order. Organizational design from the systemic mental model should provide more comprehensive, integrated, and anticipatory templates.

Note that one hypothesis assumes improved employee satisfaction through reduced conflict. For some time management development personnel have tried to improve the relationships between people in organizations. But often those who are in conflict while on the job get along well outside of the organization. It seems rather fundamental to observe that possibly the organization itself is generating the disharmonious behavior. If this is the case, then let us stop trying to teach people to be nice to each other and commence to change those organizations which generate conflict. Guest (1962) observed: "Getting people to cooperate with one another is not something that can be taught. A willingness to cooperate evolves from a change in the total system ..." [p. 96]. Although experience with the project work of pilot student groups provides subjective affirmation of these first hypotheses, their complexity and the long term nature of their implications necessitate considerable time and effort for testing.

Question 2.

What is the effect on managerial performance of using the living systems mental model?

Hypothesis 2.1. Managers who utilize the systems mental model will feel more competent when faced with problems of organizational design.

Hypothesis 2.2. Managers who use the systems mental model will indicate less departmental myopia.

Hypothesis 2.3. Use of the systems mental model by the manager will result in his greater use of staff specialists trained in systems science.

Hypothesis 2.4. Managers who use the systems mental model will consider a broader range of parameters in problem analysis, and use different criteria in deciding among alternatives.

Again the reader must be forewarned that little "hard" data exists for evaluating the effects of the pilot programs on the graduates. Two survey instruments were administered to the pilot groups as pre- and post-tests. Although changes in perception can be identified between the pre-test and the post-test, the absence of a suitable control group precludes attributing the perceptual changes to the program itself. Rigorous evaluation of the effects of systemic training is planned. At a subjective level it can be said that the pilot graduates apparently obtained the systems mental model and seemed to have acquired tools for working with or communicating from this point of view. Having some framework for visualizing and categorizing the complexities of the organization plus the possession of schematic tools for manifesting the framework should relieve some of the uncertainty of organizational design.

Adoption of the systems mental model presumes a reduction of the departmental myopia that often stagnates an organization. As soon as one begins thinking in terms of the whole system, departmental boundaries and concerns become subordinate to larger goals and problems. One can't see the organization as a system and at the same time limit one's vision to the boundary of one's own department. If myopia is reduced as suggested by the hypothesis, then one might conclude that the system mental model would serve as an excellent orientation for upper echelon managers. In the executive suite a perspective of the entire organization is mandatory yet traditional mental models have provided little conceptual structure for such thinking.

It can be asserted that as managers begin to perceive the systemic nature of organizational problems, they will then seek the assistance of those staff specialists who are trained in systems science. Not only will those skilled in system analysis be sought, but their proposals will be understood since both the manager and the staff specialist will have some similarity of viewpoint. Further, the newer analytical tools will lose their mysticism and become as common as statistical quality control, standards engineering, and economic forecasting.

Finally, the general systems or living systems mental model would logically affect the problem analysis and decision making of the individual manager. In analyzing problems, it is held that the manager will utilize a broader initial base and consider the problem in a broader context. In gathering data regarding the problem situation, the systems oriented manager will not limit his search to the immediate point of difficulty, but rather, will investigate the interaction of the apparent problem with its environment. When choosing among alternative problem solutions, the systems oriented manager will evaluate each alternative on its ability to solve the immediate problem and as well its effect on the systemic context within which the problem exists.

Question 3.

Of what value is the use of the systems mental model to those responsible for managerial education and development?

Hypothesis 3.1. Use of the systems mental model by management specialists will result in more effective training strategies.

Hypothesis 3.2. Use and application of the systems mental model in managerial education will engender organizational systemic change as well as individual change.

The complexity of contemporary organizations increases the difficulty of using training in its most beneficial manner. In addition to demands for the competent design and execution of classroom activities, the trainer bears responsibility for a strategy of training that will obtain optimum results. The strategic plan of any training effort must identify that point in the organizational fabric where the application of educational technology will have the greatest benefit. Understanding the full systemic nature of an organization can help the trainer identify the point where he can have the greatest effect. Good training strategy would avoid trying (a) to teach a new behavior pattern which the system won't support, or (b) to use education to obtain system change while focusing educational efforts upon those who have no power to change the system. Unfortunately, however, there are numerous examples of training efforts which violate these strategic principles. Consider the effort to teach hourly employees a sensitivity to the need for quality in their output. Considerable training time has been used to motivate employees to produce near perfect parts. These efforts, however, are rapidly nullified when management, under the pressure of meeting schedules, decides to accept or ship goods of lesser quality than that expected of the production man. The behavior of the system, its management, inspectors, customers, technology, et al., establishes the accepted quality norms, and classroom or propagandistic educational efforts to teach some other standard are a waste. To ask the hourly man to behave in a fashion that is out of touch with the reality of his work environment is to ask him to accept and live in a state of occupational mental illness. Thankfully, they ignore such efforts. A parallel example has occurred in management training. Numerous efforts have been made to teach managers to be more participative and democratic in their relationships with subordinates. Managers are sent off their jobs to conferences where many different forms of educational technology are used to enhance the interpersonal skills and performance of the participants. Guest (1962, p. 96) noted that there is a basic assumption behind many human relations training programs throughout industry that it is possible to teach human relations skills apart from the socio-technical matrix in which such skills are to be exercised. The general result of these efforts is largely zero because upon returning to the job, the manager responds to the world around him and adjusts his behavior to that pattern demanded by the system. The manager has returned to Rome and will behave as a Roman. Some would contend that this proves the need for training the top management in democratic/participative skills so that from the very top a receptive climate is established and subordinate efforts flourish. This contention is short-sighted. Indeed the subordinate responds to and will reflect the socio-cultural atmosphere of his work place. But further, he responds to the technology, work pace, and impersonal factors around him. Rigid procedures of accounting and high speed, linear, fully integrated production lines have overwhelmed the most pleasant personalities and stultified the most humane atmospheres.

Underlying erroneous training strategies such as those described above is the assumption that individual change means, or inevitably results in, organizational change. Or as observed by Katz and Kahn (1966): "Scientists and practitioners have assumed too often that an individual change will produce a corresponding organization change. This assumption seems to us indefensible." (p. 450). The weakness of such an approach should have been apparent long ago. The organizational systems we have allowed to emerge are so massive and entrenched that they overpower any individual.

The management education or development specialist who utilizes the systems mental model should first be able to better understand the problems of the organization as a result of his broader framework. Second, he should be able to construct training strategies which are congruent with the system and with the nature of the problem. Third, when systemic change is required he is equipped to obtain such changes. By being an agent of systemic change, the trainer is able to become an organizational surgeon rather than continue to treat peripheral symptoms.

Question 4.

Earlier it was stated that a theory of management must await a theory of organization; given the living systems framework as a theory of the organization, what is the emergent theory of management?

Hypothesis 4.1. The manager is the systems component who performs critical organization level information processes for the purpose of retarding the production of entropy.

Miller identified 19 subsystem processes which must be performed to maintain a living system. Nine of the subsystems are involved in the metabolism of (processing of) information. They are input transducer, internal transducer, channel and net, decoder, associator, memory, decider, encoder, and output transducer. It is suggested that a general framework or concept of management can be built from the view that the manager is an information processor and the service performed by the manager or administrator component for the organization, is to control the production of entropy. The outward manifestations of entropy control is the degree of order or organization in the system. (This is not to be confused with the formality of classical management theory.) Counteracting entropy results in a greater state of organization which is to say that the goal of the manager is negentropy. Negentropy is attained by the manager's execution of certain information processes in the system, particularly decision making. Reviewing the nine information processes found in living systems, the ways in which managers execute these subsystems processes can be exemplified.

Input transduction. Managers read industry and trade publications for competitor, market or other environmental information.

Internal transduction. Managers intuitively recognize the need to keep their finger on the pulse of things. Some functional staffs, for example accounting and quality control, perform some of this process for administrators in larger organizations. Sensitivity training techniques have been used to build openness in organizations thus facilitating internal transduction by the managers. "In-basket" exercises used in management development programs often contain or consist entirely of information to be transduced for consumption within the system.

Channel and net. Managers spend considerable time receiving and sending information on to others. When their output equals the input they received, they are performing the channel and net process.

Decoder. The chief engineer generally decodes the contents of technical literature before passing it on to the organization's staff members.

Associator. The association process is generally dispersed to the individual level in most organizations and it is the managers who frequently concern themselves with the performance of this process for the organization. Organizational alterations resulting from experience can be seen as association at the organization level.

Memory. The associations made by managers are often stored in their memories. These often constitute unwritten procedures or policies. The value of experienced personnel stems from this process.

Decider. Miller (1965 a, b) observed:

> The decider is the executive or administrative subsystem which controls the entire system, causing its components and subsystems to coact [p. 357]. A group's decider often is its leader - a committee's chairman, a squad's commander, a family's father - but the process also may be dispersed to all members, who decide jointly. Organizations commonly have multiple echelons with recognized chiefs [p. 358]. In systems with multiple echelons, the decider ... is so organized that certain decisions (usually certain types of decisions) are made by one component of that subsystem and others by another. These components are hierarchically arranged. Each is an echelon [p. 217].

Roberts (1963, p. 100) stated that the decision making process is viewed as a response to the gap between the objectives of the organization and its apparent progress toward those objectives. From the framework given here we would say that one objective of an organization is negentropy.

Encoder. Miller (1965, p. 359) noted that information must be transmitted out of the system in a code which can be understood by other receiving systems, or they cannot interact. Advertising material is very carefully encoded,and must usually pass the scrutiny of managers.

Output transduction. Some managers are entrusted with "speaking" for the organization. Other managers do not have this freedom and sanctions are applied when they attempt to perform this process.

In addition to performing information subsystem processes, adjustment processes can be initiated by the manager's decisions. Miller (1965b, p. 364) noted that there are three general purposes for a system to employ adjustment processes; (a) the governing or control of relationships among its subsystems; (b) the governing or control of the system as a whole; and (c) the governing or control of relationships between the system and the suprasystem. There are seven adjustment processes available to the manager. One category of adjustment process relates to the internal processing of information. In his study of "Plant Y," Guest (1962, p. 77) noted that lateral information exchanges had been blocked because of the plant managers repeated emphasis on having information flow vertically through the separate and formal channels. Under a new plant manager a new atmosphere of in-plant relationships was established and personnel were encouraged to communicate across functional or departmental lines. The operational improvements under the new plant manager appear to be a partial result of better internal communications including, of course, feedback. And, as Miller (1965a) noted, "feedback couplings between the systems parameters may cause marked changes in the rate of development of entropy" [p. 197].

As the size of organizations varies the information processes performed by the manager varies. A small firm consisting of an entrepeneur and two workers is a living system at the organization level. Within this organization the eleven information processes must be performed (or obtained from the outside). In a firm of this size the workers perform the processing of matter and energy which result in marketable commodities or services. The information processing is defined as the boss's job. In the small firm the entrepreneur typically does his own order taking (input transducing); auditing of worker time and production (internal transducing); communication of instructions to workers (channel and net); interpretation of consumer acceptance studies (decoding); connecting of worker name with output performance (associating); storage, filing or remembering of pertinent facts (memory); choice of supply sources (decider): preparation of advertising materials (encoding); and, speaking for the organization (output transducing). Additionally the entrepreneur establishes and retains the right to alter the charter of the organization (reproducer) and he sorts the incoming mail for information which he feels is important for posting (boundary). Thus in the small firm the manager may perform many or all of the information processes required by the system. In larger firms the manager's range of activities is more limited. For the most part individuals or groups are delegated the responsibility of performing many information processes. At the highest echelons of the firm the only major process performed by the manager for the entire system may be decision making. Reduction of the number of information processes performed usually accompanies more difficult or greater demands for deciding, the one system process which the manager component cannot totally relinquish without giving up his role as a manager. In firms with more than one echelon of management, those lower in rank do not spend their time making the major decisions of policy. At lower echelons the manager's (supervisor's) job consists largely of the information processes of internal transduction and channel and net. Note, however, that managers or supervisors of this echelon still make organizational level decisions. Traditionally management stops where decision making stops. Bottom echelon supervisors are valued in that they perform a boundary function (at the group level, not organization level) between the management group and the worker group (union).

It is the position of this paper that the organization may be conceptualized as a living system. Further, the information processes that must be performed at the organization level provide the basis for a concept of management. That is, the manager can be seen as an information processor or in Shull's (1962) words, "management can be viewed as a total hierarchical or intelligence and control system" [p. 133]. Boulding (1965a) said, "He [the manager] is a receiver of messages from the receptor[s] of the organization and his job is to transform these messages into instructions or orders which go out to the effectors" [p. 27].

The information processing approach to management, stemming as it does from the living systems framework, should provide assistance to academic and inhouse management educators. Several benefits might be expected.

First. A manager in an organization deals with individuals, groups, the organization itself, and the larger society. To deal effectively at each of these levels the manager must be equipped with the knowledge of many traditional disciplines. By treating each level as a living system one accomplishes the major economy of providing one conceptual framework that is applicable to all four levels.

Second. Courses designed to teach the student to perform adequately as a manager can be viewed against the backdrop of information processing. For example, given a specific course content, the kind of information process taught can be identified and the types of information subjected to that process can be classified.

Third. Living systems theory provides a framework against which all of the manager's education can be evaluated. The educational administrator can ask himself: (a) Are my students being given an integrative framework for classifying and categorizing the knowledge they have been given (and will obtain)? (b) Does the curriculum deal with each element of living systems at each of the applicable four levels? (c) Have each of the organizational level information processes been treated sufficiently to enable this neophyte to contribute to the organization he chooses to join? In some cases the questions of how much of any level or topic must be researched, but at a minimum, the omissions can now be detected. Information processes provide a plan for the design of curricula rather than necessitate the continuance of responding to omission, or designing relative to other (unplanned) curricula. As well, the role of traditional disciplines and of the newer interdisciplinary efforts can be determined by their contribution to understanding some segment of the framework.

Fourth. In-house and continuing education and training of managers will benefit from adoption of this framework. Present day managers have been subjected to a wholesale conglomerate of courses, programs, and seminars. Undoubtedly the first valuable service to be rendered by the adoption of this framework would be to integrate these many learning experiences by showing their relative contribution to understanding of the various systems levels. As well, proposed educational activities for managers could be evaluated against this framework for their ability to cover previously untouched ground. Finally, the in-house management educator would obtain a confidence based on understanding the organization and management that heretofore has been lacking and has thus paved the way for numerous "final answer" panacea seminars and programs. In both academic and continuing or in-house education efforts, the result is order arising out of chaos. In this case, the salvation is a conceptual system that provides a usable mental model of the firm and subsequently a central theme for understanding what a manager does for an organization.

ACKNOWLEDGEMENTS

The author wishes to express his deep gratitude for the insightful comments of Professor Richard F. Ericson and Milton D. Rubin, who read earlier versions of this paper.

REFERENCES

Bakke, E.W., Concept of the Social Organization, *Yearb. Soc. Gen. Sys. Res.*, *4* (1959) 95-120.

Beer, S., *Decision and Control*, London, John Wiley & Sons (1966).

Bennis, W.G., Leadership Theory and Administrative Behavior, *Admin. Sci. Quart.*, *4* (1959) 259-301.

Berelson, B., and Steiner, G.A., *Human Behavior*, New York, Harcourt, Brace and World (1964).

Boguslaw, R., *The New Utopians*, Englewood Cliffs, N.J., Prentice-Hall Inc. (1965).

Boulding, K.E., *The Image,* Ann Arbor, Mich., Univ. of Michigan Press, (1956) (a).

Boulding, K.E., General Systems Theory - A Skeleton of Science, *Management Science,* (1956) 197-208 (b).

Buck, R.C., On the Logic of General Behavior Systems Theory, in H. Feigl and M. Scriven (Eds.) *Minnesota Studies in the Philosophy of Science, Vol. 1. The foundations of Science and the Concepts of Psychology and Psychoanalysis,* Minneapolis, Univ. of Minnesota Press (1956) 223-238.

Cardullo, F.E., Industrial Administration and Scientific Management, in C.B. Thompson (Ed.), *Scientific Management,* Cambridge, Mass., Harvard Univ. Press (1914) 49-103.

Carzo, R., Jr., and Yanouzas, J.N., *Formal Organization - A Systems Approach,* Homewood, Ill., Irwin-Dorsey (1967).

Frederick, W.C., The Next Development in Management Science: A General Theory, *Academy of Management Journal, 6* (1963) 212-219.

Forrester, J.W., *Industrial Dynamics,* Cambridge Mass., The MIT Press (1961).

George, C.S., *The History of Management Thought,* Englewood Cliffs, N.J., Prentice-Hall (1968).

Gibson, J.L., Organization Theory and the Nature of Man, *Academy of Management Journal, IX,* No. 3 (September 1966) 233-245.

Guest, R.H., *Organizational Change: The Study of Effective Leadership,* Homewood, Ill., Irwin-Dorsey (1962).

Katz, D., and Kahn, R.L., *The Social Psychology of Organizations,* New York, John Wiley (1966).

Koontz, H., The Management Theory Jungle, *Academy of Management Journal, 4* (3) (1961).

Meier, R.L., Explorations in the Realm of Organization Theory, *Yearb. Soc. Gen. Sys. Res., 4* (1959) 185-200.

Miller, J.G., Living Systems: Basic Concepts, *Behavioral Science, 10* (1965) 193-237 (a).

Miller, J.G., Living Systems: Structure and Process, *Behavioral Science, 10* (1965) 337-379 (b).

Miller, J.G., Living Systems: Cross-level Hypotheses, *Behavioral Science, 10* (1965) 380-411 (c).

Rapoport, A., Mathematical Aspects of General Systems Analysis, *Yearb. Soc. Gen. Sys. Res., 11* (1966) 3-11.

Roberts, E.B., Industrial Dynamics and the Design of Management Control Systems, *Management Technology, 3* (2) (1963) 100-118.

Shull, F., Jr., The Nature and Contribution of Administrative Models and Organizational Research, *Academy of Management Journal, 5* (1962) 124-138.

Sloan, A.P., Jr., *My Years with General Motors*, Garden City, N.Y., Doubleday (1963).

Taylor, F.W., The Principles of Scientific Management, *Scientific Management*, Hanover, N.H., Dartmouth College (1912), in H.F. Merrill (Ed.), *Classics in Management*, New York, American Management Association (1960).

von Bertalanffy, L., The Theory of Open Systems in Physics and Biology, *Science, 3* (2) (1950) 23-29.

von Bertalanffy, L., General Systems Theory - A Critical Review, *Yearb. Soc. Gen. Sys. Res., 7* (1962) 1-19.

von Bertalanffy, L., *Organismic Psychology and Systems Theory*, Barre, Mass., Clark Univ. Press (1968).

Waldo, D., Organization Theory: An Elephantine Problem, *Yearb. Soc. Gen. Sys. Res., 7* (1962) 247-260.

Zalkind, S.S., and Costello, T.W., Perception: Research implications for Administration, *Administrative Science Quarterly*, September 1962.

THE CHANGING HOSPITAL ORGANIZATIONAL SYSTEM: A MODEL FOR EVALUATION

Frank Baker*

ABSTRACT

This paper presents a model for the study of a changing mental hospital organizational system and its developing community relations based on the concepts of open-systems theory. Since a hospital's goals of becoming a community mental health center emphasize the importance of increasing interaction and interdependence between the hospital, the community, and other caregiving organizations, with an increasing interpenetration of traditional boundary conditions, it is important to focus on more than intraorganizational processes. The open-systems model developed here points out the importance of doing final-outcome analysis in relation to process-oriented analysis focusing on the intermediate outputs of a system, but it also examines the exchanges and transactions across the boundary of the hospital organizational system.

1. INTRODUCTION

As the concepts of community mental health gain wider acceptance, many mental hospitals are attempting

---

* Research Associate in Psychology, Department of Psychology Harvard Medical School at the Laboratory of Community Psychiatry, 58 Fenwood Road, Boston, Mass. 02115. Preparation of this paper was supported by NIMH grant NH-09214.

to take central roles in newly developing community-oriented health programs and changes are being attempted in the organization and services of these institutions. This paper describes an open-systems organizational model that is being employed as a conceptual framework in the study of Boston State Hospital as it undergoes an evolutionary metamorphosis from state mental hospital to community mental health center (Schulberg, Caplan, and Greenblatt, 1968; Schulberg and Baker, 1968).

In attempting to evaluate the functioning of a large mental hospital within a comprehensive community mental health program, the usual problems of adequate program evaluation are compounded. In most cases the demonstration of effectiveness in such outcome measures as length of hospital stay, discharge rates and readmissions is dependent upon not only intentional variation in one program variable but upon multiple variables interacting with each other in interdependent ways. In addition to developing indicators of certain outcome characteristics, it is particularly important to study the process by which the organization searches for, adapts to, and deals with its changing goals.

Social science research on mental hospitals has focused principally on *intraorganizational* processes with an implicit assumption that the organizational problems of the hospital can be analyzed by reference exclusively to its internal structure and functioning. The classical organization models employed in this type of research have been based for the most part on this closed-systems approach.

The community mental health center idea emphasizes the importance of increasing interaction and interdependence between the hospital, the community, and other caregiving organizations, with an increasing interpenetration of traditional boundary conditions. For example, since one of the primary aims of the community mental health center is continuity of care, permeability is required not only between the various subparts of the center but also between it and other agencies in the larger caregiving network.

In the evaluation of the functioning of the hospital as a community mental health center, it is necessary to look beyond the hospital-based program to the community's total mental health network. One must consider *extraorganizational* and *interorganizational* processes as well. Three areas for focus in this type of evaluation are:

1) the intraorganizational processes of the changing state mental hospital;

2) the exchanges and transactions between the changing hospital and its environment; and

3) the processes and structures through which parts of the environment are related to one another.

2. OPEN-SYSTEMS ORGANIZATIONAL THEORY

The concept of the "organizational system" is a commonplace in organization theory today, and Scott (1964) argues that modern organization theory differs primarily from classical and neoclassical organization theory in its acceptance of the premise that the only meaningful way to study organization is to study it as a system. Katz and Kahn (1966) have recently written a book which takes an open-system approach to the study of the social psychology of organizations. They attempt to spell out open-system concepts as a framework for organizing the research on complex organizations, even though much of that research was generated from either a closed-systems approach or a fragmented or non-systems approach to theory.

The importance of the environment and of interorganizational relations in the conceptualization of the functioning of an enterprise has become the subject of increasing attention by organization researchers and theorists (e.g., Etzioni, 1960; Levine and White, 1961; Emery and Trist, 1965).

In England, workers at the Tavistock Institute of Human Relations have been active in developing organizational systems concepts, much influenced by Bertalanffy's work (1950) which first revealed the importance of a system being open or closed to the environment (Emery and Trist, 1960; Miller and Rice, 1967; Rice, 1963; Rice, 1965).

The open-system model of the Tavistock group assumes that: 1) organizations are defined by their primary task; 2) organizations are open-systems i.e., an organization admits inputs from the environment, converts them, and sends outputs back into the environment; and 3) organizations encounter boundary conditions which rapidly change the characteristics of the organization. The conceptual model for research on the mental hospital in transition that will be presented here is based on these same assumptions.

3. PRIMARY TASK

In order to analyze the internal activities of a mental hospital, it is necessary to consider the organization's directed impetus toward some goal in relation to its community. The systems model is often contrasted with the more traditional "goal model". However, the open-systems model is neither divorced from the goal approach nor is it in opposition to the goal approach. In order to survive, the organizational system must fulfill the function of achieving goals which define overall objectives and it must further develop sets of subgoals to be accomplished by subsystems such as roles and departments. The distinction between the two models for the researcher is that the systems model points out the importance of doing not only final-outcome analysis but also process-oriented analysis focusing on the intermediate outputs of a system.

Rice (1958) first introduced the concept of "primary task" to discriminate between the varied goals of industrial enterprises. He defined the primary task as the task that an institution had been created to perform. In a later book (Rice, 1963), Rice recognized the difficulty of treating the organization as if it had a single goal or task, and he cited the teaching hospital, the prison, and mental health services in particular as examples of institutions which carry out many tasks at the same time.

Like most complex formal organizations the mental hospital has multiple goals and performs many tasks simultaneously. A mental hospital admits patients, provides therapy and custodial care for patients, attempts to place patients in the community, provides employment for mental health workers, looks after its employees, conducts research, keeps records and accounts, and if it has educational programs, provides professional training for nurses and residents. Usually, there is no settled priority of goals for an organization and one of the goals becomes primary in the hierarchy at a given time according to the balance of forces then operating.

Multiple goals are not necessarily compatible and involve competition for scarce resources by those subparts of the organization which are more committed to one goal over others. For example, the organization that best fits the task of training residents is not necessarily the same one that best fits the task of returning patients to the community as quickly as possible. Sofer (1961) has shown that the organizational structure required for research programs in a mental hospital does not easily fit with the rest of the hospital.

Difficulties may arise in defining the one primary mission or task of an organization because certain goals are denied and consequent ambiguities may confound the researcher's choice of appropriate criteria for judging task performance. Traditional studies of organizational effectiveness based on success in goal achievement have depended too much on what organizational spokesmen have said. As Etzioni has pointed out, the goals which an organization claims to pursue (public goals), "fail to be realized...because...they are not meant to be realized," (1960, p. 260). It is too easy to confuse the convenient public fictions of official pronouncements with the actual functioning of a system. The stated purposes of an organization in its charter or other official papers or the reports of leaders may give a very distorted picture of the functions of the enterprise. Changes in organizational objectives as stated by officials do, nevertheless, provide a good beginning point for finding out about changes in the system which have taken place.

One of the ways the environment affects the setting of goal priorities in the mental hospital is through ideological penetration - the values which come in from the outside. Mental health facilities considered as organizations grew out of a complex interaction of cultural and technological systems. Cultural values or belief systems determine within broad limits what the goals of organizations will be. Currently, a new belief system is developing among mental health professionals which has been labeled the Community Mental Health Ideology.

As Community Mental Health Ideology penetrates a hospital, it can affect the priority of goals set for the hospital. For example, the ideas of community mental health imply extension of the hospital's goals of patient care and treatment to include improvement of the Mental Health of a population through increasing attention to programs of primary and secondary prevention. These new goals may produce changes in the task hierarchy of a hospital. A Community Mental Health Ideology Scale has been developed which is being employed in the study of the relation of this attitude to a program development (Baker and Schulberg, 1967).

4. INPUTS

Basic to the conception of an organization as an open system is the assumption that any enterprise, or parts of it, can be characterized as admitting inputs from the environment, converting or transforming them, and sending outputs back into the environment. The inputs to an organizational system include people, values, economic resources, physical facilities and technology - the variables which, as they are operated on in various ways, determine the outputs. The subsystems variables, the internal structures, attitudes, and activities account for the conversion or transformation of the inputs into outputs. The result of the internal processing is a set of output variables which are usually used in defining the effectiveness of the system. These outputs from one system in turn become inputs for other systems. These three sets of variables constitute the very core of the open-system organization theory.

The major inputs to a community mental hospital consist of people. The hospital takes in those defined as mentally ill and it is these patients that correspond to the "raw material" which is processed in an industrial enterprise. Just as the qualitative and quantitative aspects of the raw material available from the environment to an industrial system are major constraints on its productivity, so it is with the characteristics of patients. The patient input to the mental hospital may be examined in terms of demographic characteristics such as age, sex, race, socio-economic status, religion, and education. For example, a large number of geriatric chronic patients provide a major constraint on what the hospital can accomplish in terms of treating its patients successfully and returning them to the community. The prehospital level of adjustment of the patient input, i.e., how sick the people are, is another constraint. Thus it is apparent that the nature of the population served by the hospital is extremely important in determining the functioning of the system. As a state hospital takes responsibility for specific geographic catchment areas with resulting changes in the input of patients to the system, the internal structure and functions of the hospital and the character of its output will be affected. As an increasing number of individuals are seen on an outpatient basis, the balance between inpatient and outpatient programs is bound to be altered.

The rate and types of patient input are also determined by hospital-community relations. The attitudes of the community served by the mental health center will determine its use of the center. This includes both the attitudes of the sick, their families and friends, and also the attitudes of the other caregiving agencies who refer patients to the hospital. The analysis of the input of patients can be approached by analyzing the paths or routes through which patients enter, are treated, and leave the hospital, and by attempting to describe and assess those various paths.

Another kind of human input to be considered is the personnel coming into the hospital system including both the fully professionally trained as well as the residents, student nurses, and others who receive professional training. The level of and type of professional skills, the personality, attitudes, interests, goals, ideologies, and habits of the personnel are important input variables constraining the operation of the hospital system. In addition to human inputs, a hospital must receive material inputs of money, supplies, and facilities. The success of a state hospital's interaction with the sources of political, legal, and financial support largely determine the limits of its ultimate output.

5. OUTPUTS

Just as the input segment of the hospital's operation is affected by its level and type of interaction with the community, so will the hospital's output be similarly influenced. The discharge of a patient from the hospital involves such community elements as his family, employer, and friends. Their receptivity or resistance will profoundly affect the hospital's output performance.

The major output of the present mental hospital is treated patients who can be restored to the community. In the community mental health center, however, the output will be more varied and theoretically also should include such products as changes in the level of health of the population that should result from primary and secondary prevention activities (Caplan, 1961).

Although much attention is being devoted to the expanded community-oriented outputs of the mental hospital, unfortunately there will be a continued need to provide custody for some element of the population in the foreseeable future. As long as some patients are defined as dangerous to themselves and others, there will be a need for protection of the patients, staff, and larger community. Because of its physical facilities, the mental hospital will be called upon to maintain some patients within its walls. This means it also will have a patient maintenance output, including all those aspects of care necessary for maintaining chronic patients within a restricted environment.

The related operating system for, on the one hand, producing rehabilitation and, on the other hand, providing maintenance and protection will come into conflict since these outcomes are in some ways incompatible. Levinson and his associates have well documented the conflict between the custodial and therapeutic orientations in the functioning of mental hospitals (Gilberg and Levinson, 1957; Sharaf and Levinson, 1957).

As an educational research institution, the hospital will produce output of trained or educated personnel and also education for the community to fulfill its goals of changing the attitudes of the community.

Another kind of output of any organization is the morale of its employees. If the hospital is unsuccessful in maintaining a reasonably high state of morale, this will feed back as a loss in the input of necessary motivation power. Like any other organization, the mental hospital has an investment in maintaining the mental and physical health of its employees.

6. SUBSYSTEMS OF THE ORGANIZATION

Analysis of the input and output activities of the open-system hospital in interaction with its community must be supplemented by an analysis of the interaction of the hospital's own internal system within its boundaries. These interactions within the system are between role incumbents and between subsystems.

A system can be examined in terms of its units of process and its units of structure. The totality of all the structures in a system which carry out a particular process is a *subsystem*. A subsystem thus exists in one or more distinguishable units of the total system and is identified by the process it carries out. Certain subsystems are particularly relevant for the study of the changing mental hospital.

Operating Subsystems. The subsystems which carry out the input-conversion-output processes are the "operating systems" (Rice, 1963, p. 18). Three types of operating systems corresponding to three aspects of the flow process through a system can be identified in the hospital system:

1) input operating systems, which function to take in patients, staff, money, and materials, etc;

2) conversion operating systems, which function in treating patients in and out of the hospital, educating personnel, producing research knowledge, etc.; and

3) output operating systems, which function to place treated patients in the community, place trained personnel, and disseminate knowledge, etc.

In a changing mental hospital new operating structures, such as geographically regionalized treatment units, consultation service programs, and home treatment services, must be examined in terms of their structure and functioning. By recognizing the principle of "equifinality" (Bertalanffy, 1950, p. 25), i.e., that identical final states may be reached from different initial starting positions and by different routes, the examination of operating systems can make clear which is the more efficient of two systems set up to produce the same outcome. This is particularly important in health programs which use a shotgun approach for providing a number of services aimed at the same targets.

Managerial Subsystem. A particularly important subsystem in the changing mental hospital is the managerial one. As the controlling or decision-making part of the organization, it cuts across all of the operating structures of production, maintenance and adaptation. The managerial functions in the traditional hospital have been the exclusive province of the superintendent, but as the organizational structure is elaborated, the functions of management become more complex and there is pressure to share the functions of management.

In the traditional public mental hospital, there are few specialized boundary roles for linking the institution with its community. In many mental hospitals the superintendent acts as the principal link. In fact, it still is a common practice for many state hospitals to have a statement such as "address all correspondence and moneys to superintendent" printed on all stationery bearing the hospital letterhead. As a hospital becomes a community mental health center, a need will arise for an increased number of boundary spanning roles. The requirements of interaction with the environment will be far greater and the use of consultants to community health and welfare resources will become an increasingly common technique for bridging the gap between the hospital and its community.

In times of change it is particularly important for management to look realistically outward as well as inward and to relate the inside and outside aspects of the organization effectively. Not only must management coordinate the environmental pressures with the internal system forces, it must deal with conflicting demands of the system's substructures. Because management finds it easier to meet conflicts as they come up from day-to-day dealing with one part of the structure and then compromising with another part, the organization may be seen to jerk along in fits and starts.

In many mental hospitals the immediate crises and pressures confronting the superintendent generally emanate from within his own system and it is easy to become enmeshed by them. The ability to withstand them and focus on subtler, less immediate pressures from the external community is necessary if the hospital is to proceed in its transition from traditional functions to the expanded ones of a community mental health center. Not only does management have to function to coordinate the substructures and resolve conflicts between hierarchical levels, it must also coordinate external requirements with the system's resources and needs.

7. ENVIRONMENT

An open-system is defined as one into which there is a continuous flow of resources from the environment and a continuous outflow of products of the system's action back to the environment. As an open-system, an organization depends for its growth and viability upon its exchanges with the environment - that part of the physical and social world outside its boundary. The environment of the mental hospital includes the community which it serves and the other organizational systems which serve as sources of legal, political, financial, technical, and professional support.

The environment is of particular importance to a mental hospital making the transition to a community mental health center since responsibility for a specific geographical community is a basic concept in its functioning. Since there seems to be general agreement that a community mental health program requires the participation of many caregiving resources, the intersystem relations of the transitional hospital are also important.

Thompson and McEwan (1958) have conceptualized organizations as ranging on a continuum of power from those that dominate their environments to those that are completely dominated. If one conceives of a community mental health suprasystem as consisting of a network of health and welfare organizations embedded in an exchange network (Levine and White, 1961), it seems that organizations must adopt strategies for coming to terms with the other organizational systems in their environment. Whether cooperative or competitive strategies are adopted has major implications for the degree of integration of services and continuity of care provided to a community.

REFERENCES

1. Baker, F., and Schulberg, H.C., The Development of a Community Mental Health Ideology Scale, *Community Mental Health Journal, 3,* 216-225, 1967.

2. Bertalanffy, L.V., The Theory of Open Systems in Physics and Biology, *Science, 111,* 23-28, 1950.

3. Caplan, G., An Approach to Community Mental Health, New York, Grune and Stratton (1961).

4. Emery, P.E. and Trist, E.L., Socio-Technical Systems; in Churchman, C.W., and Verhulst, M. (Eds.) *Management Sciences: Models and Techniques, 2,* London, Pergamon Press (1960).

5. The Causal Texture of Organizational Environments, *Human Relations, 18,* 21-32, 1965.

6. Etzioni, A., Two Approaches to Organizational Analysis: A Critique and a Suggestion, *Administrative Science Quarterly, 5,* 257-278, 1960.

7. Gilbert, D., and Levinson, D.J., Role Performance, Ideology and Personality in Mental Hospital Aides; in Greenblatt, M., Levinson, D.J., and Williams, R.H. (eds.), *The Patient and the Mental Hospital,* New York, The Free Press, 197-208, 1957.

8. Katz, D., and Kahn, R.L., *The Social Psychology of Organizations,* New York, John Wiley and Sons (1966).

9. Levine, S., and White, P.E., Exchange as a Conceptual Framework for the Study of Interorganizational Relationships, *Administrative Science Quarterly, 5,* 583-601, 1961.

10. Miller, E.J., and Rice, A.K., *Systems of Organization,* London, Tavistock Publications (1967).

11. Rice, A.K., *The Enterprise and Its Environment,* London, Tavistock Publications (1963).

12. *Productivity and Social Organization,* London, Tavistock Publications (1958).

13. Schulberg, H.C., and Baker, F., The Changing Mental Hospital: Is It Really Changing? paper presented at the 20th Mental Hospital Institute, Washington, D.C. (October 1, 1968).

14. Schulberg, H.C., Caplan, G. and Greenblatt, M. Evaluating the Changing Mental Hospital: A Suggested Research Strategy, *Mental Hygiene, 52,* 218-225, 1968.

15. Scott, W.S., Organization Theory: An Overview and Appraisal, *Journal of Academy Management, 4,* 7-26, 1964.

16. Sharaf, M.R., and Levinson, D.J., Patterns of Ideology and Role Definition Among Psychiatric Residents; in Greenblatt, M., Levinson, D.J., and Williams, R.H. (eds), *The Patient and the Mental Hospital*, New York, The Free Press, 263-285, 1957.

17. Sofer, C., *The Organization from Within*, London, Tavistock Publications (1961); Chicago: Quadrangle (1961).

18. Thompson, J.D., and McEwen, W.J., Organizational Goals and Environment: Goal Setting as an Inter-Action Process, *American Sociological Review, 23*, 23-31, 1958.

SOME NORMATIVE IMPLICATIONS OF A SYSTEMS VIEW OF POLICYMAKING

Yehezkel Dror*

ABSTRACT

Normative general systems theory can provide a main approach to the improvement of public policymaking and serve as a basis for policy sciences. For these purposes, the public policymaking institutions and their operations should be viewed as a complex system, which can be analyzed and improved with the help of basic general systems theory ideas and concepts. Especially significant is the distinction between system outputs and component output, which leads to two main conclusions: (a) a variety of alternative changes in different components can result in similarly better policies; (b) in order to have any positive impact on the overall policy output, changes in the policymaking-system components must reach some minimum threshold.

Possibilities and needs for changes in the policymaking-system can be illustrated by the following improvement proposals: (1) explicit strategy decisions; (2) explicit learning feedback; (3) better consideration of the future; (4) extensive analysis; (5) encouragement of creativity and inventions; (6) improvement of one-person-centered high-level decisionmaking; (7) development of policy professionals; (8) development of politicians; (9) establishment of policy sciences as an

---

* Yehezkel Dror is at the Rand Corporation, 1700 Main Street, Santa Monica, California, and the Hebrew University of Jerusalem (on Leave).

academic interdiscipline and profession; (10) radical changes in school teaching of civic and current affairs subjects; and (11) explicit and systematic meta-policymaking.

Realization of such improvements based on a normative systems approach can result in important, though limited, advances in the quality of policymaking. They can be intellectually and politically feasible if intense efforts are made.*

1. INTRODUCTION

Systems theory in its general forms can be used to analyze and explain behavior and to provide a unifying and general theoretic framework for comprehending in common terms a larger number of heterogenous phenomena. Another main use of general systems theory is to provide a normative approach to the analysis of behavior and phenomena. As developed in "systems analysis" and "systems engineering," the normative approach tries to use, explicitly or implicitly, general systems theory concepts and frameworks in order to improve the operations of a given system or to design new and better systems.

In its conceptions and ideas, normative general systems theory seems to hold considerable promise for conscious social self-direction. This promise is of special importance for contemporary humanity, because of the increasing need for better public policymaking as a main mode for dealing with more acute and difficult social problems. But, at present, social systems and their

---

* Any views expressed in this paper are those of the author. They should not be interpreted as reflecting the views of The RAND Corporation or the official opinion or policy of any of its governmental or private research sponsors.

broader components are in the main excluded from re-examination and improvement with the help of system approaches. Some promising work has been done in the utilization of systems ideas for analyzing and explaining social behavior, but very little is available on normative applications of general systems theory to broad, non-sub-sub-sub-optimized social issues and systems.

The characteristics of social systems seem too diffuse, their dimensions too complex and many of their events too arbitrary (in the technical sense of being unpredictable and unexplainable by either deterministic or stochastic concepts) to fit into any "model" which is formalized enough to permit systems analysis, systems control and systems design* by available techniques. Application of the normative system orientations to social issues is regarded by many social scientists and most policy practitioners as either quite useless, or such application, in the opinion of most systems analysis professionals, must wait till social science becomes more mature and delivers the "hard data" needed for a vigorous systems approach.

It is with these views and the resulting scarcity of useful normative applications of general systems theory to the social arena that I disagree. It seems to me that general systems theory can make great contributions to social improvements, but in order to do so we must learn to distinguish between its core ideas and its secondary apparatus. What is useful for normative application to social macro-phenomena are the basic ideas of general systems theory including: the concept of a "system," the distinctions between system behavior and additive component behavior, the concepts of interaction and

---

* I prefer these terms to the concept "systems engineering," which I think should not be used in reference to social systems. This concept is too technically oriented and has connotations that are too strongly amoral (though not immoral) to fit social phenomena and their improvement requirements.

feedback dynamics, adjustive and homeostatic behavior, system-environment exchanges and interdependencies and others. What is less useful for normative treatment of social systems, at least in the foreseeable future, are (1) some philosophical assumptions of parts of general systems theory, such as the issues of entropy vs. negative entropy, and (2) especially some of the main tools of normatively applied systems approaches, such as quantitative models, optimization techniques and computer simulations.

This paper concerns an effort to use some simple general systems concepts normatively in order to explore approaches to the improvement of public policymaking. My purposes in doing so are: (a) to illustrate the possibilities of utilizing a simple general systems approach for improving complex social systems; (b) to stimulate work on one of the most important contemporary needs, namely, the improvement of public policymaking; and (c) to try to lay some foundations for a new interdiscipline of policy science, based in part on general systems theory.*

2. A GENERAL SYSTEMS VIEW OF PUBLIC POLICYMAKING

Using a very simple version of systems theory, we regard public policymaking (and, with some changes, other types of policymaking) as an aggregative process in which a large number of different units interact in a variety of partially-stabilized but open-ended modes. In other words, public policy is made by a system, the public policymaking system.

---

* Some theoretic foundations of such a policy science, based on a systems approach to policymaking, are presented in my book *Public Policymaking Reexamined* (San Francisco: Chandler Publishing Corporation, 1968).

This system is dynamic, open, non-steady-state, and includes a large variety of different and changing multi-role components interconnected in different degrees and through a multiplicity of channels. It is closely interwoven and overlapping with other social macro-systems (e.g., the productive system, the demographic-ecological system, the technological and knowledge system and the cultural system), and it behaves in ways which defy detailed modelling.*

Nevertheless, even a very simple systems perspective of public policymaking, which is feasible with available knowledge and the present state-of-the-art, leads to two important improvement-relevant conclusions:

a. As public policy is a product of complex interactions between a large number of various types of components, similar changes in the output (or similar "equi-final states") can be achieved through many alternative variations in the components. This means, for our purposes, that different combinations of a variety of improvements may be equally useful in achieving equivalent changes in the quality of policymaking. This is a very helpful conclusion, because it permits us to pick from a large repertoire of potentially effective improvements those which are more feasible under changing political and social conditions. This view also emphasizes the open-ended (or, to be more exact, "open-sided") nature of any search for improvement-suggestions: there is, in principle, unlimited scope for adventurous thinking and invention. Therefore, any list of such proposals should

---

* A much longer time perspective may permit modelling of the evolution of human institutions in historiosophic or biological terms, but is irrelevant for improving the policymaking system. But such models may become important for some future long-range policy issues, such as genetic improvements through molecular engineering, space expansion policies for the human race, and problems of total environmental control techniques.

be regarded as illustrative and not definitive.

b. A less optimistic implication of a systems view of public policymaking is that improvement efforts must reach a critical mass in order to influence the aggregative outputs of the system. Improvement efforts which do not reach the relevant impact thresholds will, at best, be neutralized by countervailing adjustments of other components (e.g., a new planning method may be reacted to in a way making it an empty ritual); or, at worst, may in fact reduce the quality of aggregative policies (e.g. through possible boomerang effect, reducing belief in capacity of human intelligence, with possible retreat to some types of mysticism, leader-ideology, etc.; or by making and implementing wrong decisions more "efficiently," and thus abolishing a basic social protective mechanism - inefficiency as reducing the dangers of foolish decisions and as permitting slow and tacit learning).

At present, many efforts are under way in the United States (and other countries) to improve public policymaking, though in a disjointed way. These efforts take a number of forms, including for instance: (a) establishment of new types of organizations devoted to improving policymaking (such as RAND, the Urban Institute, and, in another way, the Center for the Study of Democratic Institutions); (b) development of new methods which try to help better policymaking (such as systems analysis, planning-programming-budgeting-systems (PPBS), and sensitivity training); (c) establishment of new schools and departments at universities devoted to "policy studies" (such as the program in policy sciences at The State University of New York at Buffalo, the program in social policy planning at the University of California at Berkeley, the programs in analysis at MIT, the new program in public policy at the Kennedy School at Harvard, and the large number of new schools for public affairs, which programs are also in part a response to student demand, with an apparent move by top students from natural science areas to social-problem-relevant studies); and (d) various efforts to increase the utilization of behavioral sciences in government.

These and similar efforts are symptomatic of increasing awareness of the need for and constitute an important beginning on the way to better public policymaking. But, if stabilized in their present form, these programs, even if their aims are sound, are of limited usefulness and some of them may even be counterproductive, because they neglect to view policymaking as a complex system, ignore many critical improvement needs, and fail - in many respects - to reach the minimum critical mass. In particular, these programs apply in the main to low-level and technical decisions; depend on quantification; require unavailable highly-qualified talent; neglect many critical decision situations (e.g. the one-person-focused decision situation); in effect, ignore the needs for creativity, tacit knowledge and adventurous thinking, and may indeed repress them through subjection to inappropriate criteria; they tend to ignore if not to disdain "politics"; fail to face the complexities of value judgment, and have no comprehensive theoretical basis nor the necessary underpinning of academic research and professional training (other than in the rather narrow areas of operations research, systems engineering, and parts of economic theory).

What is needed, therefore, is a broad systems approach to the improvement of policymaking, by which a sufficiently large variety of improvement suggestions can be identified so as to provide a sub-set of feasible alternative improvements sufficiently large to reach the critical mass and to achieve a substantial impact on aggregative policymaking. The probable effects of any proposal must be "guestimated" (guessed-estimated) in terms of system-effects and, in most instances, a synergistic set of improvements is required. This applies to the illustrative suggestions for improvement presented in the section which follows. These suggestions are mutually reinforcing and should be implemented in sets including at least some parts of several of them.

3. SOME SUGGESTIONS FOR IMPROVING PUBLIC POLICYMAKING*

Improvement of public policymaking must, as explained, proceed in view of all main dimensions of the public policymaking system. In particular, improvements are required with respect to: (a) process-patterns; (b) structure; (c) personnel; (d) knowledge; and (e) on a broader level, "policy culture." In all these dimensions, improvements should strengthen rational-analytic capacities as well as extra-rational capacities (such as creativity, tolerance of ambiguity, propensity to innovate, and levels of aspiration). To illustrate, the following concrete proposals, dispersed over the aforementioned systems dimensions, are suggested:**

(1) Explicit strategy decisions. Special structure and process-patterns should be established to engage in basic strategy decisions, as distinguished from more-or-less *ad hoc* policymaking. Such strategy decisions include formulation of longer-range policy goals, establishment of main postures, determination of attitudes toward risk and similar "master-policy" decisions.

(2) Explicit learning feedback. Special structures and process-patterns should be established to engage in the systematic study of past policies, the drawing of future-oriented conclusions from those experiences, and the injection of these conclusions into contemporary policymaking.

(3) More comprehensive and sophisticated consideration of the future. Special structures and process-patterns should be established to encourage broader consideration of the future in contemporary policymaking.

---

*Some of these illustrations were presented by me before the Center for the Study of Democratic Institutions and benefited much from discussions with its Fellows.

**For reference to more detailed discussions of some of these ideas, see the bibliographic note at the end.

This includes, for instance, dispersal of various kinds of "future study" organizations, units, and staff throughout the social guidance cluster, and utilization of alternative images of the future and scenarios as standard parts in all policy considerations.

(4) Policy analysis as an integral part of policymaking. This involves (a) development of policy analysis as a method for better dealing with complex, largely non-quantifiable issues; and (b) establishment of policy analysis units (of different scope, size, and complexity) throughout the social guidance cluster, so as to change somewhat the patterns of policy discussions and policy formulation.

(5) Encouraging creativity and invention in respect to policy issues. This involves, for instance, no-strings-attached support to individuals and organizations engaging in adventurous thinking, avoidance of their becoming committed to present policies and establishments, and opening up channels of access for unconventional ideas to high-level policymakers. Creativity and invention should also be encouraged within policymaking organizations by institutionally protecting non-conventional thinkers from organizational conformity pressures.

(6) Improvement of one-person-centered high-level decision making. Even though of very high and sometimes critical importance, one-person-centered high-level decision making is very much neglected both by research and by improvement attempts. This in part is due to difficulties of access, on one hand, and to dependence of such decision making on the personal characteristics and tastes of the individual occupying the central position and the consequent difficulties in improving such situations, on the other. Nevertheless, one-person-centered high-level decision making can be improved, because some needs of better decision making - as explained previously - can be satisfied by a variety of means, some of which may often fit the desires of any particular decision maker. Thus, information inputs, access to unconventional opinions, feedback from past decisions, etc. can be provided by different channels, staff structures, mechanical

devices, communication media, etc. -- which provide sufficient flexibility to fit arrangements to the needs, tastes, preferences, and idiosyncrasies of most, if not all, top decision makers.

(7) Training and development of policy analysts and other policy professionals. Nearly all the improvement suggestions require persons with high moral, intellectual, and academic qualifications to serve as the professional staff for policy analysis, policy research, future studies, etc. Training of such professionals at universities and their continued development (e.g., through rotation between more detached and more applied research) is essential. Also urgently needed is a professional organization of policy analysts and other policy science professionals to promote research and training, to support recognition, to encourage mutual learning and processing of experiences, and to attempt to deal with very difficult and sensitive issues of professional ethics and qualification requirements.

Better policymaking requires also better utilization of social sciences, of law, of life sciences, and other disciplines. Preparation of graduate students in these areas for playing a role in policymaking -- both in staff positions and as independent free-thinking citizens -- requires significant changes in many of the contemporary graduate studies curricula.

(8) Development of politicians. The idea of improving politicians is regarded as taboo in Western Democratic societies, but this is not justified. Politicians can be improved within the basic democratic tenets of free elections and must be improved to increase the probabilities of good policymaking. Setting aside the more diffuse proposals on how to encourage entrance into politics of more persons whom we regard as "desirable" and how to vary the rules of the game to permit better judgment by the voter, the following proposal is offered: elected politicians (e.g., members of a state legislature) should be granted a sabbatical leave to be spent in self-developing activities, such as travelling abroad and studying. Suitable programs should be established at

universities and special centers for active politicians to spend their sabbaticals in a useful and interesting way.

(9) Establishment of policy science as a distinct area of research and study. Implied in most other improvement suggestions, and indeed fundamental for every effort to understand and improve the public policymaking system, is the need for more knowledge concerning policymaking. Also, taking into account the needs of preparing and developing policy professionals, and in view of the organizational characteristics of most universities -- recognition of policy sciences as a distinct area of research and study seems essential.

(10) Radical changes in the teaching of "good citizenship" and current affairs subjects in schools. In the longer run, better preparation of the citizen for his roles in influencing policies and policymaking are of critical importance for the adjustment of democracy to an age of increased knowledge and better multi-directional communications. A first step to meet urgent needs is a radical change in the teaching of all "good citizenship" subjects in the elementary and high schools in the direction of developing individual judgment capacities, learning information search and evaluation habits, and increasing tolerance for ambiguities, and a readiness to innovate. Intensive use of new teaching methods, such as gaming and projects, and full exposition of contradicting points of view may be helpful in moving in the desired directions. But what is really needed is a far-reaching reform of the teaching of all subjects (and of all teacher preparation), including early introduction of pupils to a system view of reality and problems.

(11) Explicit and systematic meta-policymaking. Basic to all improvement-suggestions, and indeed to the whole approach presented in this paper, is the requirement for explicit and systematic policymaking on how to make policy and on how to change and redesign the policymaking system, which is *meta-policymaking*. It is the very limited, sporadic and somewhat accidental nature of present attention to the main features of the public

policymaking system which poses the most urgent need for far-reaching change. Spontaneous adjustments, incremental change, and learning by the shock-effects of radical mistakes may have been adjustment mechanisms adequate for survival in the past, though at a quality level well-characterized by Milton Rubin by his reference to Walter Pitkin's book *A Short Introduction to the History of Human Stupidity*.* In the future, the natural adjustment capacities of the public policymaking system are not adequate. They must be complemented and, in part, replaced by the application of human intelligence and knowledge through explicit and systematic meta-policymaking. This requires think-organizations which specialize in the monitoring and evaluation of the public policymaking system and the creation of improvement proposals. In addition the quality of public policymaking must become a main concern for academic research and study, public interest and political action.

4. IMPLICATIONS FOR THE FUTURE

The eleven improvement suggestions, are only some illustrations of needed, mutually reinforcing improvements in the public policymaking system. These suggestions, especially the last, seem to raise two additional and more basic questions: (1) what can be hoped for in the way of better policymaking even if we fully apply a systems view to the improvement of the policymaking system and implement these and similar reform suggestions; and (2) is it realistic to expect any impact of a systems view on policymaking reality and do any such intellectual ideas have any political feasibility in the foreseeable future?**

---

*Milton D. Rubin, "The General Systems Program: Where Are We Going?" p.15.

**I am grateful to Milton Rubin for pointing out the omission of this very important question in an earlier version of this paper.

Our view of policymaking as the function of a complex and non-deterministic system should help us to avoid any form of *hubris* and to warn us of misplaced overconfidence in the human capacity to consciously shape (or misshape) his own future. My own feeling is that for the next ten to twenty years, some avoidance of "minimin"* in policymaking would be a great achievement, and overall improvement of public policymaking by -- in a qualitative sense -- "ten percent" would be tremendous progress which would constitute a radical change in the evolution of social auto-guidance.

Limiting our aspirations to such modest levels of change in the quality of policymaking -- which are radical in comparison with the past but do not approach some of the apocalyptical views on the possibilities of complete transformation of policymaking with the help of "science" -- we remain within what I think is a potentially feasible range from the point-of-view of our knowledge, given one essential condition: very intense efforts must be made to develop policy sciences as a general systems theory-based interdiscipline, and policy analysts as new action-oriented professionals.

The question of political feasibility is harder. As mentioned previously, the great range and equifinality of useful alternative improvement possibilities increases the probability that some of them may be or may become politically feasible. There also seems to be a growing awareness of the need for some quantum jumps in the qualities of public action which may open the way for significant improvement. In particular, crises often provide far-reaching opportunities for change, given the condition that well worked-out ideas are available, though this is a very costly way to get urgently needed improvements accepted and a highly erratic and unpredictable one.

---

*I propose "minimin" as a new term, by which I refer to the worst of all bad alternatives -- in part-contrast to the theory of games concepts of maximax, maximin, minimax.

No less important in the longer run is the impact on feasibility of good ideas, especially when consistently and convincingly presented by persons who regard it as a moral duty to struggle for their acceptance. If we really believe that a normative general systems approach can contribute to so critical a process as public policymaking, it is up to us and all who share our opinion not only to consider political feasibility as a fact*, but to attempt to shape it, and to make what is absolutely necessary also feasible. Hence, the general systems-oriented policy scientist is faced both by a fascinating intellectual issue and a great challenge.

*For an examination of the concept of political feasibility and maps for predicting it, see "the Prediction of Political Feasibility", Yehezkel Dror, Vol 2 (1969).

BIBLIOGRAPHIC NOTE

Good illustrations of attempts to analyze complex social systems with the help of a general systems framework are K. Deutsch, *The Nerves of Government*, N.Y., The Free Press, 2nd edition (1966); D. Easton, *A Systems Analysis of Political Life*, N.Y., John Wiley and Sons (1965); W. Buckley, *Sociology and Modern Systems Theory*, Englewood Cliffs, N.J., Prentice-Hall (1967); and F.K. Berrien, *General and Social Systems*, New Brunswick, N.J., Rutgers University Press (1968). An excellent collection presenting the present state-of-the-art of normative systems approach is E.S. Quade and W.I. Boucher, (eds.), *Systems Analysis and Policy Planning*, N.Y., American Elsevier (1968).

The growing interest in public policymaking, including some attempts to apply to it a behavioral or normative systems approach, are illustrated by six books, all of them published since 1968: C. Lindblom, *The Policy-Making Process*, Englewood Cliffs, N.J., Prentice-Hall (1968); R.A. Bauer and K.J. Gergen (eds.), *The Study of Policy Formation*, N.Y., The Free Press (1968); J.M. Mitchell and W.C. Mitchell, *Political Analysis and Public Policy*, Chicago, Rand McNally (1969); A. Ranney (ed.), *Political Science and Public Policy*, Chicago, Markham (1968); F.E. Rourke, *Politics and Public Policy*, Boston, Little Brown (1968); and G. Vickers, *Value Systems and Social Process*, N.Y., Basic Books (1968). The growing importance of social self-direction and its underlying basis is well presented in A. Etzioni, *The Active Society*, N.Y., The Free Press (1968).

My basic view of public policymaking and its improvement from a systems point of view is presented in *Public Policymaking Re-examined*, San Francisco, Chandler (1968). Some of the improvement suggestions are discussed in my following papers or illustrated by them: "Policy Analysts: A New Professional Role in Government Service," *Public Administration Review, XXVII*, (3) 197 - 203 (September 1967); "Proposed Policymaking Scheme for the Knesset Committee for the Examination of the Structure of Elementary and Post-Elementary Education in

Israel -- An Illustration of a Policy Analysis Memorandum," *Socio-Economic Planning Sciences, 2,* 1969 (in print; earlier version RAND paper P-3941, October 1968); The Improvement of Leadership in Developing Countries, *Civilizations, XVII,* (4) 435-441 (1967); Some Requisites of Organizations Better Taking Into Account the Future, in R. Jungk and Y. Galtung (eds.), *Mankind 2000,* Oslo, Norwegian Universities Press 286-290 (1969); and The Role of Futures in Government, *Futures, 1,* (1) 40-46 (September 1968) (earlier version RAND paper P-3909, August 1968).

SECTION V

THE ANALYSIS AND EVALUATION OF A SCIENTIFIC FIELD

Williams points out in the first paper of this section that the unprecedented growth of American science since World War II has increasingly emphasized the need for more careful planning for the Federal support of science in relation to national goals. This need has been sharpened within the last few years by the reduction of available funds. Dr. Ivan Bennett has stated the situation with terse clarity: "Science, including biomedical science, can no longer hope to exist, among all human enterprises, through some mystique, without constraints of scrutiny in terms of national goals, and isolated from the competition for allocation of resources which are finite. Unless we biomedical scientists are prepared to examine our endeavors, our objectives, and our priorities, and to state our case openly and clearly, the future will be difficult indeed."[1] In other words, science is only one component out of many in a national system of activities for satisfying national needs, and its support must be justified in terms of its contribution to the whole.

At present, no satisfactory method exists for period-

---

[1]Research in the Service of Man: Biomedical Knowledge, Development, and Use. A conference sponsored by the Subcommittee on Government Research and the Frontiers of Science Foundation of Oklahoma. Senate Document 55, U.S. Government Printing Office, Wash. D.C.:1967. pp. 7-8.

ically surveying the scientific fields with which a government agency is concerned, in order systematically to identify the areas of research activity within these areas, and set rational priorities on them. Administrators of research programs need such a survey system not only to make more satisfactory decisions between competing demands for funds, but to handle the overwhelming amounts of information which are needed, and to offer a more exact public accounting of research expenditures, particularly where basic research is concerned.

In response to this need, a new area of study has grown up within the general field of research-on-research, which has come to be called the analysis and evaluation of a scientific field. It is with this area of research that this section is concerned.

Williams' paper places the problem in national perspective, in order to show the importance to the Federal government of various forms of evaluation of various levels of scientific activity from single projects to broad areas of science, in relation to other parts of the national budget. The second paper, by Bailey, describes in detail one promising technical solution to the problem of modeling a scientific field as a system of research problem-areas.

While neither of the authors of these papers is a specialist in general systems theory or in systems analysis, they agree that one obstacle to the development of explicit general system models has been the lack of a formal or rational procedure for articulating the anatomy and functioning of a system, and a usable taxonomy or comparative classification of systems. If general models are possible, such a procedure should as a minimum (1) specify the system environment, (2) identify components and their various aggregations into sub-systems, (3) measure the dynamic interrelationships between these entities and (4) identify and evaluate inputs and outputs against specified criteria. These papers report attempts to develop methodology which satisfies these requirements with respect to a scientific field. They are contributions to general systems theory insofar as they can be generalized to systems in other fields, either by the use of the explicit formulation of their theory structure in generalized terms, or by isomorphic translation of their structure to the other fields.

Milton D. Rubin

ALLOCATIONS FOR THE DEVELOPMENT AND EMPLOYMENT OF SCIENTIFIC RESOURCES: A CONCEPTUAL FRAMEWORK AND SURVEY *

Charles W. Williams, Jr.**

ABSTRACT

This paper sets forth a conceptual framework from a general systems point of view of the context within which one must view the analysis and evaluation of allocation decisions for support of scientific fields. A brief historical summary is provided as a basis for understanding how the context of decision-making has become so broad and complex.

The framework suggested is classified into levels of allocation choice, parameters for aiding decisions for allocation and evaluation, and a scientific opportunity structure. The status of methodological development and current research within the categories of this framework is discussed together with some recommendations on where further work is needed and how this type of methodological research relates to general system concepts.

---

* Views expressed herein are those of the author. They do not necessarily reflect any official position or viewpoint of the National Science Foundation.

** Charles W. Williams, Jr. is Staff Associate, Office of Planning and Policy Studies, National Science Foundation, 1800 G Street, N.W. Washington, D.C. 20550.

1. BACKGROUND

The analysis and evaluation of a scientific field as a claimant for resources must be fitted into the more complex picture of general allocations for the development and employment of scientific resources. Questions of effective allocations of such resources have, in the mid-sixties, come to the forefront as some of the most critical issues of science policy. Why should this be so? A few reasons come immediately to mind:

1) A developmental period reaching back to Colonial days has now reached a scale and scope that have created new "climates" for decision.

2) There has been a relative shift in the locus of the decision centers within which complex allocation choices are made. This shift has resulted in a greater degree of inter-dependence between and among those decision centers.

3) Significant advances have emerged in the managerial and administrative sciences which are being called forth to deal with the greatly increased scale and complexity that result from more sophisticated levels of scientific and technological capability.

A brief summation of these trends will be helpful in setting the context for our deliberations.

During the earlier years, the foundations of "real" resources (i.e., skilled manpower, instruments, institutional structures and the theoretical and experimental levels of knowledge) were steadily evolving. Although more adequate financial support would have undoubtedly increased the rate of such evolutions, they continued for many years at a scale which did not levy heavy demands for financial resources. Barely thirty years ago the idea that science is a national resource was articulated. This was followed closely by World War II and the immediate urgency to amass all of our resources on a national scale. By the time victory had been won the value of science as an indispensable foundation for achieving national goals was well demonstrated. It was also proven that large scale organization could be applied to scientific endeavors and indeed was essential to many. Compressed into less than a decade, these conceptual and historic events were a clear "threshold" leading toward the present prominence of contemporary allocation issues. The twenty year delay may be accounted for by a number of factors.

The concentration of priorities for science as an important instrument of national security prevailed throughout the early Cold War years. Consensus was made relatively easy by the immediate benefits of exploiting the dramatic leaps in more effective nuclear power, missile and related new military technologies, and the urgent realization of the type of "military capability gap" which would have resulted from the failure to give adequate priority to these goals. The medical sciences which also emerged as an area clearly fruitful for absorbing rapid growth were of a nature which made consensus for support readily attainable. These areas, in conjunction with the continued growth of already well established agricultural research, made up the tight concentration of goals and priorities seeking large scale employment of scientific resources and research. Such "scientific and social opportunities" were coupled with a relatively small base of *real* scientific resources. Therefore, financial support was not the most basic constraint. Rather, it was used as an instrument to "call forth" a larger, more adequate foundation of trained manpower and facilities. By the mid-sixties the scientific resource base had grown to require relatively large scale financial support. The shortage of growth in such support became an important (perhaps the most important) constraint to continuing scientific development. Meanwhile, the complexities of a technological advanced society with diverse specializations were becoming evident. Some of the social, economic and ecologic side effects of science and technology became too evident to be pushed aside. The public perception of science's role had reached at least a degree of sophistication where solutions to many problems were regarded as possible only through expanded research for greater knowledge and technological capabilities. But demands were also rising for this technical exploitation to be carried out in a manner which would not entail high undesirable social or ecological costs.

During the earlier years, neither the economic nor the scientific "margins" for support of the rapidly developing scientific areas were felt to be approached - though of course such "margins" were far from precisely perceivable. Accordingly, almost any investment in competent research was "good" allocation of financial resources. But as the structures of knowledge became more advanced, those areas which had enjoyed priority attention under a relatively narrow set of national goals began to approach "the margin" when compared to other scientific areas in earlier stages of evolution that were ripe for development. (Note: Of course none of these terms can be applied very rigorously. By its very nature science is an open ended venture. "The margin" is not a real perceivable point. But the terms are useful in at least a very crude sense).

Heavier costs of scientific activities coupled with the importance of these activities in the attainment of national goals caused a relative shift from private toward public (especially Federal Government) decision centers for the allocations to support development and employment of scientific resources and capabilities. This, in turn, carried the requirement for public dialogue and for more elaborate rationales to serve as instruments for gaining the consensus necessary for making broad national investments of public funds.

Some of the advanced methods of the administrative sciences were directed to be implemented in nearly all Federal Executive Branch operations with the requirement to develop formalized systems of Planning, Programming, and Budgeting (PPB).* The agencies covered by this requirement included all of those heavily engaged in the support of scientific activities. Employment of such formalized concepts for science and research required (and still requires) significant revisions to the then existing planning and programming methods and the adaptation of PPB applications to accommodate the special considerations needed for scientific and research processes. As we continue to search for such improvements, one of the most vexing problems is the need for adequate theories and techniques to measure the direction, magnitude and effectiveness of the nation's scientific and research activities.

The mid-sixties have, therefore, seen a search for new motifs for viewing national goals and the role science and technology play therein. This has been accompanied by increasingly competitive social demands upon national resources which are at least a partial factor in the declining growth of support for science. A system of such complex national dimensions has its own built-in inertias which make it difficult to "stop" and "start" in whimsical fashion. The "pipeline" pressures evoked during the years of growth lag behind the drop in support rates and therefore create - for a while - still more competition for financial resources. These developments have converged at a time when science is reaching into more and more areas of the physical and social universes and has become an essential foundation upon which advanced societies must depend for survival, progress and the resolution of staggering human and social problems.

---

* See BOB Circular 68-2, July 31, 1967, Planning, Programming, Budgeting (PPB).

As the competition for resources has become keener, deficiencies of the methods which provide us the instruments for allocation become both more evident and more critical. For in large measure, allocation decisions for resource development and use are determinants of conducive or inhibiting environments within which are evolved the viability and directions of scientific research for both the acquisition of knowledge and its effective application. Our methods for reaching those decisions have not kept pace with events. The methodological gap includes inadequate conceptualizations, perceptions, techniques and procedural processes. Indications for the foreseeable future are that the importance, scale and complexity of these allocation problems will continue to grow rapidly. The consequences of error will become increasingly more alarming.

2. SOME CONCEPTUAL FRAMEWORKS

Allocation begins far ahead of research performance through processes in which many individuals and decision centers share. For example, career choices of young persons are an important part of national allocations. These individual choices are significantly influenced by such early factors as: the quality of secondary educational experiences and undergraduate development; the youth's view of the social role of science; the economic stability and incentives of a scientific career; and the availability of the essential high quality higher education. Plans of individual institutions (academic, business, and non-profit) are important ingredients of allocation processes. The complex evolutions of our national goals and human aspirations are partially determined by and are largely determinants of science resource allocation. Programs of governmental executive departments and agencies and of the Congress significantly shape the kinds of scientific resources America will have and the directions of their employment. Public and private budgeting and funding decisions are critical elements as are the various mechanisms through which support is linked to specific projects and performing groups. These and other factors both contribute to and react to the indispensable individual and collective creative imagination, insight and competence of scientists, engineers, administrators and political leaders. These are but a few of the many, many ingredients of, and participants in, science allocation processes.

Many attempts are made to seek "rational" allocation through adaptations of economic techniques. These are, of course, important economic questions. But to see such methods as adequate for "objective," "rational" or even "effective" resource allocations is to delude oneself in drastic oversimplifications. The questions deal with such elements as knowledge, processes, technical training, individual development, the search for solutions to problems or ways to achieve a wide variety of human goals - many of which have no "economic market" in the usual sense. Thus far, such multifarious elements have not yielded to reliable measurement, quantified expression or clear description of cause/effect relationships. They are largely matters of values - human, social, political and economic - formulated individually and collectively among widely diverse persons and organizational/institutional settings.

There are many ways to set forth conceptual frameworks for analysis of allocation problems. Table 1 shows three such frameworks that will be useful in this respect: (1) Levels of Allocation, (2) Parameters for Aiding Decisions for Allocation, and (3) Opportunity Structures for Allocation. The major section of this paper will be structured in accordance with the various levels of allocation, as described in Table 1.

Of course the various "cells" of each of the frameworks in Table 1 are highly interdependent in all directions. The inter-relationships vary widely as to the degree of directness and indirectness as well as to the intensity of the effects. Our abilities to perceive these relationships analytically are still quite limited. We have not developed adequate comprehensive descriptive models. Many of the models that we have are of recent vintage and are relatively crude. However, there is a growing amount of work going on which encompasses the categories depicted. It will be useful to discuss some of this work and its limitations, some current and prospective needs and some observations for improvements that might be sought in future research of this type.

There is now a great deal of evidence which points to a fact that perhaps is not yet well articulated. Although the respective levels and frameworks represent inter-dependent allocation processes they also require different techniques. This means that we cannot transfer methods among categories and that a balanced strategy of methodological research is needed. Moreover, the various categories cannot be linearly approached in one, two, three fashion. The combined effects of these intrinsic characteristics are significant and complex.

3. DISCUSSION OF STATUS OF CURRENT WORK

Allocations Among Competing Projects (Level 8). From the viewpoint of the first framework (Levels of Allocation), most of the systematic work and methodological developments have been at the 8th level. This is the point at which funds are actually linked to specific efforts and performance. Carefully designed systematic procedures are used to reach informed and balanced judgments which will help assure that the projects are: 1) scientifically meritorious and relevant; 2) well formulated and planned for effectiveness and cost efficiency; and, 3) to be carried out by competent people. The mechanisms include the peer rating process, advisory councils drawn from various institutions and professionally competent program directors in the disciplines or areas involved. A variety of formal and informal inter-agency groups are constantly being formed and reformed in order that overlaps and duplications may be minimized. Significant indirect contributions are made by the maintenance of effective scientific information systems which afford the "idea exchange" through which one can know the "state of the art" and what is being done by whom. The rapid and wide dissemination of such information has a great efficiency influence and contributes both to more rapid formulations of opportunities and to minimizing undue duplication of effort. The scientific ethos of seeking the highest degree of generality and the simplest form of explanations for the phenomena involved are important self-organizing forces which function to minimize inconsequential activity. But efforts that seek a reliable and valid method to express these and other elements as numeric measures or even in quantified language have not been generally successful. It is, however, desirable that we achieve such methodological tools even though their development will likely be very difficult and slow.

Many of the efforts at this "8th level" have centered on the latter stages of applied research and development. These largely evolved from the defense and space arenas where developmental commitments among alternative weapons systems entailed large scale expenditures over a number of years. A great deal of work has also centered upon product research and development within industrial operations. These efforts have yielded many of the important developments of modern decision theory. Much has been done toward quantifiable parameters which permit at least some capability to "optimize" *within* operationally defined objectives. An example is the now familiar concept of mathematical models of complex technological systems and their use in evaluating trade-offs between costs, time, and performance. But despite these advances, there is much work still to be done. Some of the more fundamental examples of this work involves progress in the area of basic research projects. We will discuss some of these efforts later.

Allocations Among the Various Areas and Fields and Within these Among Sub-Areas and Sub-Fields (Levels 6 and 7). The system of organization for the administration of research support and its close relationship to budgetary progresses are important instruments for achieving at least some continuity in allocations with respect to many of the scientific areas and fields. Perhaps a brief description of how this operates within Federal agencies will suffice to make the point clear. For example, particularly in the case of support for basic research type efforts, the organizational structures of the universities where most of such work is done can be related to fields or interdisciplinary areas which have clearly evolved and are somewhat definable in relation to research projects. These aggregations are also essentially "mirrored" in the organizational structure of supporting agencies. This means that the system has some "tempering" of allocation distribution among the fields which will militate against extreme fluctuations that might otherwise be attached to a straightforward priority approach. This idea of distinction between allocations and priorities (a special motivation for allocation) is very important. A dramatic example of its importance was pointed up by the wartime experience with national priority systems. Early in World War II we learned that if only a system of priorities operated, demands were such that some strategic materials were totally consumed on top priority programs and that this then virtually eliminated many less direct requirements that were essential to social cohesiveness and civilian needs. If not satisfied to a minimal degree, these matters quickly shift into the top priority category, to the overall detriment of the national military efforts. Some sense of "balance" is necessary within which proportional ratios can emphasize priorities to varying degrees but not exclusively. Much of our budgetary system is based, as is our political and social system, upon concepts of "advocacy." The organizational correlations with disciplines and the way research projects or programs are actually "packaged" is some assurance of representative "advocacy" within the budgetary allocation system. Since budgets are ultimately distributed to offices, sections, divisions or other organizational components, this assures at least some minimal level of support that gives a degree of stability and consistency not otherwise likely.

A great deal of pioneering work for assessing the needs and opportunities of several fields has been started. These studies have generally centered on certain aspects of the "opportunity framework" and have dealt with the relationships between those opportunities perceived and the needs of the particular field for support. Some of them have been limited to academic research needs, and therefore cannot be taken as a statement of total national need. The NAS Committee on Science and Public Policy has been a leader in these studies and has issued or is now working on the following reports:

The Plant Sciences: Now and in the Coming Decade, issued 1966.

Chemistry: Opportunities and Needs, issued 1965.

Physics: Survey and Outlook, issued 1966.

Ground Based Astronomy: Ten-Year Program, issued 1964.

Digital Computer Needs in Universities and Colleges, issued 1965.

The Mathematical Sciences: A Report, issued 1968.

The Mathematical Sciences: Undergraduate Education (Supplement), issued 1968.

Report on the Social Sciences (being Developed in Cooperation with the Social Sciences Research Council), planned to be issued Fall 1969. (The report will cover Anthropology, Economics, History, Political Science, Psychology, Sociology, Geography, Linguistics, and Psychiatry).

A number of interesting methodological innovations was employed in these reports. Though significant difficulties are yet to be overcome, these efforts represent important foundational work upon which to build. Some of the improvements needed are: 1) a solution to the problem of keeping the analyses up-dated – particularly in rapidly developing or changing fields; 2) extending the coverage and analysis to include total national needs; 3) developing more effective mechanisms to incorporate the recommendations into the various decision processes throughout the many decision centers that must be involved if the recommendations are to be carried out; 4) dealing with the various strategies and consequences associated with alternative support levels should the recommended levels not be realized; 5) addressing the very difficult problems of comparative evaluations for allocations between or among the various fields; and 6) making sure that the attention resulting from "the contemporary studies of a particular time" does not unduly bias the overall allocation process in favor of "greasing the squeaking wheel" to the detriment of other less visible but equally important needs.

A study recently completed by the RAND Corporation* for the NSF deals with the feasibility of applying conventional PPB techniques to a field of science from a Federal-wide perspective. This particular effort centered upon the field of chemistry and was an initial feasibility study to determine whether or not a full scale pilot study for the field would be warranted. Such a pilot study would have been considered worthwhile if it could be regarded as likely to aid in developing operational applications of PPB methods which would have improved the present system of programming, budgeting and evaluation. Attention was focused upon basic research. The report made a number of suggestions for improvement. But perhaps its most significant findings relate to the difficulties which make a straightforward transfer of PPB methods not feasible because a great deal more study is needed to clarify essential relationships. And, in addition, basic theoretical advances in economic analysis are needed as a foundation

---

* *An Exploratory Study of Science Resource Allocation* by S.M. Barro, the Rand Corporation, Santa Monica, California.

upon which to build appropriate quantitative decision models.

The report contained a number of observations about present allocation processes. These observations are of course selective rather than comprehensive. They are based upon analysis limited by the modest scale of an initial feasibility probe. There is a great deal of disagreement as to the validity of some of the observations contained in the report. However, the points pertinent to this paper may be summarized as follows:

1) The present system bases its rationale on needs as determined largely by the requirements felt to be necessary for satisfactory growth and development of a scientific field. There is, according to the report, inadequate analysis of what areas should receive least attention should the allocation of financial support be less than that needed and called for.

2) The report emphasized that the crucial allocation questions relate to inter-field distribution and balance. Under present procedures for formal allocations of support among projects, the questions of inter-field allocation tend to be relegated to a subordinate position because the budgeting distributions are largely accepted as operational ceilings. (Of course, this must be considered in the light of the budgetary processes discussed previously).

3) The present rationales over-rely upon the linkage of basic research to educational benefits.

4) The present orientation toward internal versus external decision criteria (i.e., criteria of evaluation internal to a discipline or to science) makes perception of relationships to social objectives of a non-scientific nature more difficult. It may also make it more difficult to "weight" decisions for oriented research. (However, this must be viewed in the light of the discussion later in this paper of allocations among departments, agencies, and social objectives).

5) The current manpower flow models are oriented more toward the internal growth and dynamics of the scientific enterprise because it is not yet clear just how to account for external requirements as the principal basis for deriving "demand" for expanding the number and balancing the skills of chemists or for scientists and engineers in general.

6) There is no apparent correlation between agency motivation for supporting research and the elaborateness or nature of proposal evaluation procedures. This point is made in connection with an assertion about the role decision-makers and their influence play upon the characteristics of the research programs. The point is that different mixtures of decision-makers seem to evolve characteristically different research programs.

7) There must be a great deal of improvement to the data systems in which are collected the statistical details about research activities. More detail must be centralized if valid quantitative models are to be developed and operated - a number of such improvements are underway.

Based upon this, other studies, and some personal work, I believe that the following additional observations should be made. PPB methods (especially as they entail economic analysis) are not easily applied to scientific activities and in some instances cannot be applied because:

1) There is no particular discernible goal structure sufficiently homogeneous to permit meaningful consistency of program categorizations that accommodate: a) demands for cross comparison; and, b) common quantitative expressions for consolidation that also meet the needs for individual orientation to the individualized goals of diverse decision centers in both the support and performance institutions.

2) There is no particular program structure which is "the best" for managerial or analytical purposes.

(This is closely related to the diffuse nature of goals mentioned in 1 above).

3) There are unresolved theoretical and methodological gaps in the theories of economic behavior of decision centers involved - particularly those centers found in governmental and academic institutions.

4) The nature of the benefits of research activities are often quite diffuse with varying intensity over time. Moreover, many of the benefits are of a non-quantifiable nature. Basic theoretical developments of a conceptual, perceptual and methodological nature are essential before valid objective quantitative models of comparative alternatives and goals may be evolved.

5) There is no consistent category of elements or criteria entering the decision processes which are sufficiently quantifiable and common to permit application of "optimization" techniques in any meaningful sense.

6) Different analytical methods or techniques are required for analysis of the resource input side of the matrix versus the output or benefit side.

7) There are no clear linkages between the input and output side of the matrix - either qualitative or quantitative.

With regard to the last two points, it is useful to distinguish between cost effectiveness and cost benefit analysis. These terms are often (if not generally) used interchangeably by economists. In general, they are seen as being the analytical linkage between the value of inputs and the value of outputs. Alternative input strategies are seen as choices relative to the same or different outputs. The perceived or calculated connection to the resulting value of such outputs is an integral part of the analysis.

However, this idea is valid only when one can discern the connecting points between inputs and outputs. In the case of much scientific research - particularly research of a basic or more general nature - such linkages cannot be clearly perceived and perhaps may not be perceived at all. Perceptions that can be delineated may be largely arbitrary, bound by artificial constraints often imposed as simplifying assumptions made in order to accommodate an analytical and quantified linkage. Such practices may be helpful in justifying the relevance of discrete courses of action within limited homogeneous contexts. But if they are then used to make comparisons and "trade-offs" between several areas there will *not* be a common denominator on which valid decisions can be made. In other words, the quantitative relationships established are not natural measurements and are therefore not free of the subjectivity found in verbal language. Such numbers may make summarization and comparison possible, without necessarily also being a common denominator. This is a particularly important point, because once the parameters become expressed in quantitative terms they are often handled in a manner which - while it may acknowledge the presence of such difficulties - operationally ignores that they exist.

It is, therefore, of value to emphasize the difference between:

(1) cost effectiveness analysis of alternative mixtures of inputs toward the achievement of a stated input objective, whether or not analytical linkages to the output benefits or objective can be charted and quantified; and (2) cost/benefit analysis that quantitatively relates the outputs to alternative inputs and to other outputs in terms that can then be used for valid comparison of "output tradeoffs" leading to "optimization" of "maximized" benefits.

Defined in this way cost effectiveness analysis may prove very helpful in evaluating input strategies for a given set of objectives even though the cost benefit calculations cannot be made.

Another type of research effort relates to the feasibility of developing quantifiable indices for parameters that can be used to reflect the "value" of basic research from the viewpoint of its scientific importance without regard to subsequent economic considerations.

An example of such basic theoretical work is offered by a project in process at Abt Associates, Inc., Cambridge, Mass. Although it may be early to make any judgments about the potentials of the project, it provides interesting insights into a rather novel approach. This project has been pursued in three phases. First, an empirical study was made of a number of selected basic research projects within Solid State Physics. From this work, three quantifiable indices that were felt to be reasonable retrospective measures of the scientific influence of a research project were developed. Attention was then turned to whether or not aggregative measures could be developed for sets of projects within, say, a sub-field (in this case, superconductivity); and also whether or not measures of prospective value could be developed that could be applied at the time of proposal evaluation.

The three indices developed for retrospective evaluation were:

1) *Project Utility Measure*. This is a number which represents the direct "content linkages" that could be identified between a "source project" (the project being evaluated) and subsequent projects (called "receptor projects")that had made use of the results of the work of the source project. The content linkages were established by expressing the results of the source project as sets of "output propositions." These propositions were then related to the receptor projects. The sets of output propositions from the source project were then correlated with a set of "input propositions" independently derived by analysis of the research reports for the receptor projects. It is significant to note that this was not

just a citation count. A citation did not qualify as a content link unless it was actually used in deriving the results (output propositions) of the receptor projects.

2) *Project Fertility Measure.* This is a relationship between the number of output propositions of the source project (as defined above) and the number of output propositions of the first generation receptor projects. The number of discrete propositions in the sets of output propositions of the receptor projects are compared with the number of discrete output propositions of the source project to derive a ratio.

3) *Project Diffusion Pattern Index.* This is an indication of the degree of the inter-dependence of the receptor projects. It is calculated by: a) determining the number of projects that were influenced by the particular project being evaluated (i.e., the number of receptor projects); b) subtract from (a) the number of generations reflected in the pattern of receptor projects; and c) divide the result of (a) and (b) by the number of "content links" reflected by the correlations of the output propositions from the source project with the input propositions of the first, second, third and fourth generation of the receptor projects.

It is felt that *if* sub-fields can be defined structurally by the propositions and queries that represent the structure for advancing knowledge within the field, then aggregative indices can be derived. And if a number of fields were structured in like manner, comparative relationships and inter-disciplinary spheres of influence may be at least roughly measured. Of course, the feasibility of developing and maintaining such definitional structures for sub-fields is still an open question.

During the second phase of the project, it was learned that the indices developed for retrospective evaluation were not transferable for prospective evaluation. In seeking to develop relatable measures for this purpose, the prospective value of a given project was redefined so that its value would be a function of the degree of change in knowledge that would be likely to ensue if the research results were consistent with the proposed questions. This could, of course, accommodate either positive or negative findings. However, it would not seek to anticipate the "random" unforeseen outcomes. The work then shifted from an emphasis upon the proportionate impact a given proposal might have upon the structure of knowledge for an entire subfield, to a relative measure of the amount of change in knowledge that would ensue from the project if successful. Probabilities of success, efficiency and competence of investigators would still require independent assessment and were not attempted within the model. Three tentative prospective measures have been set fourth and are still being evaluated. Again, it is still much too early to pass judgment upon the validity of any of these techniques.

Allocations Among the Various Geographic Regions Within the U.S. and Among Nations (Level 5). The international dialogue on the technological gap and the special considerations required as a part of the U.S. balance of payments deficits are aspects of international allocation criteria external to science. There are, of course, many other examples ranging from considerations of security classifications to what institutions and cooperative relationships are necessary in order to induce adequate environments for the conduct of research - especially that of a social science nature. The international year programs are still other forms of mechanisms - largely of an internal scientific nature - that are influential determinants of allocations among nations.

If the decline in the rate of growth for U.S. support of research and development continues, it will be likely to exert increasing pressures for decisions within the scientific community to give greater priority to U.S. scientists, even if "the best" potential performers happen to be in another country. Of course, these complex problems are far beyond this cursory survey and conceptual discussion. They will, however, be likely to become more critical aspects of science allocation decisions in the future.

Feelings have increased that there is a close relationship between the location of scientific research activities and regional economic development. This is buttressed by the growing appreciation of the cultural advantages also accruing to those regions where leading universities are located and where such institutions possess high quality scientific and research capabilities. The studies which such concerns have called forth have all shown that there are marked non-uniformities in the distribution of centers of scientific excellence. This perception is, in turn, leading to increased pressures for working out mechanisms that will - at least in the long run - redress much of this imbalance. Though more balanced distribution is not being sought at the expense of the existing centers, from a practical standpoint growing tensions must be anticipated unless a rather large upward shift in the growth of support for science occurs. It is now felt that any basic changes in the geographic balance of scientific activities will occur only as the result of conscious efforts and special criteria. As these goals are expanded, they can be expected to exert increasing influence upon allocation issues with which we are concerned in this discussion.

Allocations Among and by Functional Departments and Agencies (Level 4). Allocation decisions at this level have an important bearing upon ultimate distribution of funding among areas and fields of science. For one of the fundamental ways that science and research activities are linked to national goals is through the organizational structure of the mission agencies of the Federal government and the functional institutions of the private sectors. Examples of this point have been evident in the shifting disciplinary mixtures of some private foundations support as they have reoriented their basic emphasis from time to time.

And, of course, the same may be said for the federal support mixture as more efforts emerge that seek to relate research to broader spectra of social goals and problems.

But as this goal spectrum associated with motivations for scientific support has increased, so have the inter-dependencies and cooperative relationships between agencies also grown. Therefore, several mechanisms have been used to temper the allocation among agencies by some more comprehensive national framework. The Marine Sciences Program now being developed by the Marine Sciences Commission is an example. Certain committees of the Federal Council for Science and Technology are another. Selective reports of the President's Science Advisory Committee and the National Academy of Sciences have addressed themselves to reccommended allocations on a national scale for scientific activities associated with particular social problems such as population, world food supplies, and the effective uses of the sea.

The National Science Board of the National Science Foundation has appointed a special commission to look into ways in which the social sciences may be more effectively brought to bear upon the solution of national problems. Their work is still in process, but it can be anticipated that at least some of the recommendations will have imbedded in them aspects of resource application.

Allocations for Proper Mix of Support and Performance Among and Within Various Institutions and Economic Sectors: Governmental, Academic, Industrial and Non-Profit (Level 3). Earlier remarks have stressed that there have been, over the past three decades, fundamental shifts in the decision centers participating in allocations among and within institutions. Though the Federal Government is now the principal supporter of science, research, and development, it is not the principal performer. Therefore, inter-dependencies have become a basic fact-of-life and they affect allocations within and among institutions in fundamental and direct ways. These features of "the system" have also become more "felt" and visible as many of the factors discussed in previous pages have emerged. With this visibility came the realization that we really did not know very much about the cause/effect relationships of already vast, complex inter-dependent processes. In turn, a variety of research and study efforts have begun. Some of these efforts seek merely to provide adequate analytical descriptions of the processes and the cause/effect relationships. Others seek to improve various features of the decision processes within and among institutions. Still others, though I think these are very few, are beginning to search such fundamental questions as: How high should the relative ratios of support from federal vs. non-federal sources be allowed to go? What kinds of policy instruments could be employed to get greater proportions of private support? How do these relationships relate to tax laws? To fiscal and monetary policy? To international relations? To successful environments for scholarly pursuits?

A few examples will help to give a flavor of the research efforts within this level of our conceptual framework.

There is growing concern about the institutional impacts of federal support of university research through the project support system. The NSF is currently supporting two complementary studies of this type. One entails the analysis of how federal funding affects a single relatively new public institution of higher education as it evolves into a large campus of a state university system. This work is being pursued at the University of California's Riverside Campus. Another more extensive study has been undertaken in collaboration with the Michigan State University. This effort seeks to analyze in detail what the impact of federal support programs for academic science has been upon all of the public institutions of higher education in the state of Michigan. On the basis of such analyses, it is expected that desirable policy modifications for mechanisms for support will be highlighted. Also, perhaps there will be some guidance for the appropriate levels of funding and proportional relationship between public and private funds. Other general policy issues are likely to be surfaced and hopefully constructive suggestions for improvement can be made.

Related studies entail the attempts to construct models of university processes in ways that permit improved "testing of alternative plans and decisions" and for deriving more effective internal allocations of university resources. As institution-wide considerations become more explicit in terms of criteria bearing upon the choice of support mechanisms (i.e., project support, departmental support, or institution-wide support), they will carry an increasing influence on allocations among the fields of science. More attention will probably be paid to the institution-wide balance and inter-relationships. And institutional bodies may insert themselves into the allocation channel as a direct recipient and distributor of research support to a greater degree than is presently seen. Perhaps some previews of the type of effects that one might expect to evolve from such arrangements have been offered by the recent experiences associated with the requirement to cut the rate of expenditures against federal grants and contracts below previously planned rates. The NSF felt that more equitable decisions could be made locally on an institution-wide basis. The process has not been free of strain. Of course, the difficulties were substantially increased because of the requirement to "roll back" on previously approved schedules of activity. Nevertheless, one gets the impression that a number of new forms of "bargaining" would enter the picture if most support were funneled through institution-wide programs.

To provide more effective tools for making such decisions, academic institutions are seeking improved methods. A number of systems analysis efforts have been undertaken to provide important theoretical and methodological foundations for planning, programming, and evaluating performance within the institution-wide framework. The studies deal with such concerns as: resource planning related to the development and utilization of facilities and faculty; balance of resources and activities; models of behavior - particularly economic behavior; improved integration of administrative systems; computer services and utilization; and regional cooperative programs among institutions. If these studies lead to adoption of more systematically effective management methods, they can be expected to have at least some effects upon present methods of allocation.

Still other dimensions to this level of science allocation bear upon whether or not support should be for performance within the supporting institution or in other institutions - the old question of whether to perform needed research internally or through extramural support. So far as I know, there are no consistent methods for making such decisions. But a number of important issues can be raised as a result of previous choices. For example, if an agency or company is planning cut-backs for certain areas of work previously supported on a mixed basis (i.e., some of which has been done internally and some extramurally), should it give preference to internal capabilities or not? Also, as the purposes for which large scale efforts were organized are fulfilled, should the efforts be dismantled, the facilities transferred to some other agency, or should new goals for their utilization largely "intact" be sought? Examples of these questions are the recent search for long range goals of space research after the Apollo program and for improved ways to employ the national laboratories such as Oak Ridge, as they move beyond the immediate concerns for which they were established.

The need for national facilities or for arrangements for broad based sharing of expensive experimental facilities has called forth novel institutional arrangements such as consortia corporations. It is conceivable that still further innovations of this nature will need to be pursued in the future.

Allocations to Basic Research vs. Applied Research vs. Development (Level 2). Whether or not one agrees that these three categorizations are the most valid ones, it seems that at least some roughly equivalent notions about the nature and motivation for research or developmental activity are relevant to the way one goes about organizing for support and performance.

One source of policy for deriving "appropriate" allocations among these functional orientations arises from relating the developmental and other directly perceivable relationships between stated agency or institutional social objectives with the capabilities that would enhance achievement of those objectives. Support for the more abstract types of applied or exploratory research and of fundamental or basic research would be determined by treating them as "overhead functions" to be funded by a percentage of the level of effort derived for the latter stages of applied research and developmental purposes.

But another approach has, for a number of years now, sought to understand the myriad processes of each of the functional dimensions and how they interrelate with one another. Some of these studies seek ways to make the connections more direct, or to shorten the time from basic research to the application of the knowledge so derived. Others seek to understand the proportional relationships and to see if generalizations can be derived that may be valid for developing criteria for "balance" under specifiable circumstances. A number of examples of such work may be cited. In 1959, Arthur D. Little completed a study for the Navy entitled: *Basic Research in the Navy*.* Among other things, this study sought to explicate the relationships between basic and applied

---

* *Basic Research in the Navy*, Vol. I, A Report to the Naval Research Advisory Committee, June 1, 1959, prepared by Arthur D. Little, Inc.

research, development, and naval missions. The more recent reports of the Committee on Science and Public Policy of the National Academy of Sciences to the Congress entitled: "Basic Research and National Goals" and "Applied Science and Technological Progress" also give attention to such concerns. The same committee is now working, in collaboration with the National Academy of Engineers, on a related study dealing with technological forecasting and its relationships to technological assessment.

Though it has a number of shortcomings, the now renowned "Project Hindsight" study of the Department of Defense was an attempt to analyze systematically the respective roles and contributions of basic research vs. mission oriented applied research to a select group of modern weapon systems.

The National Science Foundation has also sought to improve insights in these matters, and by doing so, to provide improved foundations for appropriate policy decisions. A number of studies have dealt with the processes of technological innovation and transfer. Others have sought to develop methods for forecasting the social implications of scientific developments. The final report of a current project which traced the key events and their inter-connections between the evolution of basic knowledge and its ultimate application in technology for five selected examples was issued recently.* The National Bureau of Standards is supporting work seeking to model innovation processes.

From research of this type, we are now clearly seeing that innovation processes are neither direct nor linear. They involve a great deal of trial and error, false directions that end in "blind alleys," spontaneous "random" events of creativity, and very complex feedback loops. Much more work will be necessary before

---

* *Technology in Retrospect and Critical Events in Science*, prepared for the National Science Foundation by the Illinois Institute of Technology Research Institute, NSF C535.

we will have valid bases for "models" that are sufficiently descriptive and reliable for us to begin to use them formally as a principal instrument of decision making. Even more fundamental work will be required to quantify the parameters of such models. And we will need many extended time correlations in order to acquire the necessary criteria to know when to use what model. Meanwhile, these efforts can contribute to our store of information and understanding of the processes. Even being able to document just how complex these processes are will be useful in helping to avoid over-simplifications that - on the surface - appear to be "objectively rational." And, of course, whatever insight can be gained will be a useful contribution to an informed "wisdom" for the inescapable fact that choices have to be made regardless of how fragmentary our knowledge for decision-making may be.

Gross National Resources Allocated to R&D (Level 1). Various studies have sought to derive criteria for determining how much of the Gross National Product should be devoted to research and development, and how much to the related evolution of scientific resources and capabilities. Such studies seek to make comparisons to the proportional ratios of GNP which are allocated by various nations to scientific research and developmental activities. The general idea is that the nation allocating the highest proportions of GNP will be most likely to have the greatest relative increase in the quality and character of their capabilities.

These types of considerations are tempered by the study of U.S. historic growth. The present proportionate allocation of about 3% may be thought to be the appropriate level because past policies leading to this ratio have helped the U.S. to emerge as a world leader in science. Under this concept, growth in R&D support would be accommodated by the anticipated growth in the GNP. It should fluctuate marginally at about the 3% level. Derivatives of this rationale permit a continuing relative growth that would be roughly comparable to an extrapolation of past relative increase curves to some still undeterminable leveling off point. This leveling off point must not be confused with the curve of growth for R&D support *per se* which is calculated on the base of past R&D expenditures. It would be possible for us to continue to allocate proportionately more of the GNP to R&D activities, so that they continue to grow but have a declining growth rate. These macro economic approaches seem to consider R&D as a consumption function only. Though there are serious gaps in our theory for systematically exploring the judgments that increasing R&D expenditures generate increased GNP, there is a feedback which increases the basis to which an allocation percentage is applied, and it is important that we not overlook this very real relationship.

There are presently no adequate quantitative relationships that accurately express the results of research and its effect upon GNP except as a consumer. The past allocations have been largely the product of political processes rather than an objective analysis and rationale for exploiting the full potentials of the scientific opportunity structures. The adoption of any arbitrary relationship to GNP as a "correct balance" could be a serious deficiency.

Other approaches seeking to derive some estimates of gross national support needs conceive of scientific manpower as the most essential determinants of expenditures. A cost per man year can be calculated, adjusted for some index to reflect the rising cost of doing research, and applied to the present or projected manpower statistics - including students.

A fairly recent refinement of approaches to questions of gross national allocation has entailed a conceptual division that separates "big science" vs. academic science of a nature that does not require large scale investments in single purpose experimental facilities. The latter category of activity needs to retain a greater degree of consistency than does the former, which is more of an initial capital investment type of decision. Only so many viable "big projects" for new facilities are likely to be competing for consideration at any given time. Though accurate data is difficult to assemble, some informal estimates have set the ratio of the "big projects" at about 1/4 to 1/3 of the total science budget. These types of projects are considered by some to be more susceptible to "sequencing" on the basis of priorities, whereas there is some minimum amount essential to maintain the basic viability of the capabilities involved in the more general type of academic research activities. There are, of course, some upper aspiration limits beyond which additional investments would not be efficient. The determination of what the amounts should be in each category takes great care.

But in all such macro attempts, the problems of inadequate perception of the relationships between science and GNP; between GNP and attainment of qualitative human goals; and the internal criteria for judging effectiveness, balance, efficiency and quality convertible to a common parameter of a macro economic model, remain largely unsolved.

There is, therefore, presently no "objective" way of quantitatively deriving what represents an optimal allocation of the nation's resources to scientific and research endeavors in total, nor to the various dimensions of the infrastructures of the activities themselves.

Opportunity Structures. A great deal has already been said about the opportunity structures and the current work related to them. However, one important point must not be missed. Most of our methods seek to improve the rationales for deciding what to pursue. Most of these rationales are set forth in a rather limited goal structure. They seek ways to achieve specified results. None really exhausts the "opportunities" in analytic detail sufficient for one to know the opportunity costs of deciding in favor of the choices made. In other words, an implicit but seldom articulated aspect of each decision to do something is what else may have been possible to do if all opportunities were sought. From one point of view, it may be just as much or more important for us to know what is not being done as it is to know what is being done. An important part of criteria that would be effective in evaluating our allocations - particularly from a field point of view - would be to understand what opportunities we have failed to capitalize upon. Very little work deals with this dimension.

About the only opportunity costs that we have identified are those that were articulated as a part of program plans and eliminated during the budgeting and allocation process. Earlier allocations excluding significant opportunity costs are never "put into the system." Furthermore, my impression is that once the "approved allocations" are made, we do not ever construct an analysis of what has been left undone. We have no method of retaining that information as a basis for longitudinal studies that will help identify in later years key opportunities that were not exploited and to associate these "lost opportunities" with the reasons for not perceiving them and then evaluate, with the wisdom of hindsight, the correctness of deciding upon the priorities chosen. It is unlikely that we will be able really to evaluate the effectiveness of allocation decision processes unless we add this dimension. To do so is of course an intellectual and administrative challenge of the highest order. But surely a discussion of general systems concepts would be remiss in omitting this dimension. Perhaps as general systems methods are advanced the necessary tools can become available

3. SUMMARY

It is evident that pressure for the determinants of allocations emerge from the scientific community at large, from governmental and institutional laboratories and mission departments, from the political community, or from program directors' and planners' ideas about fruitful and appropriate opportunities. These varied groups have no consensus of either the theory of methods for quantifying the relative "values" of their ideas, needs, judgments, or opinions. Accordingly, quantitative comparisons of potential "benefits" is very difficult. Present techniques do not offer reliable "objective" representations useful in the normal sense of the term.

Until meaningful cause/effect inter-relationships between scientific and other national endeavors are developed and until these are reduced to a calculus, quantifiable parameters have value for indicating various aspects of decision choices but they have limited reliability as criteria for evaluating either the effectiveness, or the efficiency of the scientific and research efforts.

We must recognize the present deficiencies inherent in our ability to quantify objectively the needed inputs to, the benefits from, or the alternatives available among the various types of scientific and research pursuits.

In any systemization relative to science resources, care should be observed both to guard against infringement of scientific freedom, and to retain the decentralized decision mechanisms which determine the specific application of the resources to research activity. It is important that we develop improved perception, concepts, theories and techniques toward this end.

Program systems (such as PPB mentioned earlier in this paper) as mechanisms for organizing human activity and affecting allocations among diverse efforts are essential. But they all have inherent limitations. It is important that practioners know the limitations and even more important they behave consistently with such knowledge. The current tendency seems to be to overextend the application of program systems. These tendencies seem particularly prevalent in those areas where there is a great deal of ignorance or uncertainty or when the programs are of such a highly specialized nature they cannot be directly perceived and "felt" nor can a consensus about them be readily attained.

The economic criterion for effective system performance is not so constructed as to permit one to determine the point at which "marginal superiority" or "leadership" or "opportunity" is, in fact, attained. The effect of constant "economizing" can retard the overall required flexibilities of a system and can force it into an overcategorization and structuring which reduces its "reaction flexibility" and adaptability.

Moreover, there are a number of diverse processes and kinds of talent which are involved in retaining the basic capabilities for response flexibility and opportunistic development. Not all of these are understood in terms easily convertible to operational definitions that fit within the formalistic structures of program analysis and that can be related to quantifiable parameters.

Yet, it is important to assure that such "wisdom" is retained within the overall context of U.S. science. This requires some form of strategy, or lacking that, must be left to unstructured "natural" evolution of very diverse sets of people in very diverse institutional and disciplinary centers which might be characterized as a "system of chance." Perhaps, the heart of strategic thinking is in avoiding such chance by improving the "rationality" while at the same time retaining the openness of the system to capitalize upon random or unforeseen events.

Programs must be supported by processes which are quite different from the program itself *per se*. The advantages and limitations of program systems need clear discussion and they need to be related to the underlying processes that make a number of alternative programs and response flexibilities possible. Both of these, i.e., programs and processes, must also be interrelated with the underlying philosophies from which they evolve and/or within which they remain operative. Some of these philosophies are explicit but many, perhaps more, are still implicit.

The job of making these interrelations clear is very important, maybe even critical. If it is to be done at all, it will be done through the concepts of general systems research. Such a prospect is surely a challenge of the highest order, a challenge that I hope will evoke a response of the most creative kind.

TABLE 1

Conceptual Frameworks for Allocations of Scientific Resources

Levels of Allocation	Opportunity Structures for Allocation
1. Gross national resources allocated to Research & Development	1. Potentials internal to each field of science which are opened by current advances within that field.
2. Allocations to basic research vs. applied research vs. development, (or other categorizations which reflect such general distinctions of effort).	2. Potentials for interdisciplinary pursuits which emerge from efforts and discoveries in different fields of science, many of which are dependent upon effective interdisciplinary pursuits.
3. Allocation of the proper mix of both support for and performance within various institutions: governmental, academic, industrial and non-profit.	3. Potentials opened by cross-disciplinary inter-relations, i.e. the developments in field A make possible new fruitful opportunities in fields B, C, ...n.
4. Allocations among and by functional departments and agencies and various social goals.	4. Potentials opened by advanced technologies (instruments, etc.) and techniques.
5. Allocations among the various geographic regions within the U.S. and also among nations.	5. Potentials for greater use of available resources which may not be fully occupied or which may be susceptible to more efficient utilization (these circumstances may arise from "imbalances" or shifting emphases in the resource allocations).
6. Allocations among the various areas and fields of science.	6. Need for expanding the resource base, i.e., manpower, and/or institutions, and/or facilities.
7. Allocations among the various sub-areas, sub-fields or programs.	7. A "compelling need for select knowledge," perhaps based upon national objectives, social welfare, or the essentials of survival. These needs may emerge from scientific advances or from causes outside science such as the demands placed upon science by the population explosion, war, or the proclamation that we got to the moon during the decade of the 1960's.
8. Allocations among competing projects.	
(Note: Important questions for each of the above levels are: a) how much total support?; and b) what should be the ratio of support and performance from and by public or private institutions?)	

TABLE 1 (Continued)

Parameters for Aiding Decisions for Allocation

Quantitative	Non-Quantitative
1. Manpower Data a. Number scientists & engineers b. Number Ph. D's (S & E) c. Experience profiles d. Ratios to: students/teachers population college graduates total Ph. D's supporting personnel	1. Human aspirational attainment 2. Cultural Value 3. General Economic Development 4. "Growth" -an intrinsic need 5. Link between research & education
2. Financial Inputs (Expenditures) a. Absolute amounts by various classifications. b. Per Capita amounts for scientists & engineers, Ph. D's S's & E's, total population, institutions, regions and various demographic analyses. c. Ratios to: GNP Federal Budget Educational Cost Industrial Sales	6. Link between development & execution 7. "Inertias" of "the System" It takes relatively long periods to change fundamental balances within the infra-structure 8. Protection of Scientific Freedom
3. Financial Outputs a. Comparative life-time earnings of scientists & engineers. b. Comparative earning capacity for longer life-health; etc. generally restricted to medical or life saving type sciences. c. Sales of "new goods & services" (very spotty). d. Various "imputed" assigned values.	9. Scientific merit (qualitative) 10. Judgment of Relevance Within same context of various objéctives 11. Judgment of Effectiveness 12. Judgment of Competence a. Personal (individuals) b. Institutional
4. Patents	
5. Nobel Prizes	
6. Institutions a. Federal/state/local government agencies b. Degree granting institutions c. Technical institutes d. Industrial Firms involved in R & D e. Operational laboratories	
7. Publications a. Absolute numbers b. Citation index, etc.	

CLUSTER ANALYSIS OF A SCIENTIFIC FIELD

Daniel E. Bailey*

ABSTRACT

This paper presents the methodology of cluster analysis of the characteristics of research grants as prepared by the Scientific Information Exchange. The principles of analysis are applied to the field of physiological psychology as represented by 930 projects and 1557 descriptive phrases applied to those projects. Results of the analysis are a segmentation into clusters of the descriptors applied to the projects and a description of the salient features of each of the descriptor clusters.

1. INTRODUCTION

The activity of scientists is patterned and structured. There are segments of the body of scientists and

---

* Daniel E. Bailey, is Associate Professor in the Department of Psychology, University of Colorado, Boulder, Colorado 80302, and is also associated with Tryon-Bailey Associates, Inc.

The author is indebted to the staff of the General Systems Research Section of N.I.H. for their assistance in this study reported in this paper, particularly Dr. Stuart Wright, Dr. Carol Steinhart and Jane Swanson. John Bauer and Tom Kibler were responsible for the development of the special programs that were used in this research -- without their contribution, the analysis would not have been possible.

there are segments of the subject matter that they study. These two statements are based on the simple observation that there are professional and scientific societies that differentiate scientists on membership terms at least, and that within a group of scientists identifying themselves by a single term, such as physiological psychologist, there are specialists. This paper is a description of an attempt to study a large scientific field in order objectively to determine certain aspects of the structure of that field. The study reported is limited to one field and one restrictive source of data. The scope of the study, however, is quite large and the data are many in number. Only general results are reported here. More complete and detailed results will be reported in subsequent papers.

The study reported was done in collaboration with the National Institutes of Health, General Systems Research Section. The objective was a partitioning of a scientific field by two modes, first of the subject matter and methodology in terms of the pattern of use of the subject matter and methodology by a relatively homogeneous population of scientists, and second of the scientists in terms of the similarity of the subject matter and methodology of the work that they do. The population of scientists selected was that group of scientists who had grants approved and funded by federal and private institutions that were reported to the Scientific Information Exchange (SIE), and who were said by the SIE to be working on a problem in physiological psychology. The population of characteristics of their work, i.e., the subject matter and methods used, were the SIE descriptors that were used by the SIE for two or more of the grant applications involved. This definition selected a total of 930 grants and 1557 descriptors of subject matter and methodology. The study reported here is a cluster analysis of the 1557 descriptors -- showing the clusters and the organization of the clusters of subject matter and methodology. The results of the study are of course limited in implication to the population of grants that were involved.

2. BASIC DATA

The data are generated in the following way. When a grant proposal is approved, agencies cooperating with the SIE send a document to the SIE containing a brief description of the project. This document is a single page containing a summary of the project proposed in 200 words or less and some data on the principal investigator(s) of the project and funding information. The SIE submits each of these documents to study by persons competent in the general area of research involved. Each project is "described" by the attribution to the project of as many terms and phrases as the reviewer needs to use to "abstract" the document. The terms and phrases range from very simple to very complicated with multi-level differentiations. For example, some of the descriptive terms and phrases in the data reported here are: metabolism; rapid eye movements; blood plasma and serum (body fluids); test animals, homo sapiens; test animals, human, homo sapiens, modern. The idiosyncrasy of the description method has an advantage in that any descriptor that a reviewer initiates is appropriate -- no artificial molding of the descriptors to a fixed list is imposed. On the other hand, the idiosyncratic use of phrases and terms causes a proliferation of terms that may be detrimental to an automated study of the data produced by the SIE system.

Our attention in this study is focused on the 1557 descriptors. However, it must be remembered that the results of the study are predicated on the specific set of 930 projects involved in the study. Another set of projects may well have produced a different patterning of results with essentially the same set of descriptors. We make no particular claims for the generality of the results within any other set of projects. The methods of the study are quite general and the question of generalization of the results is empirical in the large.

Within the 930 projects the frequency of occurrence of descriptors is generally quite low. The number of descriptors used with various frequencies is shown in

Table 1. Ordinary techniques of cluster and factor analysis are not readily adaptable to data of this kind.

TABLE 1

Distribution of frequencies

Frequency	2	3	4	5	6	7-15	16-30	31-465
number of descriptors with the frequency	367	213	152	116	79	297	156	174

In the first place, the number of variables, 1557, is greatly in excess of the limits of computer programs currently being used for the more prevalent forms of analysis. Second, the extremely low frequency of the descriptors makes the problem of describing the degree of similarity of the descriptors very difficult. In much of the work on these kinds of data the presumption is that the relative frequency of occurrence of a descriptor (presence of the attribute on a project) is in the proximity of 0.5. In order to deal with these problems, and other technical problems related to them, a special technique was devised. The details and mathematics of the technique will be described elsewhere in subsequent reports. The outlines and details of the technique sufficient for presentation of the results are presented here.

3. PRELIMINARY DATA PREPARATION

The paired-comparison similarity of two descriptors was defined as the relative frequency with which the descriptors were used for the same projects, with respect to the frequency with which the two descriptors were used in the collection of projects. Specifically, the index of similarity of the jth and k^{th} descriptors was defined by S_{jk}:

$$S_{jk} = I_{jk}/U_{jk}$$

where

I_{jk} = number of projects on which descriptors j and k were both used

and

U_{jk} = number of grants on which either (or both descriptors were used.

The matrix formed by this procedure is a 1557 by 1557 matrix of pairwise similarity measures. However, the statistical properties of this matrix are not appropriate for factor analysis or cluster analysis, even if the magnitude of the matrix were not a block to the application of the analytic procedures. Treating the 1557 by 1557 matrix as a score matrix, as is often done with pair-comparison matrices, suggests taking the correlations of the columns of the matrix as a statistical measure of the similarity of the descriptors properly defined for cluster and factor analysis. This matrix is given by:

$$r_{jk} = \frac{\sum_{i=1}^{N} s_{ij}s_{ik} - (\sum_{i=1}^{N} s_{ij}^2)(\sum_{i=1}^{N} s_{ik}^2)}{\sqrt{(\sum_{i=1}^{N} s_{ij}^2 - (\sum_{i=1}^{N} s_{ij})^2)(\sum_{i=1}^{N} s_{ik}^2 - (\sum_{i=1}^{N} s_{ik})^2)}}$$

Although the computer programs that were used for the cluster analysis of descriptors could begin with the correlations defined by the equation above, they are limited to correlation matrices for 90 variables, i.e. 90 by 90 correlation matrices. Other techniques are available for as many as 5000 variables if each variable is located on a set of 15 or fewer reference dimensions. The procedures that were adopted are predicated on the availability of these two types of computer program. The first step was to do cluster analyses on several sets of 90 variables sampled from the 1557 descriptor-variables. The results of these analyses were merged to obtain reference dimensions and the second type of analysis was performed on all 1557 descriptors.

4. CLUSTER ANALYSES OF 90 VARIABLE SUBSETS OF DESCRIPTORS

The Tryon-Bailey cluster analysis computer system, BC TRY (Tryon and Bailey, 1967 and Tryon and Bailey in press), was used to cluster-analyze 90 descriptor-variable samples from the 1557 descriptors. The system is limited to dealing with 90 variables at one time for the standard cluster analysis technique. Since the most general information about the projects, and about the organization of the descriptors in the set of projects, is contained in the descriptors with the larger number of applications to the 930 projects, the descriptors that were selected for analysis were the descriptors with the largest frequencies of use (excluding the descriptor "physiological psychology" and certain composite descriptors such as "other"). The 270 descriptors with the highest frequencies were selected in three sets of 90 each. Each set was submitted to cluster analysis in the BC TRY System. The variables (descriptors) that were not well described by the cluster analysis results in the three separate analyses were combined into sets of 90 variables and cluster analyses were performed on these sets. From the results of these analyses a set of 90 variables were selected to represent the clusters that emerged from the separate analyses. These 90 variables were then submitted to cluster analysis and produced 15 clusters that were taken as a reasonable representation of the independent dimensions of variation present in the 1557 descriptors. A number of other clusters appeared in the process of the several analyses but their statistical properties were not clearly satisfactory in one way or another. For example, the statistical reliability of some composite variables represented by the clusters in this class was insufficient to warrant confidence in their reproducibility in another sample of projects.

The clusters that were finally selected all had satisfactory statistical characteristics. A list of the number of descriptors in the clusters and a cluster title suggested by the descriptors is given in Table 2. The title attached to each of the clusters is based on an interpretation by the author of the meaning of the descriptors included in the cluster.

TABLE 2

Defining clusters for reference dimensions

cluster	number of descriptors	title suggested by descriptors
1	4	vertebrate physiology
2	5	pharmacology; psychological effects
3	11	central nervous system physiology
4	6	autonomic nervous system response
5	5	developmental psychology
6	5	neuro-endocrine effects
7	6	learning and memory; conditioning
8	4	nervous system; metabolism and chemistry
9	6	sensory physiology; vision
10	7	hunger and thirst
11	3	use of rodents
12	3	clinical psychology; physical disorders
13	3	learning
14	2	clinical psychology; psychoses
15	4	sensory physiology; touch and smell

5. DEFINING THE REFERENCE DIMENSIONS

The final clustering of all 1557 descriptors is performed by a program that takes data that are values of the objects (descriptors) on up to fifteen reference dimensions. These reference dimensions may be defined in any way, providing that each object is given a value on each of the dimensions. The usual source of these scores is from factor estimates in factor or cluster analysis. The scaling of the scores on the dimensions is arbitrary and may be selected at the option of the user of the procedure. The cluster analysis is concerned primarily with relative degrees of homogeneity of groups of objects and relative distances between groups of objects, within a vector space defined by the reference dimensions.

In the present study the standard factor and cluster scoring techniques were not applicable. Consequently, a new procedure was devised. The procedure evaluates each project (not descriptor) on the degree to which the reference clusters (descriptors) are represented, relative to the number of projects on which each descriptor was used. The measure is simply the frequency within a reference cluster that the descriptors were attributed to the project. These values are interpreted as weights indicating the degree to which a project represents the descriptor reference cluster. The second step is to apply these weights to each of the descriptors in a simple averaging fashion. If a descriptor is present on a project, the weight (frequency) of that project on the reference cluster is accumulated to a total score for the descriptor on the reference cluster. These totals are divided by the number of projects on which the descriptor is used. In this way, each of the descriptors is scored on each of the reference clusters -- forming scores for each descriptor on each of the reference clusters defining the reference dimension. The scoring procedures may be interpreted as following. Each project is given a weight indicative of the relative importance of the reference cluster in the project. When a descriptor is scored on the reference cluster the pattern of weights on the projects and the pattern of projects to which the descriptor is applied are matched. If the projects to which the descriptor is applied are projects with large weights and the projects to which the descriptor is not applied are projects with small weights, the descriptor would appear to have "more" of property represented in the reference cluster than a descriptor with a different pattern of projects. In order to correct the score for the number of times a descriptor is used the sum is divided by the number of times the descriptor is used in the 930 projects.

Stated formally, the scores on the fifteen reference dimensions are formed by the following expressions. The data form a matrix of 930 by 1557 with an entry of 1.0 where the descriptor is applied to a project, and an entry of 0.0 where the descriptor is not applied to a project. The matrix is called V. The reference clusters are denoted $C_1, C_2, \ldots, C_{15}$. The weight on the kth reference cluster for the ith project is given by

$$W_{ik} = \sum_{C_k} v_{ij} \qquad i = 1,\ldots,930 \quad k = 1,\ldots,15$$

where the index j is defined on the members of the k cluster. The score of the gth descriptor on the kth reference dimension, defined by the kth reference cluster, is given by

$$x_{gk} = \frac{1}{n_g} \sum_{i=1}^{930} w_{ik} v_{ig} \qquad g = 1,\ldots,1557 \quad k = 1,\ldots,15$$

where n_g is the total frequency of use of the gth descriptor on the 930 projects.

6. CLUSTER ANALYSIS OF THE 1557 DESCRIPTORS

The BC TRY System of cluster analysis includes a program that is designed to cluster large numbers of objects when the objects are described as points in Euclidean spaces of up to 15 dimensions. The procedures for this clustering are essentially those of pattern matching. If two objects have the same pattern of scores on the 15 dimensions, they are located in the same place in the 15 dimension space. If two objects are relatively near each other in the 15 dimension space they will have relatively similar patterns of scores on the 15 score dimensions. Consequently, with suitable limits of tolerance for difference in co-location or score pattern, we can identify the clusterings of objects by assigning them to regions or sectors of the space.

A preliminary stage of the clustering procedure is to cast variables into core clusters based on arbitrary sectioning of the cluster score space defined by the scores obtained from the procedures just described. In general, the number of sectors is determined by how many broad categories are used on each of the dimensions, how many dimensions there are, and how many objects are available. Cutting each of the 15 dimensions into three segments produces 3^{15} sectors in the 15 dimension space (in excess of 14.3 million sectors). The next stage is to sort the individual objects into the sectors in the score space depending on their specific pattern of scores. The program includes options regarding the number of variables falling into a sector sufficient for the program to decide that the variables form a cluster; say p or more variables are sufficient to define a cluster. For each sector containing p or more variables a core cluster is defined, giving a total of m clusters in the entire space. Having selected this set of core clusters, the arbitrary sectors in the score space are discarded and the individual variables are compared with the centroids (averages of the member objects on each of the fifteen cluster scores) of each core cluster. The individual variables are added and excluded from a new set of core clusters on the basis of the Euclidean distance of the variable from the core cluster centroid. The simple Euclidean distance between the variable and the centroid of a cluster is divided by the square root of the number of dimensions in the score space to adjust the distances for the number of dimensions. The quotient is called the root mean square. If the root mean square is greater than one standard deviation of the scored dimensions the variable is excluded from the cluster. If the root mean square is one standard deviation or less, the variable is included in the new cluster. In this way, the centroids become condensation points for the entire set of variables. New clusters are formed about the old centroids. However, there is a good deal of room for drift of the centroids of the newly defined clusters from the initial core clusters. Clusters can merge if they are composed of variables flanking a division between two adjacent arbitrary sectors on the initial allocation. Clusters also can be split to merge with other clusters. The process of replacing clusters defined at a given iteration by the centroids of the clusters and reassigning variables to clusters is repeated until no cluster membership list changes. The centroid of each of the final clusters of variables is a vector in the score space and as such is an average representation of all of the variables in the cluster. These vectors are plotted as a profile that shows graphically the score pattern of the centroid of the cluster.

In the analysis of the 1557 descriptors a total of 39 descriptor clusters were developed. A list of the number of descriptors and cluster titles suggested to the author by the descriptors in each cluster is given as Table 3. Also indicated in Table 3 is a measure of the degree to which the variability of reference dimension scores is partitioned by the descriptor clusters, homogeneity for each cluster. A homogeneity of 0.000 indicates that the scores within the cluster are as variable as the scores over all the 1557 descriptors. The homogeneities reported in Table 3 are averages over the 15 reference dimensions for each of the 39 clusters of descriptors. Experience with this measure suggests that homogeneities above 0.8 are indicative of a well-defined and self-consistent cluster of variables with respect to the reference dimensions. The reported values indicate a satisfactory clustering of the descriptors, with possible weaknesses in clusters 2, 23, and 24, and more likely weaknesses in clusters 13, 22, and 39. Inspection of the homogeneity of these clusters for the individual reference dimensions indicates that the weakness of the clusters is isolated to relatively few reference dimensions. For cluster 2, the only reference dimension on which the degree of homogeneity is strikingly low is reference dimension number 9, sensory physiology. This indicates that the descriptor cluster, clinical psychology (assessment and evaluation) descriptors, is non-differentiated on the descriptor dimension dealing with sensory physiology (vision). The descriptors have approximately as much variability on that dimension as all of the descriptors have without respect to the descriptor clustering. That is, we could expect a descriptor in cluster 2 to have a high score on dimension 2, just as much as we could expect it to have a low score. Similar analyses of the other descriptor clusters having small homogeneities indicate that the low average homogeneity of descriptor cluster 23 is attributable to a low homogeneity of the cluster on reference dimension 12; likewise for cluster 24 and dimensions 1, 6, and 8; cluster 13 on dimensions 10 and 15; cluster 22 on dimensions 3, and 11; and cluster 39 on dimensions 1, 3, and 11.

TABLE 3

Descriptor clusters in the 1557 descriptors

	Homogeneity cluster	Number of descriptors	Title suggested by descriptors
1	0.893	78	instrumentation -- radiation
2	0.776	16	clinical psychology -- assessment and evaluation
3	0.889	34	performance training
4	0.876	18	evaluation and measurement
5	0.902	61	instrumentation, methods, subjects -- non-physiological
6	0.896	91	instrumentation, methods -- physiological
7	0.805	33	microelectric study of sensory systems
8	0.823	16	assessment of metabolic disorders
9	0.817	29	learning -- brain and physiology
10	0.824	24	clinical psychology -- physical disorder
11	0.873	44	clinical effects of metabolic disorder
12	0.803	26	memory mechanisms in rodents -- physiological

TABLE 3 (Continued)

	Homogeneity cluster	Number of descriptors	Title suggested by descriptors
13	0.739	36	motivation and instinct -- feeding
14	0.889	56	instrumentation -- sensory
15	0.861	15	sensory -- touch, taste and smell
16	0.909	30	vision and hearing -- organs and neural mechanisms
17	0.824	40	CNS metabolism
18	0.804	38	autonomic nervous system
19	0.868	31	learning -- conditioning
20	0.819	38	endocrine system
21	0.803	35	CNS enzymology
22	0.759	41	child development -- normal
23	0.780	34	child development -- disease and retardation
24	0.777	46	sex
25	0.828	42	instrumentation -- GSR
26	0.837	28	instrumentation -- autonomic, clinical
27	0.839	26	instrumentation -- autonomic
28	0.907	70	instrumentation -- neurological
29	0.915	55	basic nervous system
30	0.855	24	CNS and peripheral nerves
31	0.854	6	psychopharmacology of learning and memory
32	0.844	32	CNS tissue structure
33	0.907	51	clinical -- general deterioration
34	0.883	36	clinical -- metabolic disorders
35	0.866	69	pharmacology -- mechanisms
36	0.916	39	pharmacology -- clinical
37	0.861	9	psychopharmacology in clinical psychology -- physical disorders

TABLE 3. (Continued)

	Homogeneity cluster	Number of descriptors	Title suggested by descriptors
38	0.822	39	ethology, ecology -- physiology
39	0.744	42	vertebrate physiology

The number of descriptors included in the study makes impractical the presentation here of a complete table of the descriptor members of the clusters. However, by selecting the most frequently used descriptors (on the 930 projects) the meaning of the descriptors becomes somewhat evident. The lists of the selected descriptors for each of the clusters are presented in Table 4. Each descriptor is listed by descriptor number and the frequency is given to the right of the descriptor. Descriptor clusters 31 and 37 have too few members to be readily interpretable, particularly in view of the low frequency (the highest frequencies are listed) of occurrence of those descriptors. However, inspection of the mean scores on the reference dimensions for those clusters provides a suggestion of meaning (see below).

TABLE 4

Selected lists of descriptors defining descriptor giving abbreviated name and frequency of the descriptor in the 930 projects

Cluster 1: Instrumentation -- radiation

151	Entomology, physiology/behavior	6
165	Entomology, physiology	12
276	Invertebrate physiology	12
383	General topics in medicine and psychology/prognosis	10

TABLE 4 (Continued)

646	Radiation effects	18
987	Test animals, invertebrates	18
1330	Physiological psychology, irradiation effects	5
1335	Physiological psychology/irradiation effects	23
1336	Physiological psychology/non-specific	31
Cluster 2: Clinical psychology -- assessment and evaluation		
94	Instrumentation/sensory techniques/ pupillography	13
1076	Clinical psychology/schizophrenia	49
1079	Clinical psychology/psychoses	72
1083	Clinical psychology/general	8
1275	Personality	8
1403	Social Psychology, family interaction/ mentally ill	6
Cluster 3: Performance training		
382	General topics in medicine and psychology/ follow-up	10
1115	Education and training	11
1150	Evaluation and measurement	13
1343	Cognitive processes/intelligence	14
1546	Project location index (geographical terms)	8
Cluster 4: Evaluation and measurement		
658	Therapy	4
1260	Performance	4
1277	Personality/other	7
1345	Cognitive processes/mediational processes	4
1357	Rehabilitation	4
1435	Therapy/conditioning	6
Cluster 5: Instrumentation, methods, subjects -- non-physiological		
44	Instrumentation	147

TABLE 4 (Continued)

401	Body temperature	22
702	Stressors	25
1039	Test animals, homo sapiens	411
1058	Clinical psychology/emotional states	46
1063	Clinical psychology/psychiatric patients	38
1151	Evaluation and measurement/tests and methods	36
1152	Evaluation and measurement in psychology	56
1254	Occupational psychology	27
1257	Performance/physiological factors	29
1262	Performance	58
1270	Personality/organization	40
1279	Personality	58
1413	Social psychology	28
1431	Stress behavior	28
1549	VA program/psychology and behavior	58
Cluster 6: Instrumentation, methods -- physiological		
141	Blood plasma and serum (body fluids)	34
211	Environment	27
348	Tests and methods	33
575	Metabolism	81
705	Stressors/physiological	35
728	Tissues/pathology	26
678	Sensory organs/visual, basic studies	37
769	Urogenital system	26
793	Biometric parameters	39
838	Substances/enzymes	29
970	Substances	285
1162	Experimental or situation variables/ response parameters	36
1167	Experimental or situation variables	102
1234	Motivation and instinct/social behavior	27
1246	Motivation and instinct	171
1334	Physiological psychology/other	82
1338	Physiological psychology	928
1351	Cognitive processes	95
1428	Stress, behavioral aspects/ specific stressors	100

TABLE 4 (Continued)

1432	Stress, behavioral aspects	124
1458	Miscellaneous psychology	104
1478	Sleep	49
1483	Miscellaneous	65
1484	Miscellaneous	130
1557	VA program	86
Cluster 7: Microelectric study of sensory systems		
53	Microelectrodes	18
164	Entomology, physiology/sensory nervous system	6
664	Sensory organs/pains	14
667	Sensory organs/propreoceptors	7
1304	Brain and CNS mechanisms	26
1371	Sensory and perceptual processes/ kinaesthetic sensitivity	16
1372	Sensory and perceptual processes/pains	11
1393	Sensory and perceptual processes	12
1453	Language and linguistics	6
Cluster 8: Assessment of metabolic disorders		
37	Chromatography	8
886	Amino acids, non-specific	6
895	Diethyltryptamine	12
1077	Clinical psychology/senile psychoses	6
Cluster 9: Learning -- brain and physiology		
1004	Test animals/birds/galeiformes	10
1012	Test animals, primates/macaca	21
1015	Test animals, primates	24
1155	Experimental variables/intensity of stimulus	18
1200	Learning and retention/retention/short term memory	15
1201	Learning and retention/retention/trace, consolidation	25
1202	Learning and retention/retention/other	35
1205	Learning and retention/retention	88
1208	Learning and retention/other	16

TABLE 4 (Continued)

1293	Physiological psychology/brain and CNS mechanisms/anatomy and physiology/learning	129
1337	Cognitive processes/animal intelligence	13
1367	Sensory and perceptual processes/audition	14
Cluster 10: Clinical psychology -- physical disorder		
80	Photography and camera	10
1064	Clinical psychology/physical disorders/ etiology, course of disease	14
1066	Clinical psychology/physical disorders/ other	51
1068	Clinical psychology/physical disorders	74
1084	Communication/clinical states	8
1480	Miscellaneous	8
Cluster 11: Clinical effects of metabolic disorder		
149	Body fluids/urogenital system/urine	26
150	Body fluids	58
328	Non-research	15
381	Diagnosis	11
387	General topics, medicine and psychology	34
570	Metabolism	16
572	Metabolism in disease	48
791	Biometric parameters/sex	15
829	Carbohydrates	11
1050	Aging/non-specific	12
1053	Clinical psychology/affective disorders	19
1085	Clinical psychology	266
Cluster 12: Memory mechanisms in rodents -- physiological		
441	Medical nutrition/states of general malnutrition	11
609	Blood brain barrier	10
1002	Test animals/birds/columbiformes	14
1024	Test animals/mammals/rodentia/muridae	20
1029	Test animals/mammals/rodentia	168
1186	Learning and retention/maze learning	18
1210	Motivation and instinct/activity	21

TABLE 4 (Continued)

1213	Motivation and instinct/exploration and stimulus learning	11
1242	Motivation and instinct/other	32
1313	Physiological psychology/dietary deficiency	10
1436	Therapy/electroshock	13
Cluster 13: Motivation and instinct -- feeding		
438	Medical nutrition/diet components/fats	18
442	Medical nutrition/states of general malnutrition	15
445	Medical nutrition	40
805	Hypothalamus	48
1220	Motivation and instinct/hunger and feeding/other	12
1221	Motivation and instinct/hunger and feeding	15
1222	Motivation and instinct/hunger and feeding/non-specific	10
1224	Motivation and instinct/hunger and feeding	46
1243	Motivation and instinct/thirst	18
1292	Physiological psychology/brain and CNS mechanisms/anatomy and physiology/feeding	22
Cluster 14: Instrumentation -- sensory		
97	Sensory techniques	39
101	Instrumentation/S	49
611	Brain injury-trauma	44
689	Sensory organs	31
710	Surgery/nervous system	20
715	Surgery	26
726	Nervous and sensory tissues, pathology	21
812	CNS and peripheral nerves/temporal	21
1305	Physiological psychology/brain and CNS mechanisms/anatomy and physiology/motor	23
1369	Sensory and perceptual processes/audition	64
1383	Sensory and perceptual processes/vision/perception of intermittent stimuli	21
1388	Sensory and perceptual processes/mixed, non-specific	25

TABLE 4 (Continued)

1395	Sensory and perceptual processes	248
Cluster 15: Sensory -- touch and smell		
668	Sensory organs/temperature	12
669	Sensory organs/touch and localization	19
690	Sensory organs/chemoreceptor	21
1370	Sensory and perceptual processes/ cutaneous sensitivity	22
1373	Sensory and perceptual processes/ taste and smell	24
Cluster 16: Vision and hearing -- organs and neural mechanisms		
670	Sensory organs/auditory, basic studies	26
678	Sensory organs/visual, basic studies	37
679	Sensory organs/auditory	47
686	Sensory organs/visual	67
693	Sensory organs	172
1302	Physiological psychology/brain and CNS mechanisms/brain - neural/audition	33
1303	Physiological psychology/brain and CNS mechanisms/brain - neural/vision	49
1307	Physiological psychology/brain and CNS mechanisms/brain - neural	114
1386	Sensory and perceptual processes/vision/ other	29
1387	Sensory and perceptual processes/vision/ misc.	26
1389	Sensory and perceptual processes/vision	109
Cluster 17: CNS metabolism		
593	Nervous system/chemistry and metabolism	50
716	Tissues, biophysical studies	17
719	Tissues, biophysical studies	39
758	Intracellular metabolism	22
872	Nucleic acid/RNA/other	17
874	Substances/nucleo compounds/RNA	18
875	Substances/nucleo compounds	22

TABLE 4 (Continued)

1300	Physiological psychology/brain and CNS mechanisms/metabolism and biochemistry	63
Cluster 18: Autonomic nervous system		
347	Tests and methods/other	16
430	Cardiovascular system	64
452	Digestive system/stomach	10
613	Nervous system	14
1028	Test animals/carnivora	29
1031	Test animals/carnivora/canidae	29
Cluster 19: Learning -- conditioning		
910	Substances	14
1165	Experimental or situation variables/ response parameters/strength	16
1171	Learning and retention/operant conditioning	83
1177	Learning and retention/conditioning/misc.	48
1179	Learning and retention/conditioning	176
1181	Learning and retention/escape and avoidance	67
1192	Learning and retention/methods	19
1194	Learning and retention/reinforcement/ negative	17
1196	Learning and retention/reinforcement/other	37
1198	Learning and retention/reinforcement	71
1211	Learning and retention	393
1416	Stress, behavioral aspects/electroshock	18
1195	Learning and retention/reinforcement/ schedule	21
Cluster 20: Endocrine system		
538	Endocrines in disease/stress	15
540	Endocrines in disease	31
542	Endocrine system	73
627	Neuroendocrine system	49
846	Substances/hormones/adrenalin	14
870	Substances/hormones	72
1328	Physiological psychology/endocrine effects/other	22

TABLE 4 (Continued)

1331	Physiological psychology/endocrine effects	67
Cluster 21: CNS enzymology		
343	Test methods/biochemical	17
521	Enzymology	18
559	Metabolism/basic studies	22
817	Isotopes/radioaction/other	11
857	Hormones/noradrenaline	15
888	Proteins and amino acids	32
958	Serotonin	11
962	Substances (84)	28
Cluster 22: Child development -- normal		
166	Child growth	17
173	Pregnancy and newborn	11
178	Children, neurological diseases	11
183	Children	51
285	Developmental physiology	13
286	Developmental biology	16
1043	Neonate (animal evaluation parameters, characteristics)	34
1046	Animal evaluation parameters/ characteristics	46
1047	Animal evaluation parameters	46
1092	Developmental psychology/infancy	25
1094	Developmental psychology/pre- and peri-natal effects	20
1097	Developmental psychology/longitudinal development	44
1107	Developmental psychology	78
Cluster 23: Child development -- disease and retardation		
8	Genetics/gene action/disease and anomalies	10
181	Children, diseases	23
311	Follow-up	7
313	Follow-up	7
548	Medical and veterinary microbiology	6
1061	Clinical psychology mental retardation/ other	14

TABLE 4 (Continued)

1062	Clinical psychology/mental retardation	15
1089	Developmental psychology/adolescence	6
1105	Developmental psychology/other	18
1132	Evaluation and measurement/test development	7
Cluster 24: Sex		
219	Pregnancy	9
229	Reproduction	22
245	Vertebrate physiology/endocrine system	14
523	Anterior pituitary, basic studies	13
525	Sex, basic studies	15
527	Endocrine system, basic studies	42
1297	CNS Mechanisms/anatomy and physiological psychology/sex	16
1327	Physiological psychology/endocrine effects	17
Cluster 25: Instrumentation -- GSR		
45	Electromyography	16
47	Galvanic skin response	63
52	Instrumentation (g)	68
1207	Learning and retention/verbal learning	12
1283	Physiological psychology/autonomic system/emotion	36
1286	Physiological psychology/autonomic system/sensation and perception	10
1288	Physiological psychology/autonomic system/other	40
1289	Physiological psychology/autonomic system	35
Cluster 26: Instrumentation -- autonomic, clinical		
83	Polygraph	12
86	Instrumentation (P)	31
87	Respiration techniques	11
88	Respiration techniques	12
608	Nervous system/autonomics/other	29
665	Respiratory system	10

TABLE 4 (Continued)

1285	Physiological psychology/autonomic system/patterns, clinical disease	17
Cluster 27: Instrumentation -- autonomic		
24	Instrumentation/cardiovascular techniques/ heart rate	51
30	Instrumentation/cardiovascular techniques	68
422	CVS/heart	22
602	Autonomics	19
603	Autonomics/cardiovascular-respiratory	29
610	Nervous system/autonomics	86
1284	Physiological psychology/autonomic system/learning	36
1290	Physiological psychology/autonomic system	145
Cluster 28: Instrumentation -- neurological		
43	Analog computer	73
65	Instrumentation/neurologic techniques/EEG	61
75	Instrumentation/neurologic techniques	263
104	Instrumentation	419
118	Medical physics	50
588	Musculoskeletal system	40
630	Nervous system	42
632	Nervous system	528
1035	Test animals, mammals; other	127
1294	Physiological psychology/brain and CNS mechanisms/anatomy and physiology/level of consciousness	37
1299	Physiological psychology/brain and CNS physiology	85
Cluster 29: Basic nervous system		
63	Neurologic techniques/CNS stimulation, other than self	73
300	Animal preparations, CNS lesions and ablation	137
302	Animal preparations	144
312	Animal preparations	194

TABLE 4 (Continued)

594	Nervous system/mechanisms of transmission	166
595	Nervous system/nerve pathways	143
599	Nervous system/basic studies	297
803	Cerebral cortex	123
815	CNS and peripheral nerves	287
1018	Test animals/primates	124
1032	Test animals/mammals/carnivora	107
1038	Test animals, mammals	463
1209	Learning and retention, miscellaneous	64
1301	Physiological psychology/brain and CNS mechanisms/anatomy and physiology	333
1314	Physiological psychology/brain and CNS mechanisms	501
Cluster 30: CNS and peripheral nerves		
69	Instrumentation/neurologic techniques/evoked potential	33
120	Mathematical biophysics/control system theory/coding and signal analysis	6
682	Sensory organs/visual/retrieval conditions	6
815	CNS and peripheral nerves	287
1377	Sensory and perceptual processes/vision/adaptation	11
1379	Sensory and perceptual processes/vision/color vision	8
Cluster 31: Psychopharmacology of learning and memory		
310	Animal preparations/split brains	10
1341	Cognitive processes/delayed response	7
1378	Sensory and perceptual processes/vision/brightness discrimination	10
Cluster 32: CNS tissue structure		
18	Instrumentation	11
143	Biophysical, chemistry and molecular biophysics	10
718	Histology and cytology	13
732	Tissues	82

TABLE 4 (Continued)

759	Cellular physiology	50
798	CNS and peripheral nerves/brain	37
973	Test animals, invertebrate	10
Cluster 33: Clinical -- general deterioration		
926	Substances (58)	18
1051	Aging	20
1052	Clinical psychology/alcoholism	21
1082	Clinical psychology/other	23
1256	Performance/evaluation	7
1449	Delusions and hallucinations	9
1450	Dreams	15
1467	Fatigue	10
1547	Project location index/North America	11
1548	Project location index	12
Cluster 34: Clinical -- metabolic disorders		
385	General topics medicine/treatments	6
567	Metabolism in mental disease	15
1072	Clinical psychology/depressive psychoses	9
1075	Clinical psychology/psychoses/ manic depressive	6
1078	Clinical psychology/psychoses	6
1441	Therapy/psychotherapy	8
1462	Chronic-degenerative	12
1466	Miscellaneous (C)	15
Cluster 35: Pharmacology -- mechanisms		
464	Pharmacology, dynamics/mechanism of action	16
470	Pharmacology, dynamics	38
481	Pharmacology, group action/neuroeffective	33
485	Pharmacology, group action/psychotropic	147
489	Pharmacology, group action/stimulation	18
492	Pharmacology, group action/tranquilizer	25
495	Pharmacology/group action	226
504	Pharmacology/tests and methods	16
508	Pharmacology	238
598	Nervous system/anesthesia	49

TABLE 4 (Continued)

905	Amphetamine	21
909	Substances (50)	53
913	Chlorpromazine	22
917	Substances (54)	59
943	Substances (74)	26
1317	Physiological psychology/drug effects/ emotion	21
1318	Physiological psychology/drug effects/ learning	49
1322	Physiological psychology/drug effects/other	60
1323	Physiological psychology/drug effects	48
1325	Physiological psychology/drug effects	207
Cluster 36: Pharmacology -- clinical		
459	Clinical pharmacology	36
460	Pharmacology, dose and response	37
476	Pharmacology, group action/energizer	13
482	Pharmacology, group action	17
502	Pharmacology and side effects	27
553	Intoxication and drug addiction	17
924	Substances (56)	19
932	Substances (66)	12
934	Lysergic acid	13
936	Substances (72)	20
955	Substances (80)	27
967	Substances (86)	23
1320	Physiological psychology/drug effects/ sensation and perception	18
1321	Physiological psychology/drug effects/ techniques of evaluation	37
1433	Therapy/chemotherapy	27
1443	Therapy of mental disorders	54
Cluster 37: Psychopharmacology in clinical psychology -- physical disorders		
941	Magnesium pemoline	4
1022	Test animals, mammals/rabbits	7
1081	Clinical psychology/other	5

TABLE 4 (Continued)

Cluster 38: Ethology, ecology -- physiology		
198	Ecology	18
238	Comparative physiology, vertebrates	12
252	Vertebrate physiology/nervous system/ sensory systems	8
260	Vertebrate physiology, sensory organs	21
994	Fish test animals	8
995	Test animals/fish	8
1006	Test animals/birds	31
1016	Test animals/mammals/primates/other	8
1232	Motivation and instinct/locomation	33
1235	Motivation and instinct/sex	35
Cluster 39: Vertebrate physiology		
230	Vertebrate anatomy	10
231	Vertebrate anatomy	12
233	Vertebrate physiology/behavior/other	36
235	Vertebrate physiology/behavior	40
251	Vertebrate physiology/nervous system/CNS	30
254	Physiology of nervous systems, vertebrates	39
268	Vertebrate physiology	75
1212	Motivation and instinctive behavior/ aggression	15
1237	Motivation and instinctive behavior/theory	10

The meaning of the descriptor clusters can be illuminated further by considering the reference dimensions defined by the reference clusters listed in Table 2. Each of the descriptor clusters from the clustering of the entire set of descriptors has a mean and standard deviation of scores on each of the reference dimensions in the score space. Each of the dimensions in the score space is defined with a mean of 50 and a standard deviation of 10. The homogeneity of the scores in the clusters of descriptors is much greater than the homogeneity of the scores across clusters, as indicated above, with a few notable exceptions. Thus, the pattern of mean values of the scores within descriptor clusters can be taken as a relatively non-variable description of the character of the descriptor clusters. That is, the scores within a cluster will all be relatively the same as the mean score within the cluster, compared with the scores for the entire set of descriptors. The profile of the mean standard scores on the 15 reference dimensions for a descriptor cluster therefore is an accurate description on the characteristics of the descriptors in the cluster, with respect to those reference dimensions. The mean standard score profiles of the clusters on the reference dimensions may be presented graphically, with the reference dimensions represented on the base of the graph and the mean standard score for the dimensions represented on the vertical axis. These graphs are presented as Figure 1. Several score profiles are presented in most of the graphs of Figure 1. In all graphs, except 4 and 14, the profiles shown are for relatively similar clusters of descriptors. Graphs 17 and 18 are for clusters of descriptors defined by pooling descriptors from two or more of the 39 descriptor clusters listed in Table 3. Each graph has the reference dimension titles shown for those dimensions with extreme mean values.

Several properties of the clusters and the reference dimensions are evident in the graphs of Figure 1. All of the graphs have relatively high points on one or more reference dimensions, indicative of salient characteristics of the descriptor cluster represented in the graphs, except for descriptor clusters 1 and 6, of graph 1. On the same graph, cluster 5 also displays only minor departures from a relatively flat profile. All three of these descriptor clusters are populated with descriptors that involve some type of instrumentation, specialized use of instrumentation, or general methodological principles of physiological psychology research. Since the reference dimensions did not include an instrumentation or general methodology dimension, it is not surprising that this flat profile is discovered. Also, it points out the basis on which the other descriptor clusters are differentiated. In all of the other descriptor clusters the pattern of reference dimension mean scores is composed mostly of scores between 55 and 45 with a selection of very high values. The flat portion of the profiles is almost uniformly between 45 and 50, below the mean value of 50 on the standard scale used for the reference dimensions. There are few score means below 40. These properties of the profiles are a consequence of the scoring device used to obtain the reference dimension scores -- only the presence of a property is used in developing the scores. As a consequence, when a property plays a strong role in the definition of a descriptor cluster it will tend to stand out as a sharp deviation in the upper direction. Put in other terms, the descriptors are either generally relevant or irrelevant to projects and there is very little tendency for a reference dimension to have a score implying the negation of a property of projects, as in "not pharmacology." This generalization does not hold true for all descriptor clusters on all reference dimensions, as is illustrated with reference dimension 3, "Central nervous system physiology, pathways and mechanisms." These patterns are all to be expected in terms of the character of the scoring procedures, which are sensitive only to the patterns of actual attributions of characteristics (descriptors) to events (projects). In reference dimension 3 the extreme of the scores are given by the descriptor clusters 1, 3, 4, 33, and 34, all composed of descriptors dealing with purely behavioral, nonphysiological, types of studies. The projects with the descriptors in those clusters are those with the fewest uses of the descriptors defining the third reference dimension, and hence the low score means. This is not to say that the projects involved are not physiological psychology projects, just that they involve relatively few of the descriptors in reference dimension 3.

Interpretation of the graphs is relatively straightforward. In Graph 1 the score profiles indicate low relevance on all reference dimensions, particularly for descriptor cluster 6, which is composed of descriptors dealing with instrumentation and methodology. This descriptor cluster undoubtedly appears as a separate cluster because of its lack of distinctive patterns of use in the collection of projects; it is the "common denominator" of the entire collection of descriptors, including the descriptor "physiological psychology." Because each of the other descriptor clusters had some common pattern of occurrence in the project, leading to a separation of the clusters from the mid-point of the score space, cluster 6 shows up as a distinct cluster. Other descriptor clusters have distinct patterns of peaks in the profile of mean scores. For example, in Graph 6, descriptor clusters 7, "microelectric study of sensory systems," and descriptor cluster 15, "sensory -- touch and smell, "show almost identical patterns with the exception of the extremeness of the level on reference dimension 15, "sensory physiology." The difference between these clusters is not apparent in the graphs themselves. Only inspection of the actual descriptor lists (Table 4) for these clusters gives a strong clue regarding the difference in meaning. The relatively lower mean scores of cluster 7 on reference dimensions 10 and 15 are coordinate with the author's interpretation of the memberships of the two clusters. Studies of "hunger and thirst" and "touch and smell" do not involve the use of same degree of neuroelectric analysis as is involved in studies of "vision and hearing."

Graph 2 illustrates another type of situation encountered in attempting to interpret the cluster profiles. All three descriptor clusters in this graph are high on the reference dimension 12, "Clinical psychology; physical disorders," whereas cluster 11 alone in this graph is high on reference dimension 14, "Clinical psychology; psychoses." Also, descriptor cluster 10 is relatively distinct from descriptor clusters 11 and 3 on the reference dimension "CNS physiology; pathways and mechanisms." In the light of these distinctions, and from inspection of the descriptors composing the clusters, the interpretation of clusters 3 and 11 as involving non-physiological attributes of clinical studies is probably justified. Cluster 3 lacks the component of the study of psychoses and is quite low on the reference dimension "CNS physiology," consonant with the author's interpretation that the descriptors deal with the clinical treatment of physical disorders through performance training.

Additional interpretative analyses of the descriptor clusters would be along the lines generally followed in the paragraphs above. These analyses are left to the reader. However, one of the descriptors is so special that it deserves special attention: "Physiological psychology," descriptor 1338, in descriptor cluster 6. This descriptor has a peculiar characteristic -- it is used on all of the projects. At first blush, it may seem that this descriptor should not have a place in any of the descriptor clusters since it is clearly related to all of the other descriptors in a non-differentiating fashion with respect to the projects. However, it also is the common denominator or the origin point of the entire set of descriptors. A descriptor, since it is a one-sided property (no negatives), can depart from the other descriptors only if it applies to a pattern of projects different than that to which the other descriptors apply. In that sense, all descriptor clusters may be seen as a departure from the unifying theme -- "physiological psychology." If these conjectures are accurate, we would expect that the profile of scores on the reference dimensions of the descriptor "physiological psychology" would be at the mean of the reference dimensions. Graph 19 of Figure 1 is the graph of the scores of "physiological psychology" on the fifteen reference dimensions. The graph is sufficiently like a flat line to warrant some confidence in the interpretation just offered.

The clusters defined by the procedure just described are next condensed successively into a hierarchical structure. The clusters with the closest centroids (smallest distance between their centroids in the 15 dimension space defined by the 15 reference dimensions) are pooled and the centroid of the pooled cluster is calculated. The next two nearest clusters, perhaps including the cluster newly formed by pooling the previous two, are pooled and the centroid of the pooled cluster is calculated. This procedure is continued until there is a single cluster with its centroid at the mean value of all of the reference dimensions (if all objects, descriptors, are included in the clusters). This hierarchical condensation can be described graphically by a dendrogram similar to those used in numerical taxonomy. In general, each of the successive stages will pool clusters that are further apart than the previously pooled clusters (this condition is sometimes violated by the vagaries encountered by the averaging process used in calculating successive centroids). The distance of the two clusters for each of the successively condensed clusters is an indicator of the degree to which the score profiles are similar or divergent. In the latter stages of the condensation procedures clusters formed in the "corners" of the score space by the previous condensations will be consolidated. These condensations will merge relatively different patterns of scores simply because they are the most similar of all the clusters defined at that point in the condensation. A graph of the hierarchical condensation, a dendrogram, is given in Figure 2. The distance between the pairs of clusters condensed at any given point in the process is indicated at the base of the graph, scaled both in terms of distances in the 15 dimension space (integer values) and in terms of the number of standard deviation units represented by the distance (standard deviation of the original reference dimension scores).

Each descriptor cluster is identified on the left of the dendrogram, with its cluster number. As two clusters are pooled, the pooled cluster is given a number, in succession beginning with 40 (39 original clusters). Pooling of clusters is indicated by joining lines extending to the right from the cluster number on the left. The pooled cluster is indicated by a line extending to the right from the point of condensation.

Graphs 17 and 18 of Figure 1 show pooled or condensed clusters. For Graph 17 the patterns of the pooled clusters are highly similar. The three original clusters are combined in the graph in successive steps, giving an average profile on larger and larger numbers of descriptors. Cluster 57 score profile represents the average of the scores of the 40 descriptors in cluster 17 and the 35 descriptors in cluster 21. The next step is to include in the average all those descriptors in cluster 32. Cluster 64 is defined by the pooled sets of definers of clusters 17, 32, 21, and 12. The pattern of variation is still quite clear in spite of the relatively large distance separating the clusters included in 64. Quite the opposite is observed in the next step in the condensation. When the condensation of clusters 62 and 64 is considered, the profile of the condensed cluster, 65, is very nearly flat. This is due to the fact that the condensations leading to the condensed cluster 59 are based on clusters that are relatively closer to the origin of the score space than are clusters such as 17, 32, 21 and 12. The clusters are nearer each other but average out to a flatter profile due to the relative closeness of the profiles to a flat (at the mean) profile. The large number of descriptors in condensed cluster number 62 simply overcomes the extreme profile points in cluster 64 to produce a flat profile. All of the other condensations in the dendrogram that represent distinctive profiles of mean scores are implied in Figure 1 by plotting on the same graph those clusters that condense. Condensations other than those represented by graphs with several clusters produce relatively flat profiles of mean scores. The limits of condensation, preserving a relatively high degree of distinctiveness of the profiles, are indicated in the dendrogram by circles drawn on the lines to the right of the condensation points that retain profile distinctiveness. Condensation cluster 48 condenses with cluster 31 at a distance represented just off of the distance scale of the graph. Attempts to make interpretations of condensed clusters beyond the level of condensation marked with the circles is probably doomed to failure; the composites represented are defined by descriptors too diverse in meaning.

7. DISCUSSION

The prima facie sensibleness of the results of the analysis is the primary validation of the success of the attempt to structure a scientific field. However, had different reference dimensions been used, the descriptor clusters could well have been different. Limitations in the program to cluster the entire collection of descriptors did not permit the use of more than 15 reference dimensions and no assurance can be given that a different set of descriptor clusters would not have been found with additional reference dimensions. However, we have made 39 relatively sensible distinctions among the descriptors and 16 higher order clusters (the 13 condensations and the 3 non-condensed clusters, 13, 15, and 37). A criterion of the empirical validity of the results developed here is quite difficult to imagine. There are no immediately available models that would provide a standard of comparison. As a consequence, some caution must be observed in generalizing the results to other samples of observations or other scientific fields. In particular, any implications of these results for the development of policy for the support or non-support of research programs are to be developed only with the greatest caution.

Two severe problems in the data were overcome by the methods used. The first of these problems is the very large size of the data set itself. This problem was overcome by the combination of multiple passes at the initial clustering to define the reference dimensions and by the large-scale pattern recognition program of the BC TRY System. However, problems related to size are primarily problems of logistics and not problems of substantive interest. The second problem is of more interest. Techniques for dealing with infrequent event data of the sort encountered in this study are not very well developed. The methods that are known deal primarily with patterns of events, or attempt to analyze the event matrix by graph-theory methods. The magnitude of the physiological psychology data set and restrictions on the pattern or graph methods make the methods ill suited to the problem. The technique used in the analysis reported in this paper takes advantage of the very large number of observations made on each of the event variables -- i.e., a large number of projects on which the descriptors were observed. The weight matrix is an attempt to capture the general patterns of individual differences among the projects in terms of the most general dimensions of characteristics. The procedures for calculation of the score matrix, in turn, result in a composite of weights on projects to which a descriptor is applied, for each reference dimension. The weight matrix is analogous to the allelic values on a chromosome and the descriptor selects the genes with certain allelic values to produce, in an additive fashion, a phenotypic value.

The selection of projects in the way that was used for this study is a point of some concern. There are quite possibly many research programs in the field of physiological psychology that are not represented in the study, for at least two reasons. First, the descriptor "physiological psychology" had to be used on a project before the project was included in the sample from the total collection of SIE projects. It is possible that had the selection been made on a more inclusive basis, such as "Physiological psychology *or* (behavior *and* (physiology *or* pharmacology *or* ... *or* nervous system))" that the collection of projects would have included a different variety of types of research and the attendant descriptors attibuted to those projects. The upshot of that would have been a different set of clusters in the final analysis. Comments made in the previous section about the appearance of the descriptor "physiological psychology" are relevant to this point. The analysis reported here is a "within" analysis compared to an analysis including a mixture of several defining criteria, such as "physiological psychology *or* pharmacology *or* neurology *or* psychiatry". The descriptor clusters would no longer center on the cluster containing the descriptor "physiological psychology". Also, the differentiations made in the study reported here would probably not be detected, in comparison with the differentiations present among the projects and descriptors in the study with the more general criterion of project sampling. The analysis reported in this paper is a microanalysis of "physiological psychology" compared to the macro-analysis that is implied by the more general coverage in project sampling. On the other hand, the "physiological psychology" study would be a macro-analysis compared with an analysis of the projects and descriptors defined by a more restrictive sampling criterion, such as "physiological psychology *and* clinical psychology". It seems to the author that a desirable study of scientific activity would proceed from the highest level of inclusion of projects and descriptors through progressively more restrictive samples of projects. Each of the successive steps in restricting the range of projects included in a study would be defined by the macro-analysis at the preceding level of restriction from the all-encompassing stage. The more detailed, and hence the least general, structure might be based on an analysis of descriptors within a set of projects as narrowly defined as one of the descriptor clusters in the data reported, such as "micro-electric study of sensory systems."

Another property of concern of the data of the study reported is that the research projects included had financial support under one or another of the agencies cooperating with SIE. Programs that did not have approved grants, for any reason, including no application and rejected application, are not included in the study. The description of the structure of a scientific field cannot be complete or accurate unless the basic data include the data excluded from this study. The data were not accessible for this study. Consequently, a more conservative and accurate title of this paper might have been "cluster analysis of supported research in a scientific field." A study of the actual activity in a scientific field would include information regarding activity not supported and activity supported by grants that do not specifically cover the work.

The results reported in this paper are descriptive only. Availability of other information about the projects and the results of the projects would add meat to the bare-bones of the description. Many types of secondary information are easily thought of -- merit, dollar cost, impact of research results on other research, type of expenditures (personnel, equipment, capital, etc.), history of support for individual projects, history of support for type of project, etc. Of particular interest would be an analysis of the differences in supported and non-supported projects, especially in terms of the degree to which the projects conformed to a specified configuration of characteristics. The utility of this kind of information is most apparent in the analysis of projects as distinct from the analysis of descriptors reported here.

Apart from the simple interest in the structure of a scientific field, a number of other potential uses of this kind of analysis are possible. Although the experienced scientist will have a certain amount of knowledge about what sorts of things his colleagues are working with, it would be unlikely for that knowledge to be systematic or highly reliable as related to empirical fact. Consequently, information like that provided in this report could be an information base to be used by scientists in order to more completely understand the structure and relationships in the germane fields. From the standpoint of funding agencies, the information provided by an analysis of a scientific field being supported could be much more valuable than it is to the scientist. Administrative and supervisory personnel in such agencies must make decisions regarding how to deal with proposals. Having a systematic description of the actual structure of a scientific field should prove beneficial from the standpoint of efficiency and accuracy of disposition of project proposals. In particular, in administering the review apparatus of the agency certain committees or panels of experts are established to deal with proposals. If the structure of the committees and panels is not in some reasonable degree of concordance with the actual structure of the scientific fields involved, the functioning of the agency, its administration, and its panels is likely to be inadequate. The concordance between the scientific field and the structure of review apparatus probably should extend so far that the coverage and overlap of review structure and field structure are at least roughly the same.

One of the more attractive potentialities in studies such as the one reported here is the possibility of following the development of a scientific field over a time period. By periodically sampling the production of research proposals the development of new combinations of characteristics of projects could be detected, as could the demise of such combinations. At the very least, repeated analyses of a scientific field would provide a basis for the development of perspective in funding agencies on the structure, balance and thrust of proposal generation in the scientific field.

REFERENCES

Tryon, R.C. and Bailey, D.E., The BC TRY Computer System of Cluster and Factor Analysis, *Multivariate Behavioral Research, 1,* (1966) 95-111.

Tryon, R.C. and Bailey, D.E., *Cluster Analysis,* New York, McGraw-Hill Book Company, in press (expected 1970).

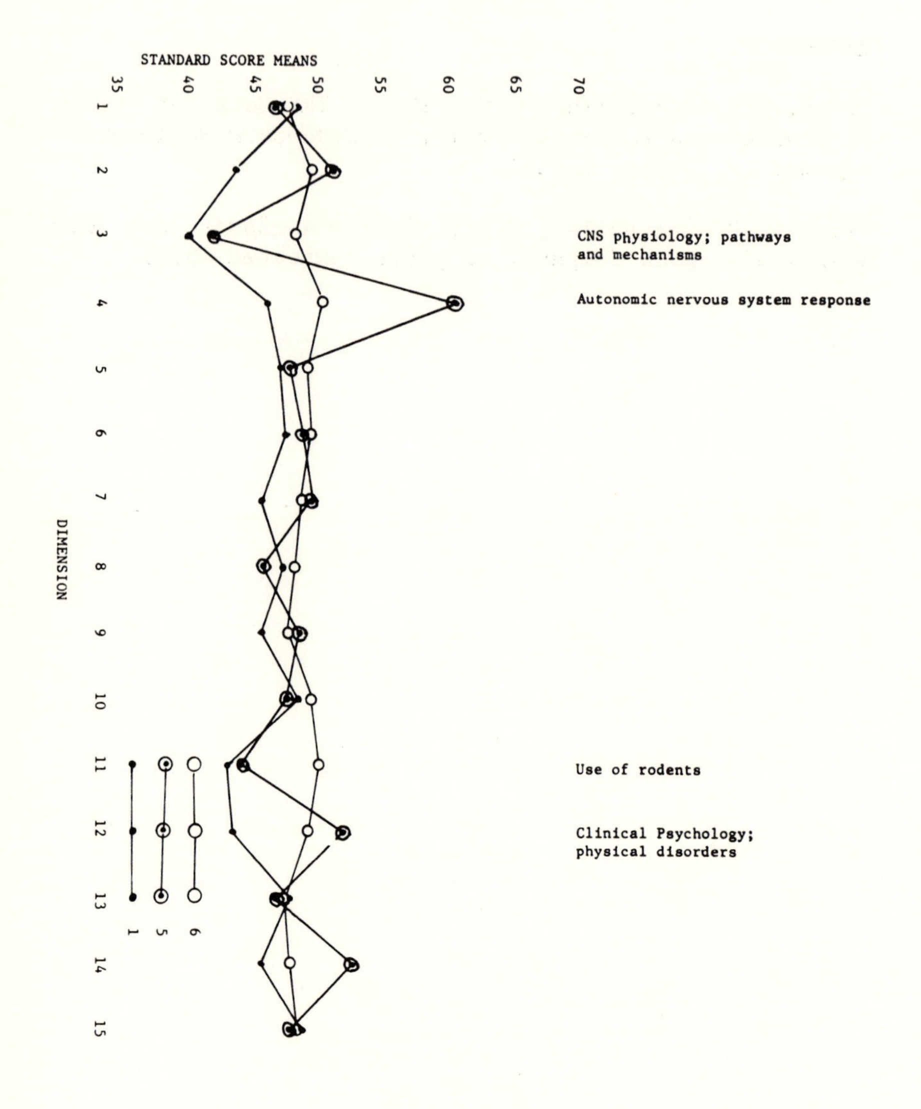

FIG. 1. (Graph 1).
Profiles of Mean Scores of Descriptor Clusters

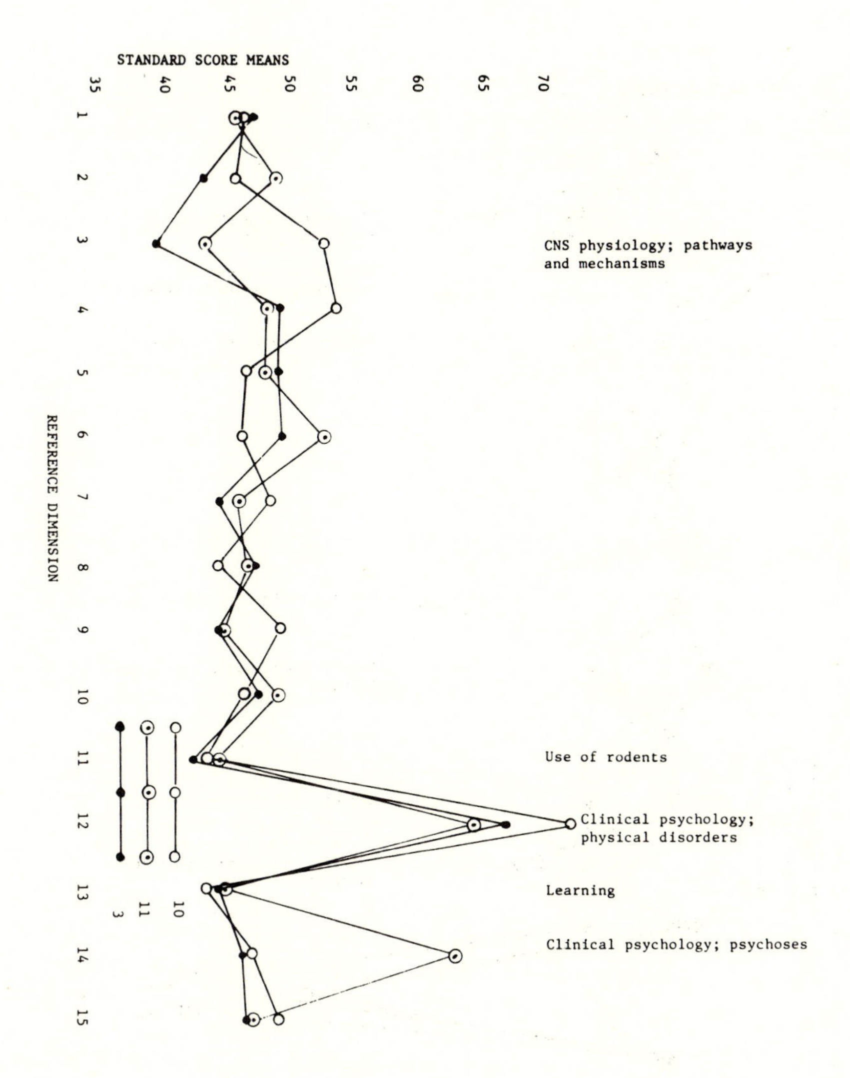
STANDARD SCORE MEANS
35
40
45
50
55
60
65
70
REFERENCE DIMENSION
1
2
3
4
5
6
7
8
9
10
11
12
13
14
15
10
11
3
CNS physiology; pathways and mechanisms
Use of rodents
Clinical psychology; physical disorders
Learning
Clinical psychology; psychoses

FIG. 1. (Graph 2).

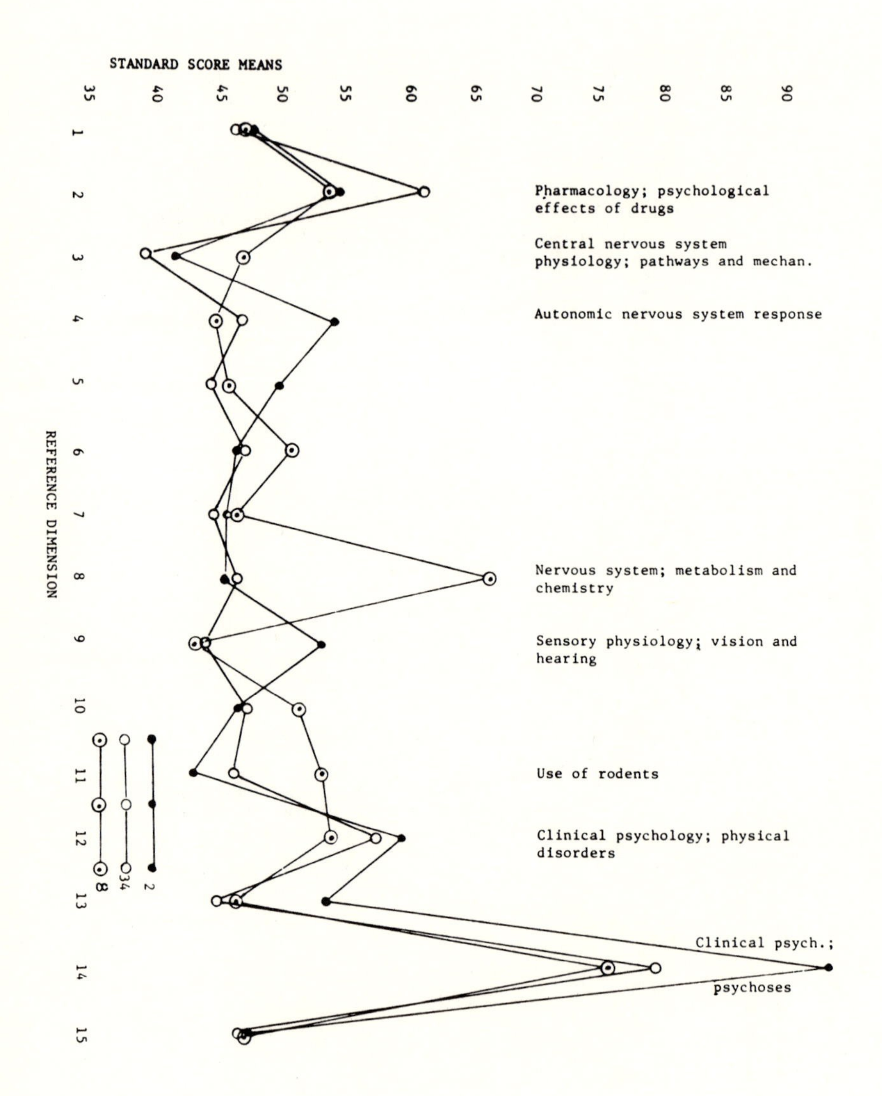
STANDARD SCORE MEANS
35
40
45
50
55
60
65
70
75
80
85
90
REFERENCE DIMENSION
1
2
3
4
5
6
7
8
9
10
11
12
13
14
15
Pharmacology; psychological effects of drugs
Central nervous system physiology; pathways and mechan.
Autonomic nervous system response
Nervous system; metabolism and chemistry
Sensory physiology; vision and hearing
Use of rodents
Clinical psychology; physical disorders
Clinical psych.; psychoses
2
34
8

FIG. 1. (Graph 3).

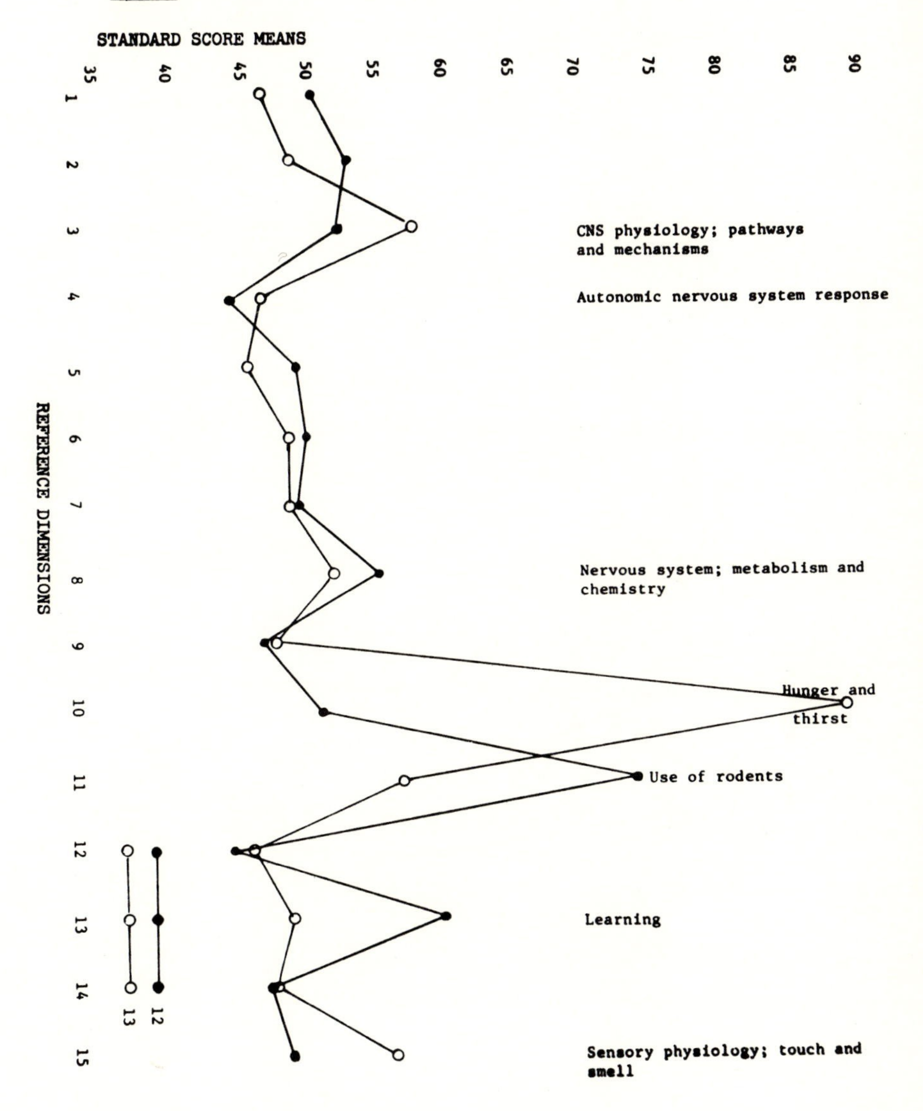
Figure 1 (continued)
Graph 4
STANDARD SCORE MEANS
35
40
45
50
55
60
65
70
75
80
85
90
REFERENCE DIMENSIONS
1
2
3
4
5
6
7
8
9
10
11
12
13
14
15
CNS physiology; pathways and mechanisms
Autonomic nervous system response
Nervous system; metabolism and chemistry
Hunger and thirst
Use of rodents
Learning
Sensory physiology; touch and smell
12
13

FIG. 1. (Graph 4).

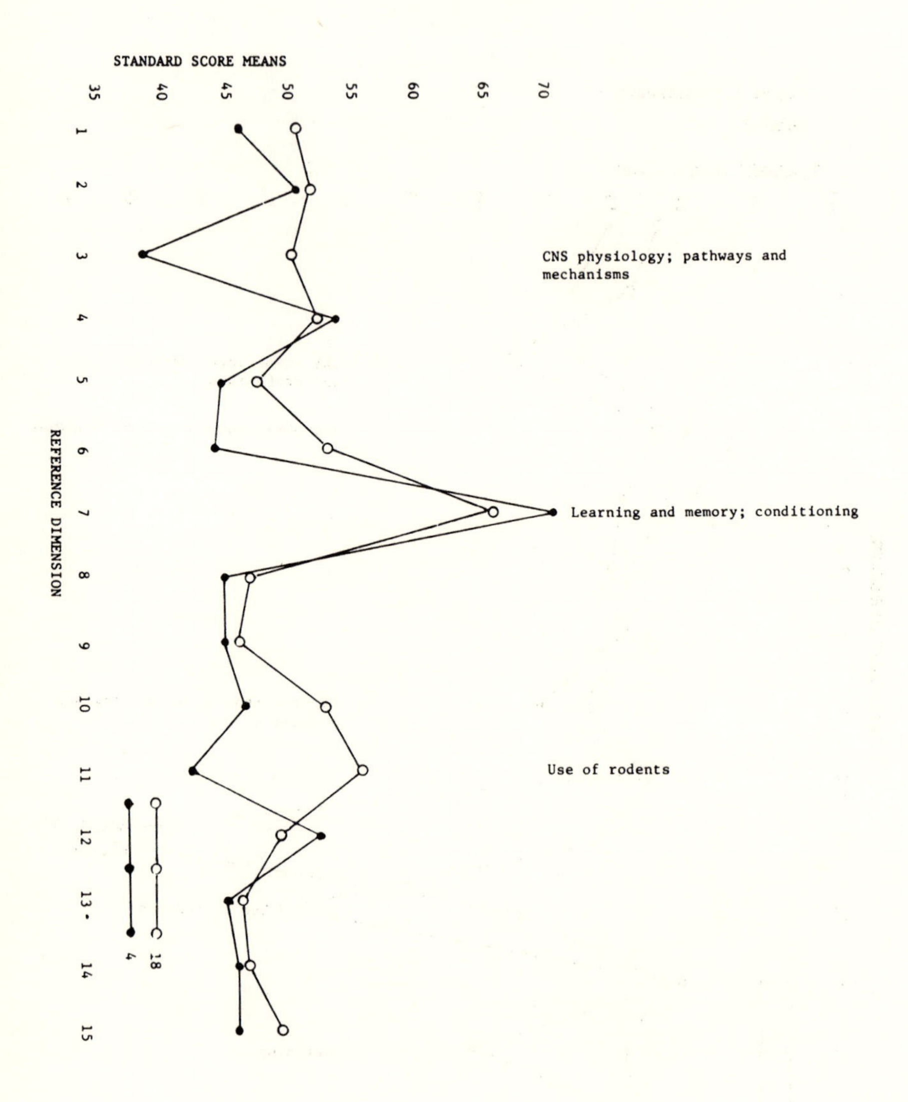
STANDARD SCORE MEANS
35
40
45
50
55
60
65
70
REFERENCE DIMENSION
1
2
3
4
5
6
7
8
9
10
11
12
13
14
15
CNS physiology; pathways and mechanisms
Learning and memory; conditioning
Use of rodents
18
4

FIG. 1. (Graph 5).

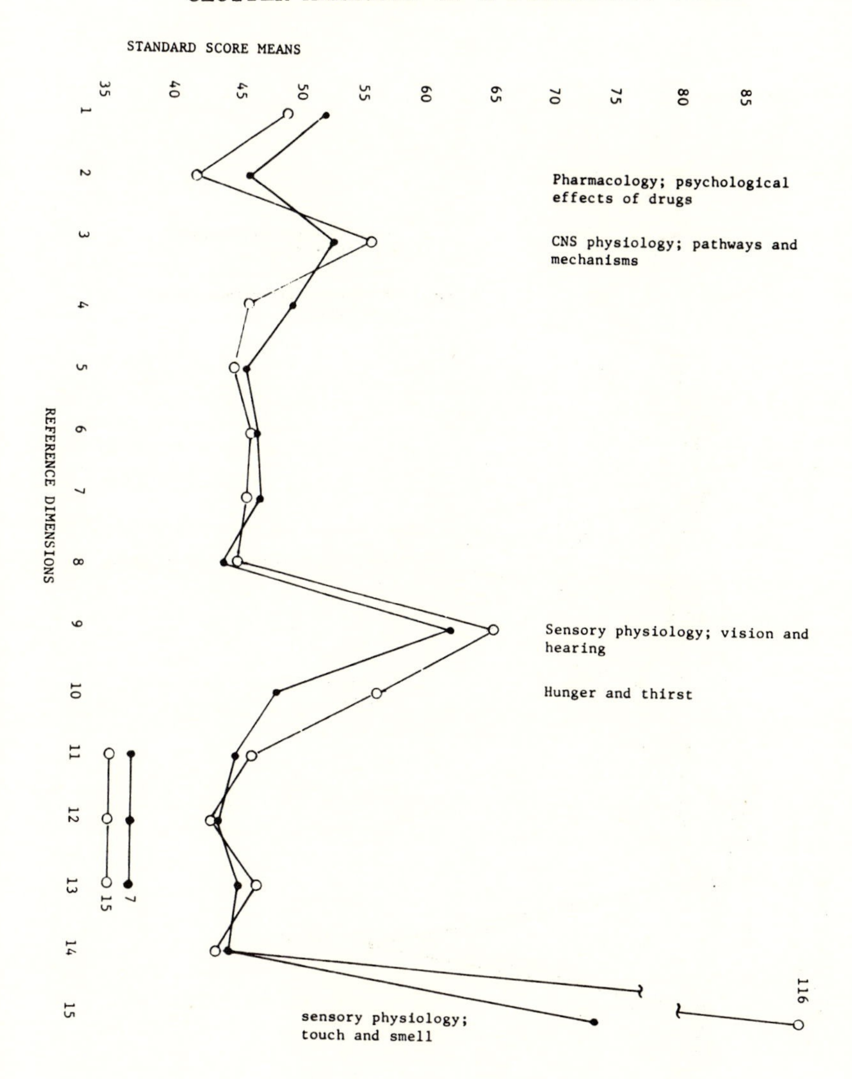
STANDARD SCORE MEANS
35
40
45
50
55
60
65
70
75
80
85
REFERENCE DIMENSIONS
1
2
3
4
5
6
7
8
9
10
11
12
13
14
15
Pharmacology; psychological effects of drugs
CNS physiology; pathways and mechanisms
Sensory physiology; vision and hearing
Hunger and thirst
sensory physiology; touch and smell
116
7
15

FIG. 1. (Graph 6).

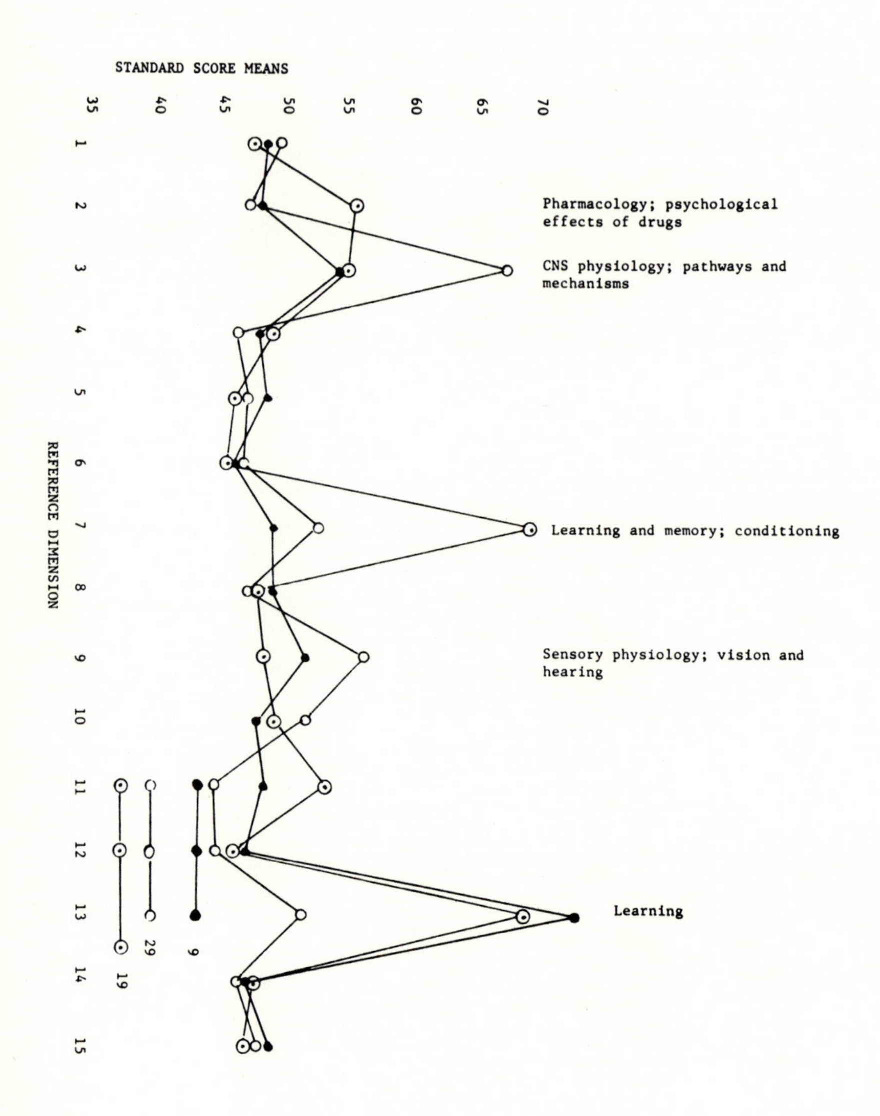
STANDARD SCORE MEANS
35
40
45
50
55
60
65
70
REFERENCE DIMENSION
1
2
3
4
5
6
7
8
9
10
11
12
13
14
15
Pharmacology; psychological effects of drugs
CNS physiology; pathways and mechanisms
Learning and memory; conditioning
Sensory physiology; vision and hearing
Learning
9
29
19

FIG. 1. (Graph 7).

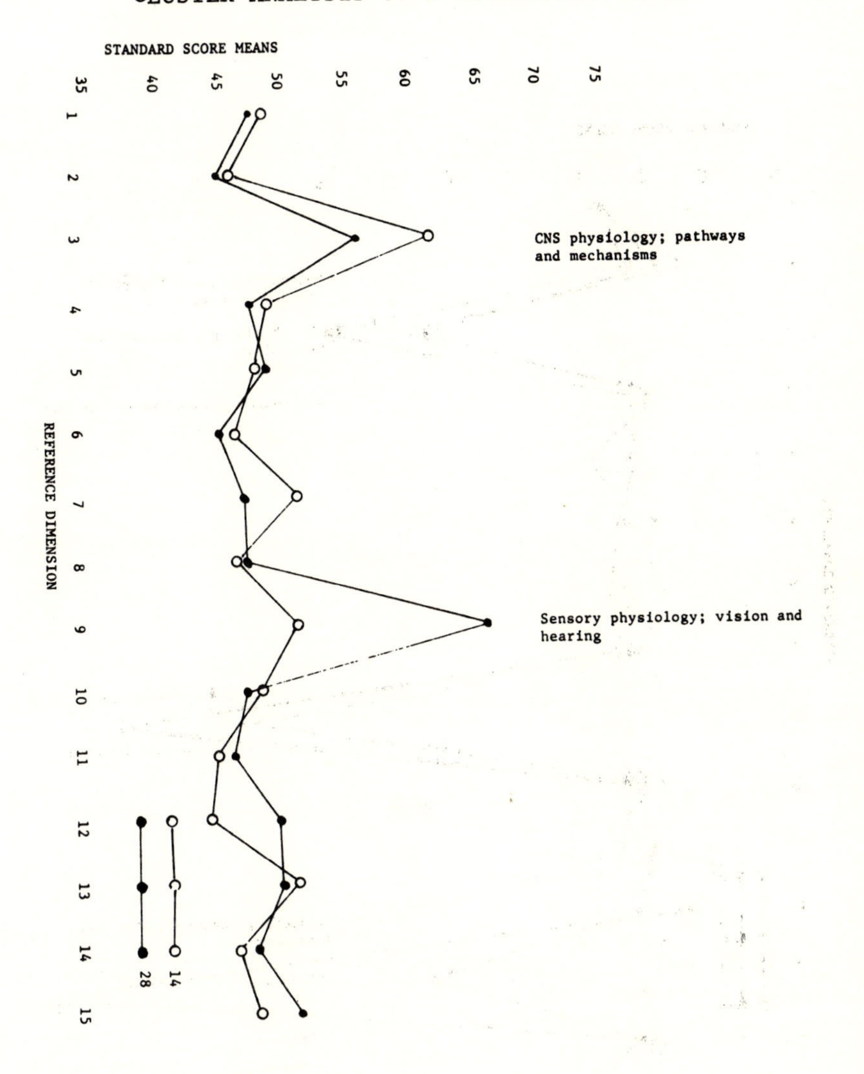
STANDARD SCORE MEANS
35
40
45
50
55
60
65
70
75
REFERENCE DIMENSION
1
2
3
4
5
6
7
8
9
10
11
12
13
14
15
CNS physiology; pathways and mechanisms
Sensory physiology; vision and hearing
14
28

FIG. 1. (Graph 8).

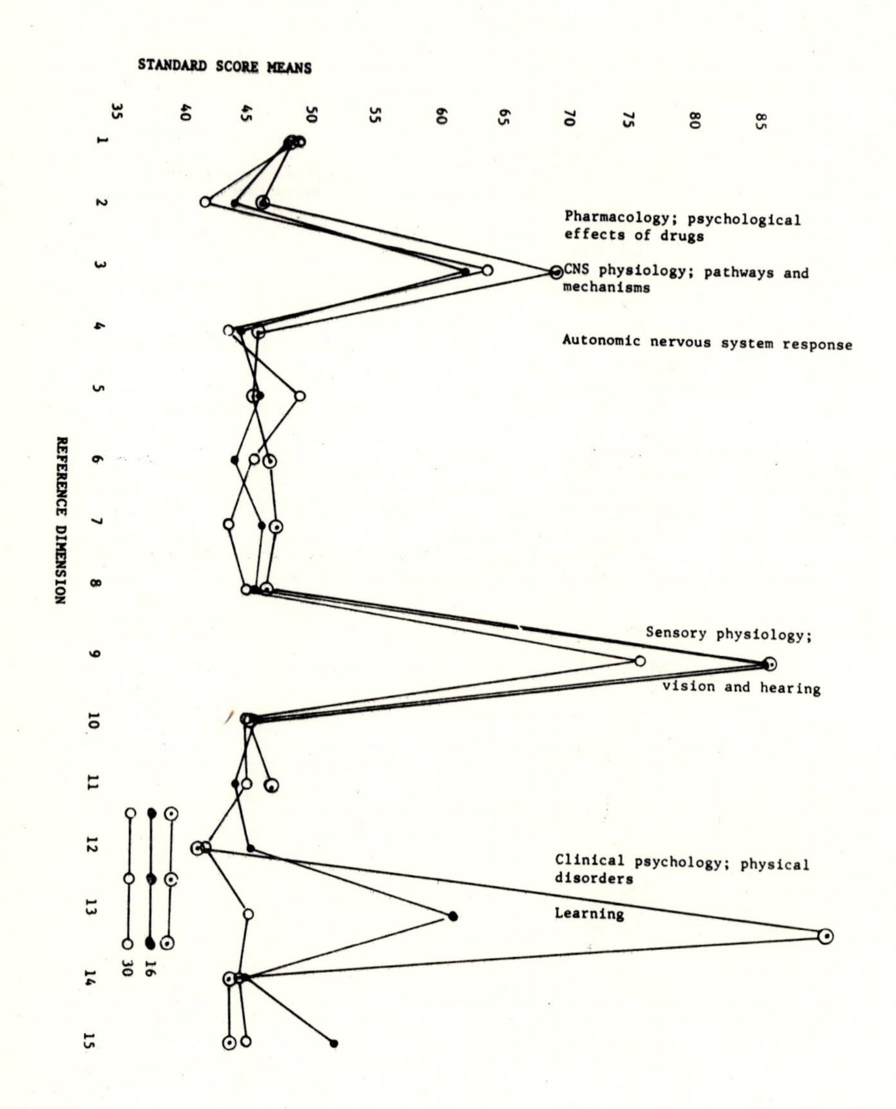
STANDARD SCORE MEANS
35
40
45
50
55
60
65
70
75
80
85
REFERENCE DIMENSION
1
2
3
4
5
6
7
8
9
10
11
12
13
14
15
Pharmacology; psychological effects of drugs
CNS physiology; pathways and mechanisms
Autonomic nervous system response
Sensory physiology;
vision and hearing
Clinical psychology; physical disorders
Learning
16
30

FIG. 1. (Graph 9).

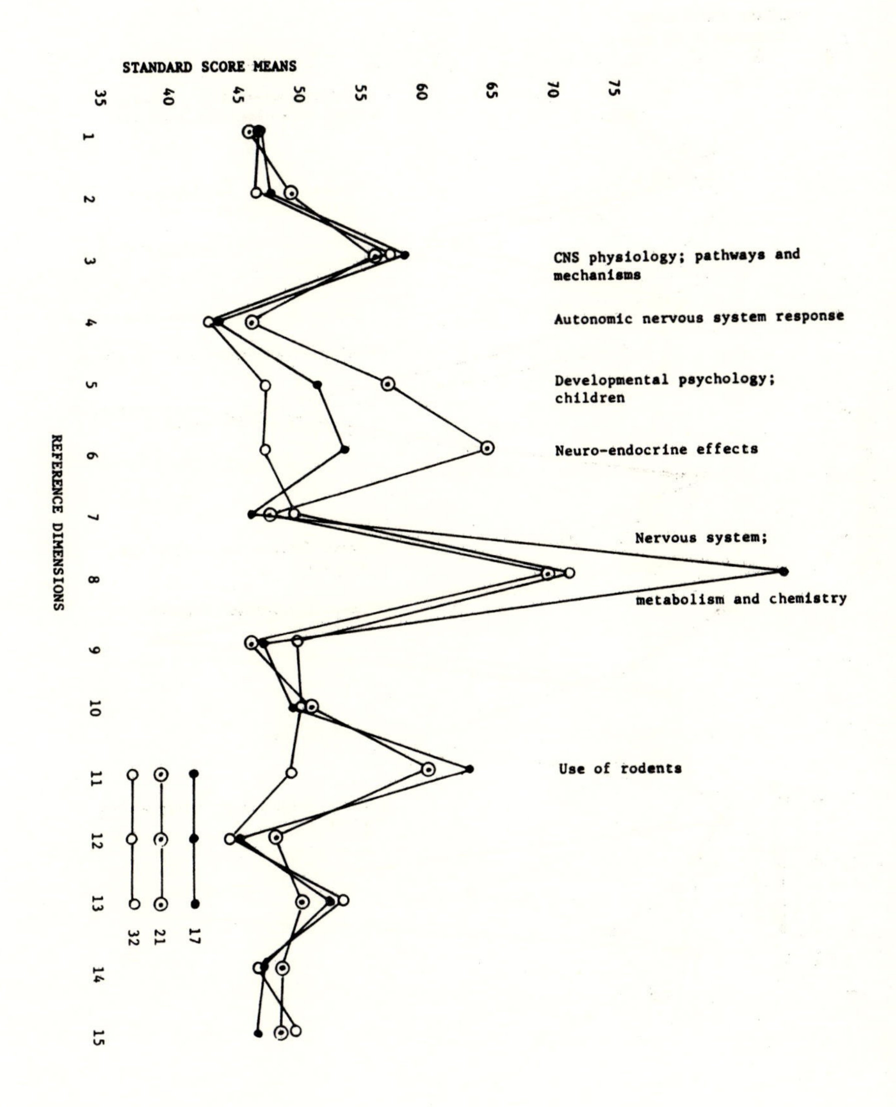

FIG. 1. (Graph 10).

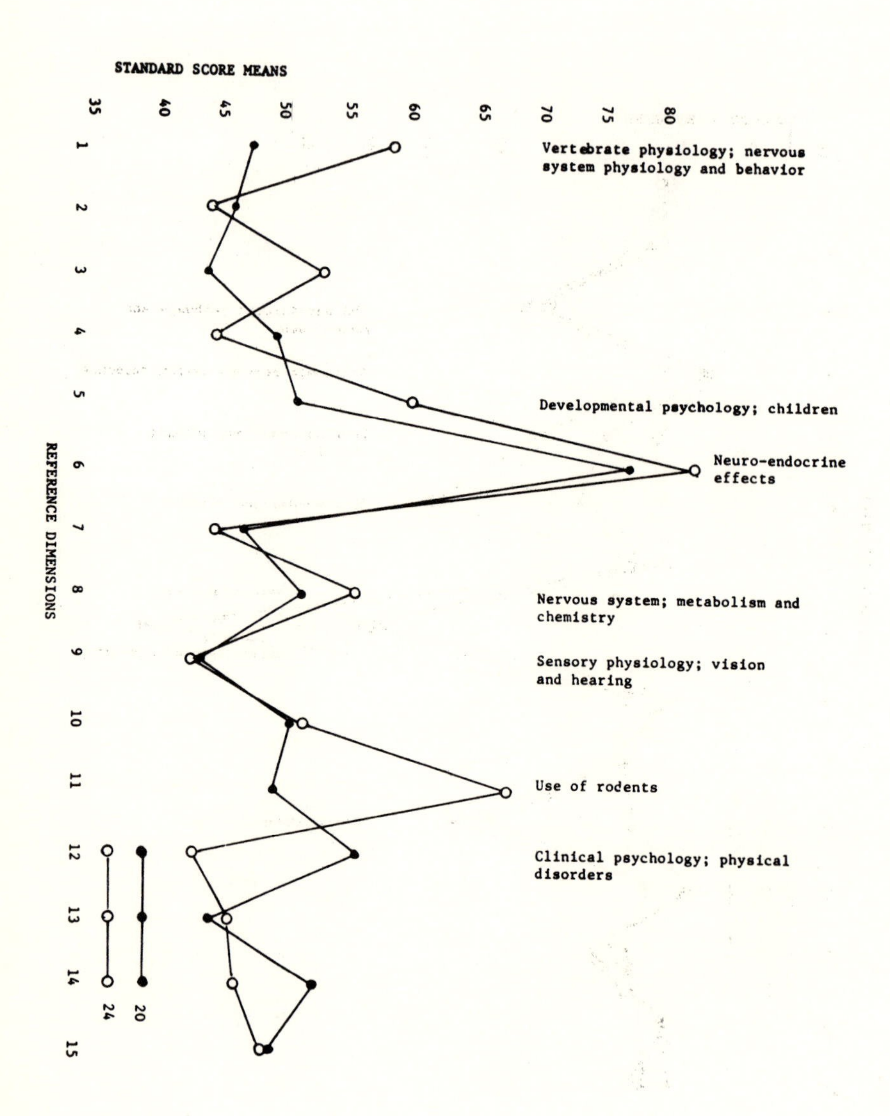

FIG. 1. (Graph 11).

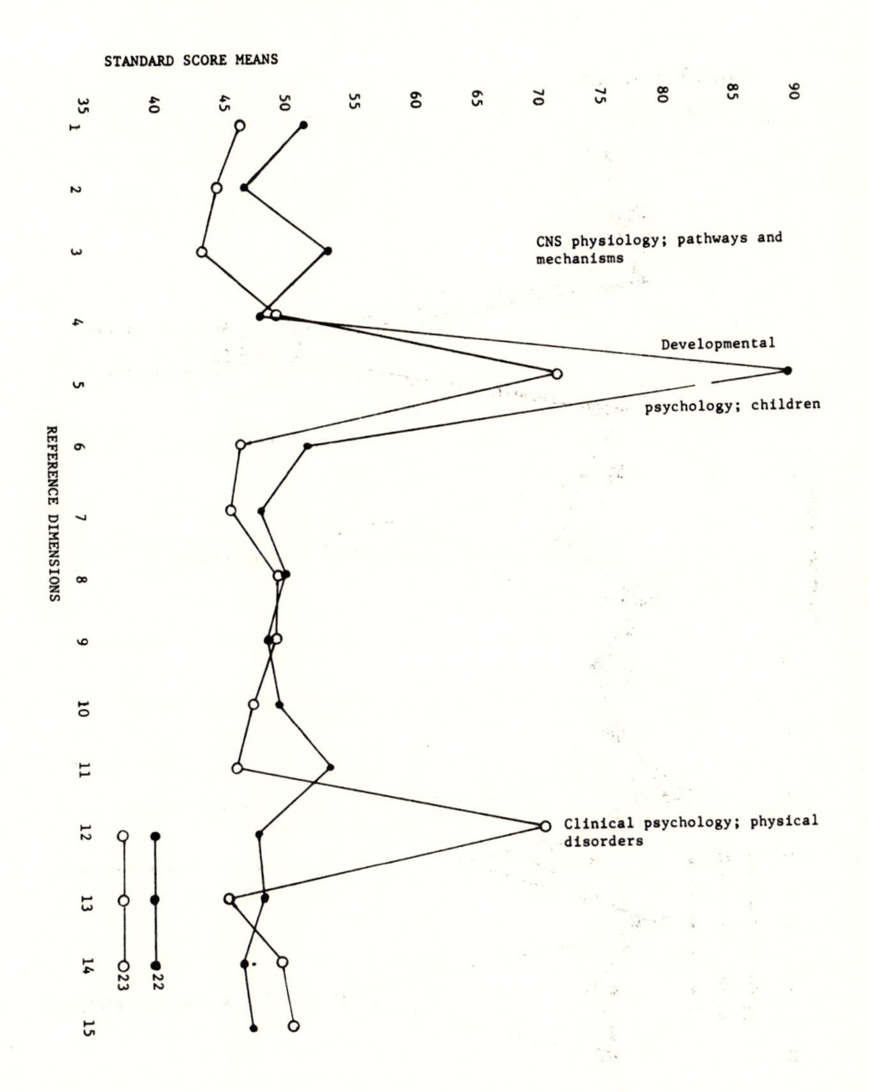

FIG. 1. (Graph 12).

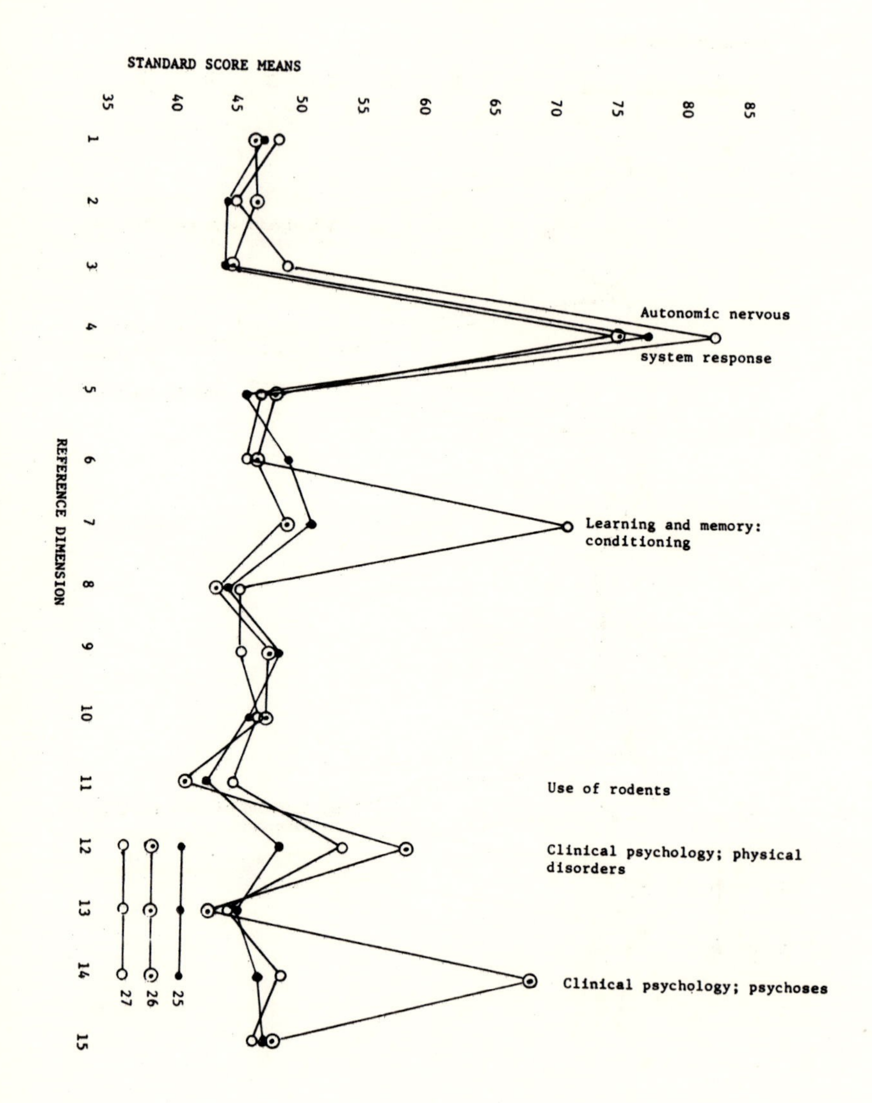

FIG. 1. (Graph 13).

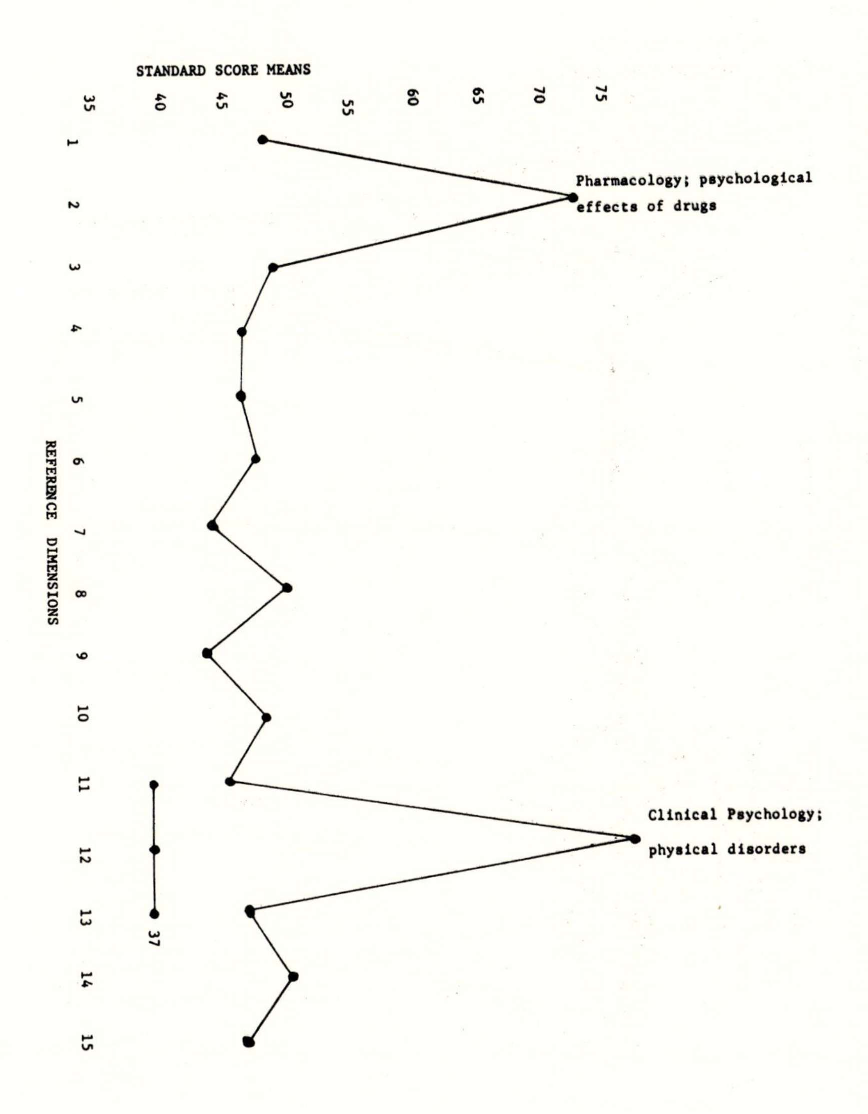

FIG. 1. (Graph 14).

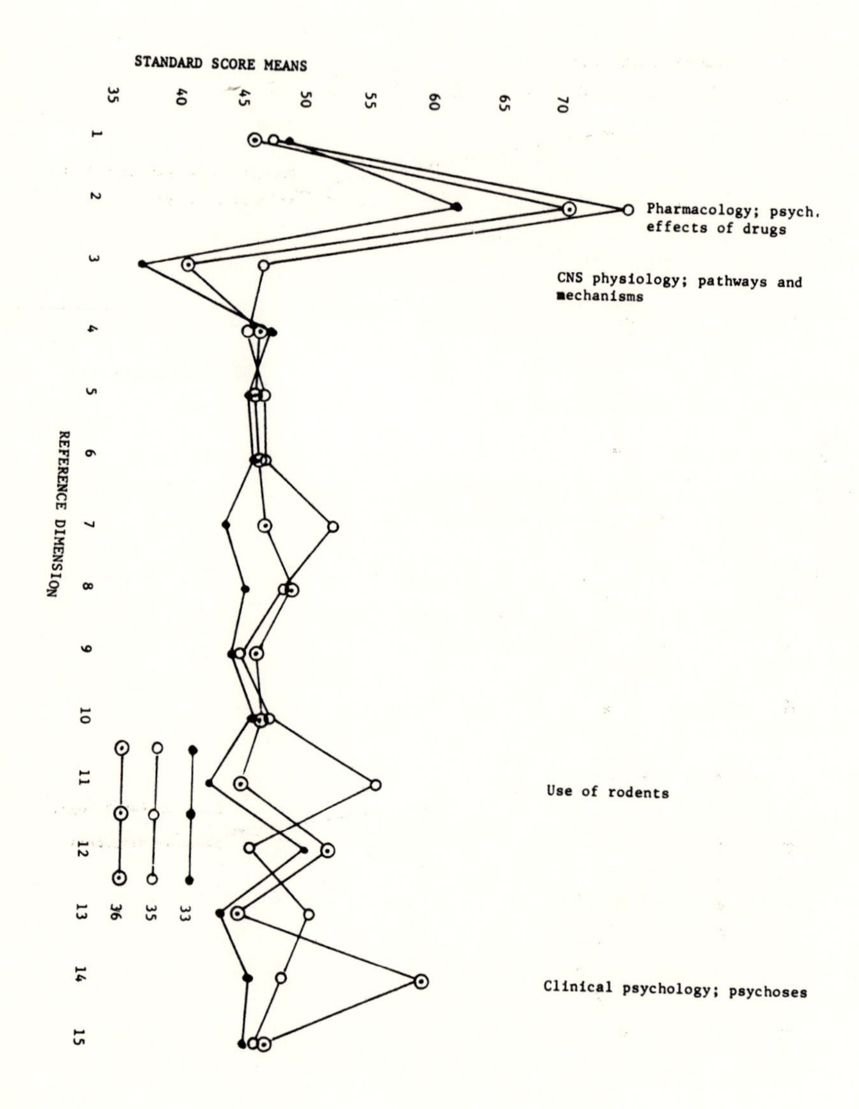

FIG. 1. (Graph 15).

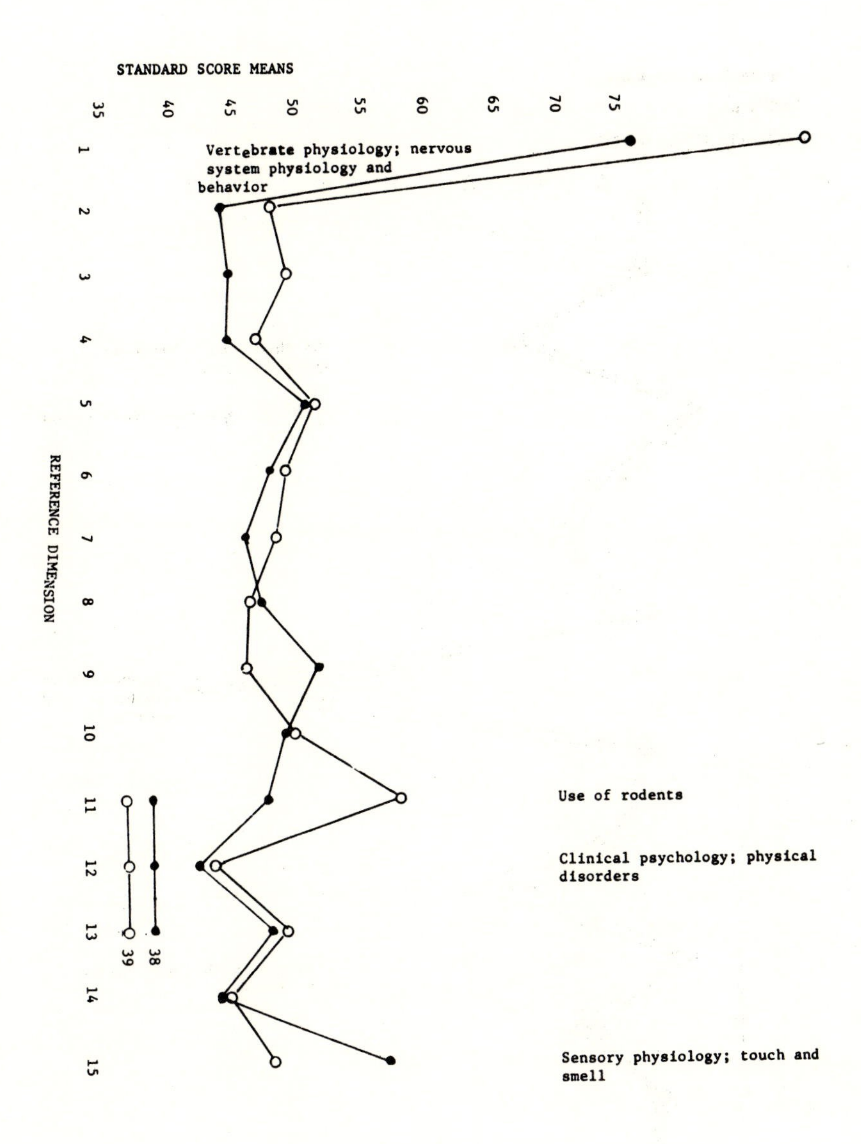

FIG. 1. (Graph 16).

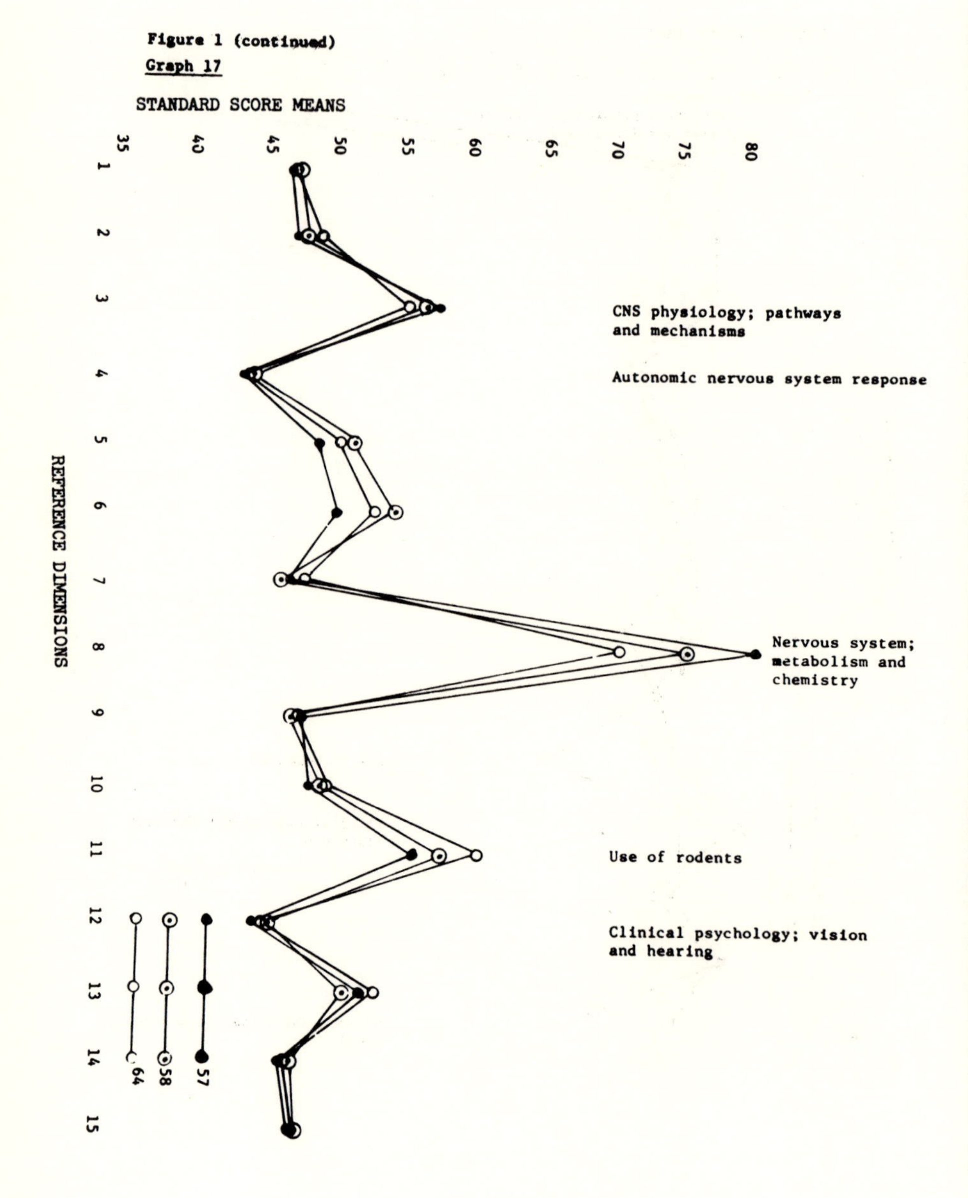

FIG. 1. (Graph 17).

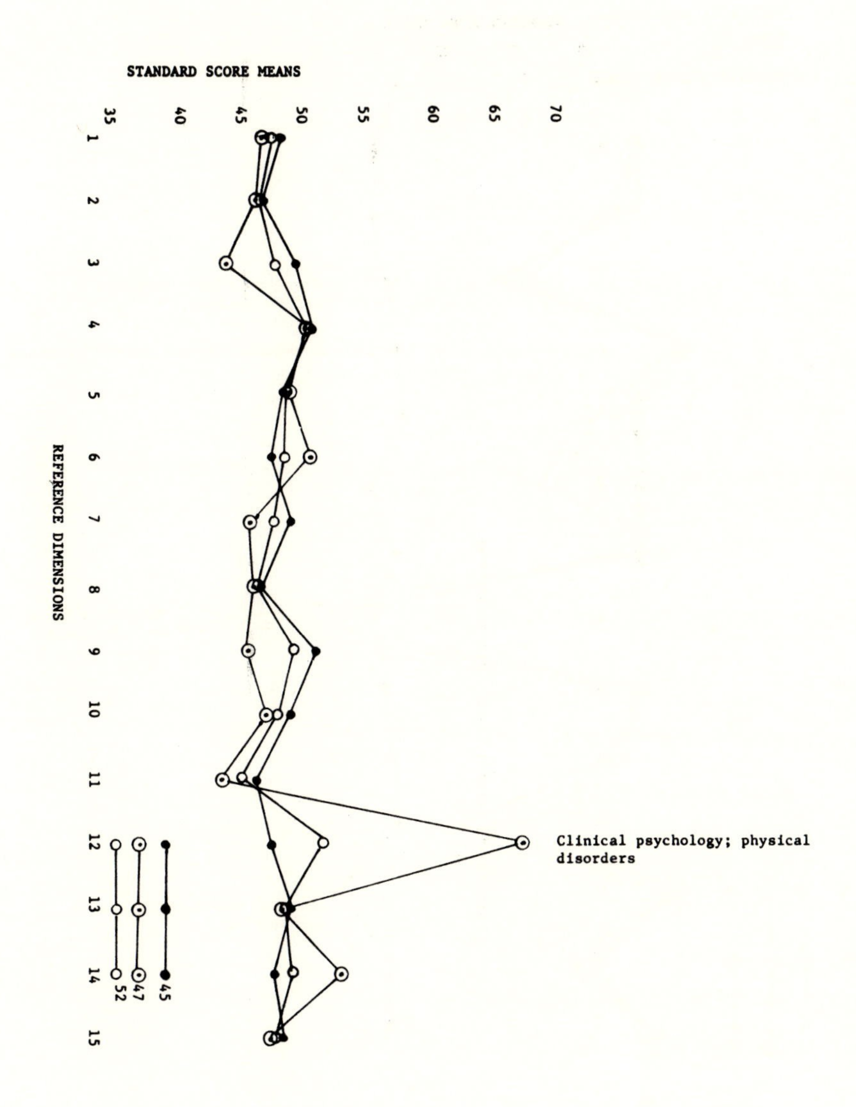

FIG. 1. (Graph 18).

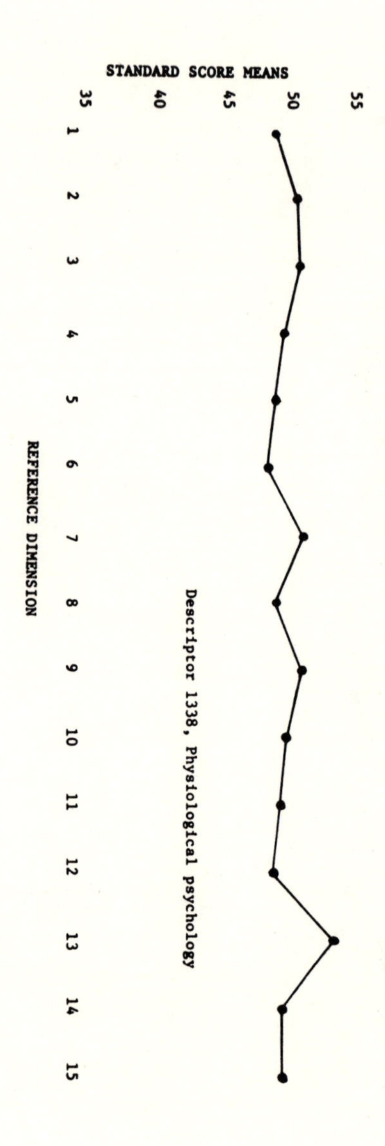

FIG. 1. (Graph 19).

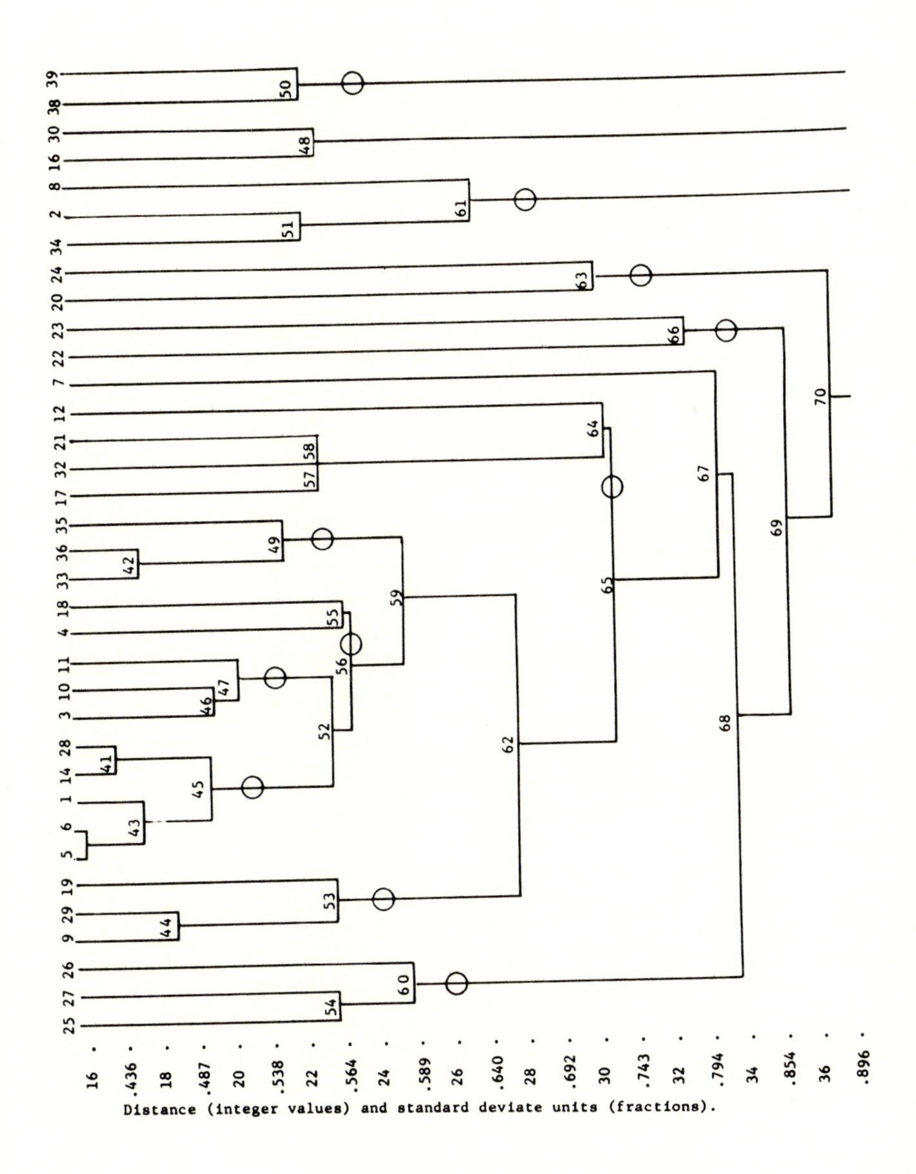

FIG. 2. Dendrogram of the descriptor clusters

SECTION VI

INTERNATIONAL SYSTEMS

The two papers in this section relate to aspects of international systems. The first, by Chase, is a mathematical treatment of the buildup of an arms race. He discusses the objectives of the buildup, whether the opponents seek superiority, damage limitation or assured destruction, and the influence of these objectives on the stability of the arms race. The second paper, by Kaplan, is a discussion of the application of systems theory to international affairs. He discusses theory in general and the use of different types of theory in explanation, and the use of comparative case studies of previous historical systems, as well as computer simulation of the international system.

Both papers are concerned with military conflict between nations. The first one is a mathematical analysis of an arms buildup between two powers, and the conditions leading to war or peace. The second one is a discussion of the use of theory and simulation in the study of relations among multiple powers.

Milton D. Rubin

ON ARMS RACES AND ARMS CONTROL

P.E. Chase*

ABSTRACT

In international relations, action/reaction processes occur in many forms and on many levels. One process always of major concern is the arms race, which has a primary effect upon both war outbreak and upon its antithesis arms control. This paper sets forth first an organized scientific approach to identifying an arms race, second a generalized model for studying the dynamics of the arms race, and thirdly relates the model to arms control parameters and then to international behavior. The study approach of arms races suggested combines the quantitative statistical techniques with the knowledge of the area specialist to determine when a competitive military build up is underway. The statistical techniques are known to work. Also the area specialist has enough "windows" into the study process to ensure that it be realistic. As to the dynamic feedback model employed in the paper, it is based on the assumptions that racing nations watch each other, compare the information about the other's capability with their own, and then adjust the perceived relationship accordingly. Upon applying the model, three new stimulating factors pop forth. Overestimation in intelligence[1] and worst case force posture planning[2] are shown to be stimulating at all times. A set of mutually incompatible military

---

* Manager System Research Department, The Bendix Corporation Aerospace Systems Division, 3300 Plymouth Road, Ann Arbor, Michigan 48107

goals[3] such as both parties seeking damage limitation is particularly stimulating. Seeking superiority is not as stimulating as seeking damage limitation. Seeking assured destruction is not likely to be stimulating in and of itself. Overestimation in intelligence (partly created by secrecy) and worst case planning are thought to be the culprits in the continuing steady state nuclear arms race between the US and the SU. Suggested control parameters are communications, trade, development of second strike weapons, ameliorating gestures, negotiating sound arms control treaties, and cultural exchanges. The relationship of these parameters to US policy is obvious.

This paper sets out to do three things with regard to arms race theory and arms control. First it suggests in recipe fashion how a thorough study of this major international relations problem might be undertaken.* Second, it describes a more general method of studying arms race theory.** Third, using the newer method it aims to identify clearly some primary conflictual behavior within and between governments that must be resolved before arms races can be converted to arms control.

As a guide let us see how we might try to answer three basic questions.

1. Is there competitive military buildup between states?

2. What are the quantitative parameters that control the arms race?

3. What is the appropriate national behavior that

---

* This portion of the paper is based upon an earlier study published in the 13th General Systems Yearbook by the author.[3]

**The model is derived in second paper by the author.[4]

will create a race, slow it down, stop it, or reverse it?

A basic study of arms races must be conducted before a complete set of formal and informal arms control measures can be derived that will preclude or control the race itself. There has been some interest in specific theoretical aspects of arms races on the part of the academic community and isolated general interest in government circles. There is a need, however, for an organized, steady attack at answering the questions posed earlier. The plan is depicted in Fig. 1 as a block diagram.

The first block is concerned with determining the rate and type of military build-up for a state. The second block is a determination if state A and state B perceive each other as a military threat. Given that there are build-ups for both states A and B, and that we have determined a mutual perception of each other as a military threat, we would then be in a position to suggest that an arms race exists between state A and state B. Let us now turn to a more complete description of each of the three steps.

The military build-up for a state might be measured by different units to show elementary variations in rate increase patterns and even wider differences in the characteristics of the race.

We may wish to plot (1) total military procurement dollars; (2) strategic military dollars; (3) the ratio of total military procurement dollars to GNP; (4) the ratio of strategic military dollars to GNP; (5) levels of armaments; and (6) incremental increases of procurement dollars, levels of armaments, and the ratios listed above as (3) and (4). Even men under arms might be a useful measure. Note that the economist might use dollars, the military man levels of armament, and the analyst ratio measures. In some cases the measurement parameters are dependent upon the nature of the data available; in others the purpose of the analysis might suggest the types of measurements.

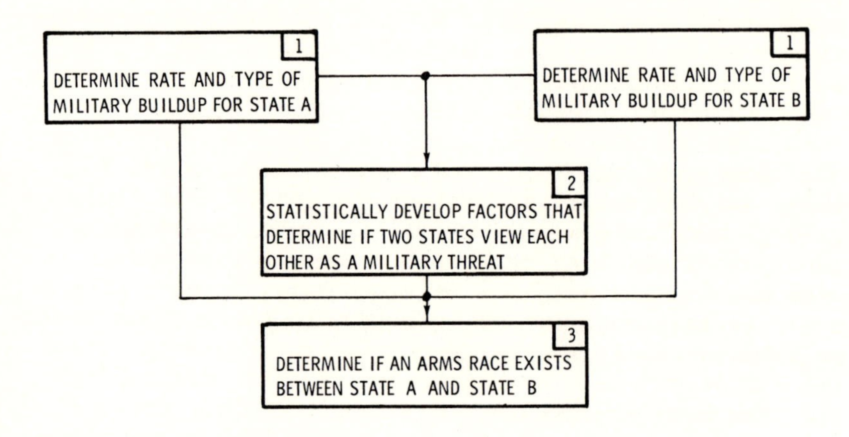

FIG. 1. Is there a competitive military build-up between States A and B?

We may now wish to plot the above parameters as a function of year for each country, and as one country versus another country, with the year noted on the subsequent graph. The plotting of the build-ups as a function of time shows the tendency towards either stability or instability. The plot of each A's armaments versus B's armaments yield interesting conclusions. A straight line (45° slope) indicates an equivalent build-up; a progression towards an asymptote implies one country has stopped racing, and a spiraling as Boulding [1] has suggested infers stability. We may wish also to take decade averages, standardize, and use serial analysis in which the yearly fluctuations about the decade mean give insight into short-range prediction - particularly if it is cyclical.

Some authors, such as Richardson [7] (pp.89, 96, 103), and in particular, Smoker [4] (pp.59, 61; [10], p.162), have made somewhat similar plots, avoiding, in particular, absolute dollars. Smoker, for example, uses arbitrarily weighted defense ratios because they are relative measures and because they conceivably avoid problems of interconversion of fiscal and yearly ratios and inflation and depreciation. A rather interesting plot has been included, as an example of U.S. total military and space expenditures versus S.U. total military and space expenditures from 1945 to 1966 (Fig. 2). The spiraling suggested by Boulding as well as the other interpretations can be readily discerned in these charts.

Another form of military build-up that is well hidden within the total procurement dollars is the qualitative build-up: the expenditures may follow a type which was given above, and yet the actual military capability of the country has improved markedly. This is called a qualitative arms race, and is evaluated by estimating weapons effectiveness. Such an estimation, of course, is known to be difficult; but certainly, for marked differences, it can be made.

An even finer grained type of qualitative arms race is the determination of increases in specific defensive weapons of one country that appear to have a marked

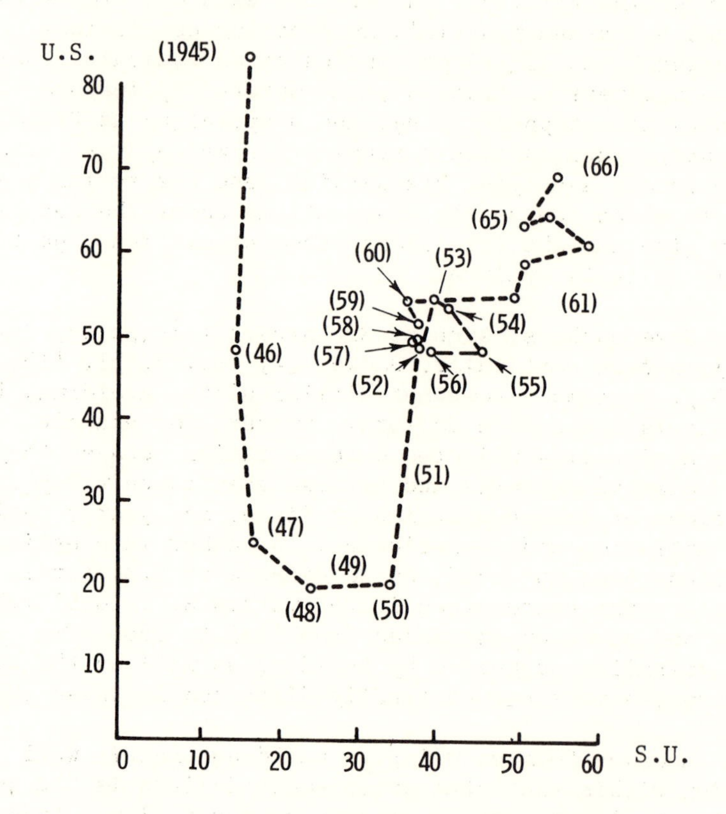

FIG. 2. Annual U.S. Military and Space Expenditures Versus Annual Soviet Military and Space Expenditures, 1945 through 1966
Source: D.N. Ivanoff, Douglas Missile and Space Systems Division, June 1966.

functional relationship with an increase of specific offensive weapons of another country. A simple example is the increase of bombers in the United States and improvement of bomber defense in the Soviet Union.

Returning to Box 2 of Fig. 1, we will list and discuss some of the ways in which countries perceive each other as a friendly ally, or more importantly, as a military threat. A simplistic notion of threat is used here. It includes capability and intent for races with conventional weapons and capability and opportunity (temptation) in strategic races. Rummel [8] in a study of conflict behavior between nations has suggested a set of conflict parameters which are a beginning point for our listing. A conflict parameter is defined as the goals or values of two or more parties which are incompatible and in which both parties are aware of the mutual incompatibility. Some of the parameters that Rummel suggests are antiforeign demonstrations, negative sanctions, protests, severance of diplomatic relations, expulsion or recall of ambassadors, expulsion or rejection of lesser officials, threats, limited military action, troop movements, and acquisitions. In addition to these, other parameters will have to be suggested when considering strategic conflict (e.g. deterrence). Statements of elites (McNamara, Khruschchev, President Johnson, and their peers) constitute one such parameter. Another may be the determination of extreme interest on the part of one country's intelligence in the research and development capability of another country, or the amount of discussion in technical magazines of research and development of a given type of weapons systems with the adversary's functionally opposing weapons type grossly identified.

Whereas the activity represented by Box 1 identified capability, Box 2 identifies the perception of the use of the capability by the other state. For example, the U.S. is an open society, and much of its capability appears in the open press. The information that is not available to the Soviet Union is how the United States intends to use its capability with regard to the Soviet Union. The Soviet Union is secretive about both capability and intent. This, of course, is a distinct assymmetry between the two countries which, in terms of arms races, may not prove to be an advantage to either. One method of determining important threat parameters is factor analysis. Here a large array of the parameter values, listed earlier for both countries A and B, is laid out, with A along the vertical and B along the horizontal. Conceivably, one or two selected parameters, which represented rapid change of procurement, would be added to this array. If it were possible to obtain a large enough sample size of arms races, the correlation values of each entry in the array could be developed. Since all entries would have common units, that is, correlation coefficients, the array can be manipulated as a matrix, and factor analysis performed. Tests could be made to see if enough variance were accounted for by this particular matrix; and if not, procurement is lagged in time; that is, delayed by one or two years behind the manifest parameter. Again, attributable variance could be determined. Hopefully, the attributable variance would be between 50 and 60% for the first one or two factors. If, again, there is not enough variance accounted for, one could rotate orthogonally both matrices individually, that is, with and without the procurement lag, and see if either accounts for a larger percentage of variance. Out of this, statistical factors are obtained that would tell if there was a strong interaction between states A and B in which they were aware of each other, and were watching closely what type of procurement and R and D the other country was conducting. These factors would then be applied to two types of pairs of states to test the prediction capability - one pair in which racing is obviously not present, and one pair in which it is just as obviously present. The process is described below in a number of steps.

1. For nation pairs having high conflict, determine if military build-up is present for year for both.

2. For nations having large weapons increase, determine if conflict with another nation is present,

using conflict parameters.

3. Determine Sets

    a. Conflict (C) and weapons increase (WI)
    b. C, and No WI.
    c. No C, and WI.
    d. No C, and No WI.
    e. C and WI One Nation.

4. Perform detailed case studies over time for set with conflict and mutual weapons increase.

5. For nation-pairs with conflict and weapons increase one nation, lag one or two years to see if other nation increases procurement. For identified lag, add nation-pairs to set a.

6. Divide into two sets

    a = e (lagged)
    b or c or d or e = e (lagged)

7. Correlate for two sets of nation pairs, using parameters of conflict and weapons procurement.

8. Perform a factor analysis for each set to describe races and non-races.

9. Select factors that contain sufficient variance.

10. Select the parameters that correlate highly with the factors, determine means and standard deviations for each interaction or non-interaction as desired.

11. Calculate the discrimination coefficient for each parameter.

12. Apply the maximum likelihood ratio to select the correct hypothesis - race or no race.

13. TEST: Select 30 cases randomly from two groups

and determine predictive ability.

14. Apply technique to present situations to see if races exist.

This approach combines the skills of the traditional, narrative area specialist, who has pertinent knowledge of the nations, the alliances, the regional organizations, and the region itself, with the analytical techniques of the quantitative political scientist. The regional specialist plays a key role in steps one (1) and four (4). The statistical methods of steps seven (7), eight (8), ten (10), eleven (11), and twelve (12) are well established for broad problems of this nature. Upon completion of these fourteen steps there should exist a thorough statistical description of arms races, a method of determining if an arms race is imminent or underway, and a partial listing of primary stimulating and damping parameters.

Fig. 3 blocks out the important steps in determining quantitative control parameters for those mutual military build-ups that were determined to be arms races. The first step (Box 4) is to specify the qualitative characteristics of the arms race. For example, the type of weapons in the race would have to be noted. That is, whether they were offensive versus defensive, offensive versus offensive, or a mix of both. We would have to note whether the weapons were strategic or tactical, the number of countries in the race, the limit on the state of readiness as to whether it was deployment, or further back in sequence, R and D procurement. Finally, of course, we would have to note the time period of the race.

The next step (Box 5) is the development of the quantitative characteristics from the data obtained earlier in step 1 in the determination of the military build-up. Some of the parameters that we might use are aggregated measures of procurement - specific measures of procurement, aggregation of weapons effectiveness, and the like. Our intention now, for each pair of countries

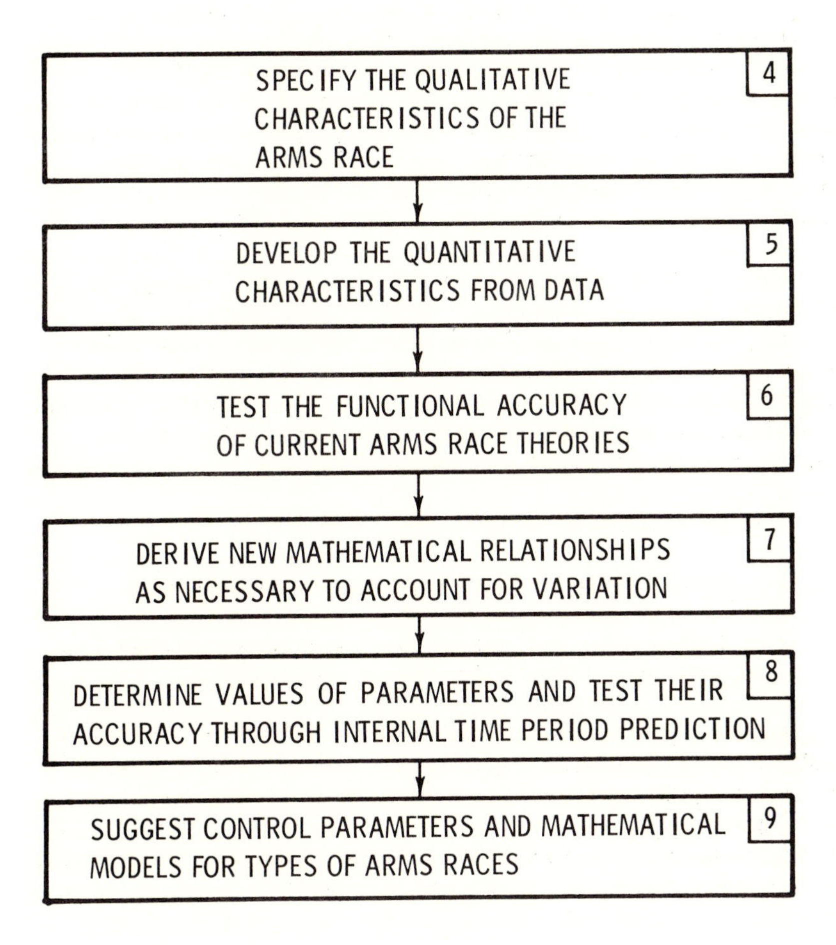

FIG. 3. What are Quantitative Parameters that control the arms race?

selected, would be to take the developed quantitative characteristics and test the functional activities of arms race theories as noted in Box 6. For example, Richardson's equations suggest that build-ups occur in a linear fashion, a relationship he warned against. If we are able to find any logical measures that give us linear build-ups, then we would certainly keep the functional factor of Richardson's equations. There are other arms race theories that we could use (McGuire [6], Caspary [2], and in particular, Smoker [10]).

None of the above theories are general enough to permit us to study the basic interaction pattern of watching (intelligence), comparing military capability (analysis of central war) and adjustment of the posture (new procurement and R and D). We will describe briefly an analytical approach derived elsewhere (4) that is general enough to serve our purposes.

The general assumptions are twofold:

1. Interaction exists between the two states.

2. Each state evaluates the unbalance between the states and adjusts in some fashion.

A variety of equation pairs can be derived from the feedback loop shown in Fig. 4. If we set the adjustment functions $G_1(S)$ and $G_2(S)$ proportional to the unbalance, the comparison functions $H_1(S) = H_2(S) = 1$, the watching functions $E_1(S) = E_2(S) = 1$, and use absolute procurement dollars as a measure, we obtain Richardson's equations.

$$\frac{dY_1}{dt} = K_{11}\ (K_{22}K_{13}Y_2 - K_{12}Y_1) \qquad (1)^*$$

$$\frac{dY_2}{dt} = K_{21}\ (K_{12}K_{23}Y_1 - K_{22}Y_2) \qquad (2)$$

---

*See list of symbols at end for definitions.

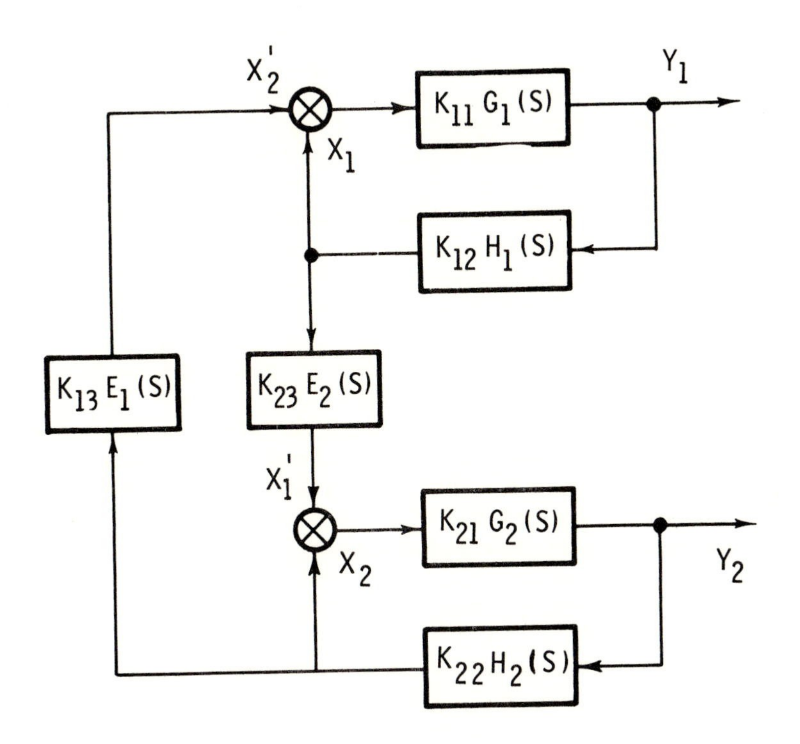

Equations

$$dY_1/dt = K_{11}(K_{22}K_{13}Y_2 - K_{12}Y_1)$$

$$dY_2/dt = K_{21}(K_{12}K_{23}Y_1 - K_{22}Y_2)$$

Stability Conditions

$$K_{11}K_{12} + K_{21}K_{22} \geq 0$$

$$K_{13}K_{23} \leq 1$$

FIG. 4. The Feedback Model

The stability conditions are

$$K_{11}K_{12} + K_{21}K_{22} \geq 0 \tag{3}$$

$$K_{13}K_{23} \leq 1 \tag{4}$$

The equations which show the correspondence to Richardson's parameters are

$$K_{11}K_{22}K_{13} = k; \; K_{11}K_{12} = \alpha \tag{5}$$

$$K_{21}K_{12}K_{23} = \ell; \; K_{21}K_{22} = \beta \tag{6}$$

Richardson's equations appear now as a special case - a pair of linear differential equations. A simple additional economic constraint, the law of diminishing returns in weapons purchasing, suggests a modification of the adjustment functions $G_1(S)$ and $G_2(S)$

$$\text{Let } G_1(S) = (C_1 - M_1Y_1)\frac{1}{S} \tag{7}$$

$$G_2(S) = (C_2 - M_2Y_2)\frac{1}{S} \tag{8}$$

where C is the cut-off of sources available for military procurement M is cost per unit of maintaining forces.

It can be shown

$$\frac{dY_1}{dt} = K_{11}K_{22}C_1Y_2 - K_{11}K_{12}C_1Y_1 + K_{11}K_{12}M_1Y_1^2 - K_{11}K_{22}M_1Y_1Y_2 \tag{9}$$

$$\frac{dY_2}{dt} = K_{21}K_{12}C_2Y_1 - K_{21}K_{22}C_2Y_2 + K_{21}K_{22}M_2Y_2^2 - K_{21}K_{12}M_2Y_1Y_2 \tag{10}$$

The equations above represent the general introduction of the economic law of diminishing return and not the artificial introduction to satisfy the intuition as performed by Caspary [2]. The analyses of this pair of non-linear equations and other pairs obtained from the introduction of other functional forms of $E_i(S)$, $G_i(S)$, and $H_i(S)$ are underway.

This more powerful approach to deriving Richardson's equations has cast in a natural way a different light upon the stability equations. The four parameters of Richardson have now been replaced with 6 parameters - three for each nation. These are now parameters of watching, comparing, and adjusting, which is the most basic division of competitive behavior. Stability equation (4) interestingly enough is a function of the intelligence parameters K_{13} and K_{23} only. Stability equation (3) is a function of the comparison parameters K_{12}, K_{22} and the adjustment parameters K_{11}, K_{21}.

Some stimulating and damping behavior can be found readily from these stability equations.

SITUATION 1. Overestimation of intelligence.

If intelligence is complete for both states and unquestionably accepted,

$$K_{13} = K_{23} = 1$$

and $$K_{13}K_{23} = 1, \text{ which is stable.}$$

We have tacitly assumed that intelligence and uncertainty are the same. Suppose that even with complete intelligence, the analyst plays safe by putting in a safety factor $a_1>1$ and $a_2>1$.

$$a_1a_2K_{13}K_{23}>1$$

Then equation (4) is unstable. Unfortunately the tendency in the intelligence community when $K_{13}<1$ and $K_{23}<1$ is to estimate in such a fashion that $a_1K_{13}>1$ and $a_2K_{23}>1$. An interesting hypothesis to test is: Whenever one nation feels a threat from another,

$$a_i\ K_{i_3}>1.$$

Corollary:

Whenever two nations feel mutually threatened,

$$a_1K_{13}>1$$

$$a_2K_{23}>1$$

SITUATION 2. Two nations seek simple superiority in a perceived outcome and can increase their arsenal.

Satisfaction with military comparison
$K_{12}>1$ and $K_{22}>1$.
Dissatisfaction with military comparison
$K_{12}<1$ and $K_{22}<1$.
For complacency $K_{12} = K_{22} = 0$.
The adjustment parameters ($K_{11}K_{21}$) are positive if the nation is able to make its desired adjustment. They are negative whenever the nation is unable to adjust. Equation (3) becomes for only one dissatisfied state: $-K_{12}K_{11} + K_{22}K_{21}<0$.

If the satisfied nation is complacent, $K_{22} = 0$ and $-K_{12}K_{11}<0$, which is unstable.

As nation 1 catches up and passes, nation 2 becomes dissatisfied, nation 1 indifferent, and the situation is still unstable.

SITUATION 3. Two nations seek complete damage limitation counterforce with respect to each other.

Three conditions exist.

A. Both nations are able to adjust.
B. One nation only is able to adjust.
C. Neither nation is able to adjust.

For condition A:

$$K_{12}<1;\ K_{22}<1;\ K_{21}>1;\ K_{11}>1$$

and Equation 3 becomes

$$-K_{12}K_{11} - K_{22}K_{21}<1$$

Condition A is unstable.

For Condition B:

$$K_{12}<1;\ K_{22}<1;\ K_{21}>1;\ K_{11}<1$$

$(-K_{12}K_{11} + K_{22}K_{21})$ may be equal to, greater than or less than 0, depending on relative size of $K_{12}K_{11}$ and $K_{22}K_{21}$.

Actually the "race" becomes one sided.

For Condition C:

$$K_{12}<1;\ K_{22}<1;\ K_{21}<1;\ K_{11}<1$$

and

$$K_{12}K_{11} + K_{22}K_{21}>0$$

because neither nation can adjust upwards. Stability exists.

SITUATION 4. Two nations seek assured destruction.

Situation 4 differs from Situation 3. In Situation 3 the nation always wants to adjust *if it can*. In Situation 4 either or both nations can be complacent even when they can adjust upward. Both Situations 3 and 4 are alike until either or both states acquire what they perceive is assured destruction. At this point

$$K_{12}>1;\ K_{22}>1;\ K_{21}>1;\ K_{22}>1$$

and equation 3 is stable.

The equation does not state that stability must result from mutual objectives of assured destruction. It only states that when obtained the nation behaves in a stable fashion.

In summary of the goals sought by each nation (Situation 2 - superiority, Situation 3 - damage limitation, Situation 4 - assured destruction), damage limitation will always be destabilizing while the nation has capability, superiority will possibly lead to instability but not necessarily so, while assured destruction will lead to stability (all other things being equal) when a nation reaches that condition.

Why then do two nations continue an arms race when both have an assured destruction capability?

1. Each nation is not satisfied with assured

destruction and wishes to obtain identifiable superiority as a position from which to bargain. Equation (3) becomes unstable.

2. Because the intelligence community would rather err "on the high side", Equation (4) is unstable and couples to Equation (3) because

3. A nation "hedges its bet" by procuring more than it needs (either in actual weapons or in research and development) and this effectively drives Equation (3) unstable. Hedging is equivalent to dissatisfaction mathematically in Equation (3).

Is it little wonder that the current arms race between the Soviet Union and the United States continues with only temporary periods of stability?

This brief discussion indicates the value of studying arms races with feedback control models in which various equations can be derived and when tractable solved explicitly. However, the value lies more in deriving the stability conditions to determine the effect of external nation behavior and internal government practices upon arms races.

We have seen how to derive arms control parameters both from a statistical analysis of past arms races and mathematical analysis of established behavioral practices. We might list some of the control parameters tentatively established by this analysis and the analysis of others in an unordered fashion.

Stimulus

Secrecy

Threatening Gestures

Overestimation in Intelligence

Selection of Mutually Incompatible Military Comparison Criteria

Worst Case Force Posture Planning

Damping

Communication

Trade

Cultural Exchange

Develop Second Strike Weapons

Arms Control Treaties

Ameliorating Gestures

In light of these control parameters, what is some appropriate national behavior that might prevent a race, slow it down, stop it or reverse it? Nations might:

1. Increase trade, communication, and cultural exchange with the Soviet Union. This is "building bridges."
2. Replace threatening with ameliorating gestures where possible.
3. Be less secretive (particularly the Soviet Union).
4. Refrain from overestimation in intelligence, worst case posture planning (hedging), and seeking either damage limitation or simple superiority.
5. Seek assured destruction. Develop invulnerable second strike weapons and forego deploying ABM's.
6. Obtain arms control agreements consonant with the five points above.

"Nuclear" arms control might result if the above practices are followed. Attendant reassurance should be gained if an invulnerable assured destruction is maintained. It would seem, however, that an important basic arms control question is: How much is the assured destruction capability eroded if the nations began practicing the other five points?

SYMBOLS

a_1 = intelligence "safety factor" for State 1.
a_2 = intelligence "safety factor" for State 2.
C_1 = cut off of sources available to State 1 for military procurement.
C_2 = cut off of sources available to State 2 for military procurement.
$E_i(S)$ = intelligence factor for State i.
$G_i(S)$ = controller (arms-level-adjustment) factor for State i.
GNP = gross national product.
$H_i(S)$ = effectiveness criteria factor for State i.
k = incentive of State 1 to accumulate arms because of the strength of State 2.
$K_{11}G_1(S)$ = functional relationship employed by State 1 to adjust its arms level in accord with power imbalance.
$K_{21}G_2(S)$ = functional relationship employed by State 2 to adjust its arms level in accord with power imbalance.
$K_{12}H_1(S)$ = functional relationship defining effectiveness against State 2 in terms of the arms level of State 1.
$K_{22}H_2(S)$ = functional relationship defining effectiveness against State 1 in terms of the arms level of State 2.
$K_{13}E_1(S)$ = transfer function for actual arms level of State 2 with respect to intelligence capability of State 1.
$K_{23}E_2(S)$ = transfer function for actual arms level of State 1 with respect to intelligence capability of State 2.
ℓ = incentive of State 2 to accumulate arms because of the strength of State 1.
M_1 = cost per unit of maintaining forces for State 1.
M_2 = cost per unit of maintaining forces for State 2.

S = Laplace transform operator
X_1 = effectiveness of State-1 arms against State 2.
X_2 = effectiveness of State-2 arms against State 1.
Y_1 = arms level of State 1.
Y_2 = arms level of State 2.
α = satisfaction of State 1 with its own arms level.
β = satisfaction of State 2 with its own arms level.
Δ_i = imbalance of military power as seen by State i on the basis of intelligence information.
X_1' = effectiveness of State-1 arms against State-2, as perceived by State 2.
X_2' = effectiveness of State-2 arms against State-1, as perceived by State 1.

REFERENCES

[1] Boulding, K.E., *Conflict and Defense*, New York, Harper and Row (1962).
[2] Caspary, W.R., Richardson's Model of Arms Races, Description, Critique, and an Alternative Mode, *International Studies Quarterly, 2* (10), Detroit, Wayne State University Press (March 1967).
[3] Chase, P.E., The Relevance of Arms Race Theory to Arms Control, *General Systems, XIII* (1968).
[4] ———, Control Theory and the Nuclear Arms Race, *Bendix Technical Journal, 1* (3) (Autumn 1968).
[5] Ivanoff, D.N., *Threat Analysis Briefing, 1,* Douglas Missile and Space Systems Division, DAC-57902, Huntington Beach, California (June 1966).
[6] McGuire, M.C., *Secrecy and the Arms Race*, Cambridge, Mass., Harvard University Press (1965).
[7] Richardson, L.F., *Arms and Insurgency: A Mathematical Study of the Causes and Origins of War*, N. Rashevsky and E. Trucco (eds.), Pittsburgh, Boxwood (1960).
[8] Rummel, R.J., Dimensions of Conflict Behavior Within and Between Nations, *General Systems, VIII* (1963).

[9] Smoker, P., Fear in the Arms Race, *Journal of Peace Research* (1964).

[10] ———, Trade, Defense and the Richardson Theory of Arms Races: A Seven Nation Study, *Journal of Peace Research* (1965).

SYSTEMS THEORY AND INTERNATIONAL RELATIONS*

Morton A. Kaplan

ABSTRACT

The application of systems theory to the field of international relations is discussed. The use of simplified models is considered, to represent complex systems such as brains in order to gain some insight into the behavior of the complex systems. A distinction is made between mechanical and homeostatic equilibrium, and the classificatory use of the latter is discussed. However, he believes in a comparative theory of systems rather than in a general theory of systems, in that different kinds of systems require different kinds of theory. The pathology of regulatory mechanisms is considered with regard to the biological as well as mental phenomena of any acting entity. He has modelled six types of international systems and has studied the kinds of equilibria in the systems. In particular he discusses in his paper the "balance of power" theory and the nature

---

* Paper prepared for delivery at the American Association for the Advancement of Science, Dallas Texas, December 28, 1968. Portions of this paper are taken from "Systems Theory," in Charlesworth, *Contemporary Political Analysis*, Free Press, 1967, pp. 150-163; "Systems Theory and Political Science," in *Social Research*, Vol. 35, No.1, Spring, 1968, pp. 30-47; "The Systems Approach to International Politics," in Kaplan, *New Approaches to International Relations*, St. Martin's Press, 1968, pp. 381-404.

of the actors, and why, after 1870 their nature was such that the "balance of power" was no longer effective. With a computer, it is possible to build the features of a theory into a program, and to explore the interrelationships of propositions concerning the strategic structure of a model of the international system. In one computer model, while a hegemony-inclined actor at first made the system unstable, it became clear that eventually he would have been eliminated.

I. Despite the use of the honorific word "theory" in "systems theory," systems theory is not a theory, although the concepts that are part of it are useful in constructing theory. Systems theory, although it provides a perspective concerning the real world, no more gives rise to specific propositions concerning it than does the infinitesimal calculus give rise to specific propositions of physics. Advice to a political or social scientist to use systems theory to solve a problem, even when it is the appropriate methodology, would advance them as far but no farther than would advice to a physical scientist to use the methods of science.

When the concepts of the systems theory are used to construct theory in substantive areas of political science, they can be most helpful; when they are used to evade substantive problems, they can be misleading or harmful. There is no such thing as theory in general; there is only theory about some specific subject matter.

II. Systems theory of the type used by the author was originally developed in neurology by scientists interested in brain behavior. The most concise and original account of systems theory occurs in W. Ross Ashby's book, *Design for a Brain*. Ashby and his colleague W. Grey Walter developed systems theory to explore analogues for certain aspects of the brain's functioning. Thus, for instance, Grey Walter designed a machine with a motor and headlights capable of plugging itself into a wall socket to recharge its battery when its headlights began to wear down and of plugging itself into a gas pump when its engine began to run out of fuel. By employing analogues of this kind, which Ashby and Walter did not mistake for the real brain system, they were able to explore how to produce simplified versions of certain observed behaviors. They used concepts previously employed but fused them into a coherent scheme of conceptions now called systems theory.

The reason for using a machine to simulate simple behaviors characteristic of only some of the brain's operations rather than the more complex behaviors of the entire system is transparent. It is very difficult to study exceptionally complex systems; the brain is a remarkably complex system with some ten billion or more cellular components and numerous and very poorly understood linkages among these components. Theoretical depth requires simplicity; many students of science would argue that Galileo's breakthrough, which launched the growth of modern theoretical science, was the choice of a simple problem that the mathematics of his age was capable of solving. In any event, it is well known that two-body problems are easy to solve, that multi-body problems are very difficult and extremely complicated to solve, and that the larger the number of bodies considered, the more *ad hoc* and the less general the solutions.

Physicists are fortunate that their simple problems can be found for the most part directly in the world of nature. Ashby and Grey Walter were not able to isolate within the brain structure the simple elements of the problems they desired to work with; they therefore built machines that were simple and that manifested some of the behaviors to be explored. The solutions achieved by such devices are heuristic in at least two senses: the machine model chosen may indeed manifest the behavior observed in reality but as the consequence of a different structure and a different set of operations. The machine model quite often employs components that have no exact counterpart in nature. This type of test for internal structure and systems properties is indirect. Were it possible to study the system directly by repeatedly assembling and disassembling it, by testing the parts independently and in sequence; were it possible to use mathematical causal analysis; we would have more confidence in our conclusions about such systems. Where such methods are not available or at least not appropriate, model building may be quite useful. The simplicity of the model is not necessarily a shortcoming.

III. Some of the most important concepts of systems theory deal with equilibium. However, it is important to point out that systems theory does not imply the desirability or the actuality of system stability. It is true that the systems that do persist usually have the greatest importance for us; this reason would no doubt justify a concern with systems in equilibrium. There are, however, prior and theoretically more interesting reasons for concern with the problems of equilibrium. Differences in types of equilibrium are important in distinguishing differences in types of systems. In addition, there is an important reason stemming from the principle of economy. Many more non-isolated modern social systems fail to persist through a time period of properly chosen duration than those that do persist. Although the parameter values that produce instability, that is, those external inputs inconsistent with a given local stability, are large in number and differ in multifarious ways from case to case, the conditions producing stability are much more limited in number. Thus, concern with local stability, at least in the initial stages of inquiry, focuses attention on a limited number of systems, a limited number of variables, and a limited range of variation. Although the problem of inquiry may still be most difficult, it is much more focused and manageable than is a concern with problems of instability. Moreover, to the extent that we can understand systems in equilibrium, we have a fulcrum for the study of systems in disequilibrium. If we can construct a theory for a system or type of system, as a system in equilibrium, we can then inquire how individual variations in the parameters will produce deviant or unstable behavior. To know why a system changes, develops, or breaks down, it is surely helpful to know why the conditions of change are inconsistent with the prior states of the system. If we cannot answer this latter question, it is doubtful whether we have correctly assigned the reasons for instability or change.

Although the physicist usually deals with systems in mechanical equilibrium, the social scientist, if he deals with a system in equilibrium, invariably finds it in some variety of homeostatic equilibrium. This is an important difference, for, although the concept of mechanical equilibrium conveys only limited information and is only of limited explanatory power, the concept of homeostatic equilibrium is merely a classification device. When a physicist says that two equal forces will cancel each other, he does not merely assume their equality from the cancellation. He has an independent measure for the equality. This is not true of systems in homeostatic equilibrium, whether they are biological or social. For instance, there is no independent measure that will establish an equality between the amount of perspiration and the lowering of body temperature, although it is true that an underlying mechanical equilibrium does exist. President Kennedy's response to the attempted rise in steel prices produced a roll-back to the old price level; there is, however, no meaningful way in which it can be said that his activity was equivalent to the reduction in price. If we wish to explain either biological or social behavior, we require a description and an explanation of the processes that characterize a particular system of equilibrium or control.

This may also help to explain why systems theory should not be expected to imply a general theory of all systems. Although general systems theory does attempt to distinguish different types of systems and to establish a framework within which similarities between systems can be recognized despite differences of subject matters, different kinds of systems require different theories for explanatory purposes. Systems theory not only represents a step away from the general theory approach but also offers an explanation for why such efforts will likely fail. Thus, the correct application of systems theory to politics would involve a move away from general theory toward comparative theory. That is, using systems theory one would search for different theories for the explanation of different types of systems.

IV. Nonetheless, classification of homeostatic equilibria into subtypes does provide some limited information that explains how an empirical theory of values may be developed. Consider an ordinary homeostatic system such as the automatic pilot in an airplane. If a plane deviates from level flight while on automatic pilot, the automatic pilot will sense this and, by the application of negative feedback, will adjust the flight pattern of the plane back to level. Consider, however, the case in which the automatic pilot has been incorrectly linked to the ailerons of the plane. If the plane now deviates from level flight, the automatic pilot will sense this. It will now made an adjustment, but the adjustment, instead of bringing the plane back to level flight, will throw it into a spin.

In principle it would be possible to build an ultrastable automatic pilot the behavior of which was not critically dependent on the linkages to the ailerons. That is, the automatic pilot could be so built that it would reject its own behavior patterns if these increase the deviation from level flight. It could then "search" for a set of behaviors that would restore the level character of the flight and, when it found it, continue to use it so long as it maintained the critical variable within the established critical limits for variation. Such an ultrastable system would have distinct advantages for survival over merely stable homeostatic systems. Even ultrastability, however, is not sufficient for complex biological survival. Ashby, therefore, applies the term "multistability" to those cases where the part functions of the system are themselves individually ultrastable and where they can therefore "search" relatively independently for those critical behaviors consistent with the maintenance of the system. It is obvious that complex social systems must also be multistable.

Multistable systems that operate at the conscious level, that is, that employ abstract generalization and universal criteria, are valuational and capable of moral behavior. When these concepts are combined with the concept of perfect information, it is possible to use systems theory to develop an empirical critique of values.

One version of systems theory attacks the problem in the following manner. The needs of the actor's system are set by its structure. The objectives of the actor's system are set by its needs in its environment as it understands that environment and its needs. The objectives of the actor's system provide its values. The objectives that, in fact, would satisfy the needs of the actor's system are valuable for it.

Is illness pathological? It may be considered pathological if resources that will restore the previous state of health are available to the ill person, for instance, if medicine can be purchased at the drugstore. Information concerning the availability of the medicine may set off a chain of events leading to the purchase of the medicine and the restoration of the state of health. However, consider the case where the medicine is not available. Is the condition pathological? It may be regarded as pathological in the sense that the morbid condition would be subject to correction if the medicine *were* available. This definition, however, is different from the first. In the first case, the input of information is sufficient to correct the condition. In the second, information is insufficient for that purpose. In either case, medicine is valuable for the actor system.

It may be objected that illness is pathological regardless of the availability of medical aid. However, from the standpoint of regulation, the system may be regulating optimally under given conditions. Consider still a third case, namely, the physiological state of brain damage resulting from carbon monoxide poisoning. Is this state pathological? Obviously the actor system cannot regulate itself to correct this condition. Was the dodo bird pathological because it was inadequate for purposes of survival? It surely was a victim of poor heredity. Carbon monoxide poisoning is not hereditary. But the consequences of carbon monoxide poisoning are as irreversible as those of hereditary endowment. Any non-pathological system, however, will attempt to avoid carbon monoxide poisoning.

Thus, from the standpoint of the regulatory mechanisms, a mechanism is pathological only if it does not produce optimal behavior under the existing or probable future conditions. A neurosis is pathological only if it regulates the personality system inefficiently when compared to the practicable alternatives. Neurotic adaptation involves the withdrawal of regulatory capacity from the metatask regulatory system. Repression, for instance, is a task-oriented mechanism. Therefore, the capacity necessary to recognize and to adapt to changed or improved circumstances is weakened. Moreover, the adaptation, once acquired, works, even if inefficiently. It permits some dangers to the system to be avoided. It uses automatic ready-at-hand responses which free the limited capacity of the regulatory system for other necessary life functions. A changed system of responses would involve many unknowns. Particularly, given the great tension of the neurotic system and the implicit, although not always conscious, recognition of its own inadequacies, it fears and resists efforts to change it or information indicating the desirability of change. Often information of which the individual has no conscious awareness is most influential in leading to action. Thus there are times when the individual feels compelled to act by forces over which he lacks conscious control. At still other times, beliefs are influenced by factors of which the individual lacks conscious awareness. This last qualification clears up a difficulty. Since it also makes some sense to say that the valuable is something that is valued, the discrepancy between a person's belief concerning his values and the valuable impugns the pragmatic adequacy of a definition based upon actor system needs. If, however, beliefs will correspond with needs under conditions of complete information--and this is a theoretically meaningful hypothesis, even if not fully testable, (except in simple models)--the discrepancy is not fatal. Moreover, if a pathological regulatory mechanism can be demonstrated in those cases in which conscious awareness of information concerning systems needs does not eliminate the discrepancy, the conclusion would be reinforced.

However, we are not dealing with Grey Walter's simple machine. We will not actually be able to carry out many of the tests required, although we are able to talk about them, I believe, in a meaningful fashion. We may be, and probably are, so tightly bound by our cultural environment that we cannot distinguish the real choices adequately. One can, however, conceive of a hierarchy of thought experiments designed to explicate the concept of the valuable. We can recognize that it may not pay the neurotic to overcome his neurosis, that the cost of change may be too high in some sense. But would we not say that at some earlier point in the neurotic's career, if he had been able to experience imaginatively alternative life patterns, he would have chosen to have been placed in a different environment so that the neurosis would not have developed? In the same way, with respect to social choices,we are bound by the history and environmental constraints of our system. But can we not state that we prefer an environment with fewer constraints where we might choose more freely from among available alternative systems? This may not be a practicable choice for those of us whose lives are so imbedded in a particular system that we cannot detach ourselves except at great cost, just as the neurotic may be unable to detach himself from his neurosis except at a cost that is too great. But surely it should be possible to recognize that some systems impose such great costs upon some individuals that the systems are unjust, at least as contrasted with theoretical alternatives, although perhaps not when contrasted with real alternatives within the given environment. This at least gives us a standard by which we can reach beyond the practical exigencies of the circumstances we are embedded in and by which we can judge the society in which we live.

Would we have such preferences? One cannot state this with assurance. But surely there is no *a priori* reason why individuals cannot have psychological and social system needs that are best satisfied if other individuals lead happy and useful lives, that is, needs that would reflect interdependent utility schedules. Since there is as much evidence that such interdependencies exist as for the contrary hypothesis, one cannot contend that the evidence alone negates it. Moreover, it can be shown, at least for certain kinds of social situations, that moral rules are required for society.

V. The writer has used systems models or theories to analyze six types of international systems, two of which have existed in the past and four of which are hypothetical. The models utilize five sets of variables: the essential rules, the transformation rules, the actor classificatory variables, the capability variables, and the information variables.

The essential rules of the system state the behavior necessary to maintain equilibrium in the system--thus they are essential. The transformation rules state the changes that occur in the system as inputs cross the boundary of the system that differ from those required for equilibrium and that move the system toward either instability or the stability of a new system. This is necessarily one of the least developed aspects of the model; fully developed, however, it would provide models of dynamic change.

The actor classificatory variables specify the structural characteristics of actors. These characteristics modify behavior. For instance, "nation state," "alliance," and "international organization" name actors whose behaviors differ as a consequence of structural characteristics. The capability and information variables require no comment here.

There are three kinds of equilibria in such systems. There is an equilibrium within the set of essential rules. If behavior occurs that is habitually inconsistent with one of the essential rules, one or more of the other essential rules also will be changed. If the set of essential rules is changed, changes will occur in at least one of the other variables of the system. Or, conversely, if changes occur in one of the other variables of the system, then changes will occur in the essential rules also. If changes occur at the parameters of the system, changes will also occur within the system, and *vice versa*.

Such models necessarily abstract from a far richer historical content. The theories therefore can be used for the derivation of consequences *only* under explicitly stated boundary or parameter conditions. For instance, the statements concerning alignment patterns of the "balance of power" model apply only at the level of type of alignment, and they do not specify the actual actors who participate in specific alignments. They specify even this broad consequence only for stated values of the exogenous and endogenous variables. These models or theories can be brought closer to empirical reality by engineering in other factors that complicate the first approximations employed in the initial models. Thus, as we come closer to reality--and this is still at a high level of abstraction--we loose generality. We begin to employ procedures closer to the step-by-step engineering applications of physical theory than to the generalized theoretical statements of physical theory.

An illustration of the way in which a theory may be engineered may be useful. The "balance of power" theory indicates that alliances should shift rapidly and that wars should be limited. One of the reasons asserted for this is that within the framework of the assumptions of the model it is more important to maintain the existence of a defeated state because of its potential importance as a future alliance partner than it is to secure the gains that could be obtained from its complete conquest. If one looks at the behaviors of the European national actors after 1870, they are inconsistent with the prescriptions of the model. Alliances tended to be long-term, to hinge around France and Germany, and the ensuing war was, according to the standards of the time, unlimited.

One possible approach to this problem would be to define every descriptively different state of the world as a representation of a different system. Thus the European system after 1870 could be called a rigid "balance of power" system. Apart from being theoretically uninteresting--for by this method we would be forced to label a system as different every time it manifests different behavior--this device is unnecessary. The "balance of power" model is a first-order approximation. The first-order approximation simplifies reality by extrapolating from considerations of internal politics in such a fashion that states are treated as if they were free to optimize according to international considerations only. When one applies a first-order approximation to reality, it is necessary to adjust for the specific boundary differences that play a role in the actual situation to which the application is being made. If one takes account of the actual temper of French politics after the seizure of Alsace Lorraine, it is evident that French public opinion became revanchist as a consequence of the seizure. Germany could neither cede Alsace Lorraine to France nor compensate France satisfactorily in other matters. As a consequence France and Germany could not be alliance partners for the foreseeable future. Thus they naturally became the poles of opposed alliances. Additionally the motivation that, according to the model, constrains the actors to accept limited objectives in war no longer operated with respect to these two states. Since neither was a potential alliance partner of the other, neither had this particular incentive to limit its aims against the other.

This example also illustrates the conditions under which one continues to employ a given theory or to adopt another theory. As long as the theory continues to explain behavior when suitable adjustments are made for the parameters of the system, that particular theory will continue to be employed even though the actual behavior is seemingly inconsistent with the prescriptions of the model under equilibrium conditions. When, however, one requires a different theory to explain the behavior, regardless of whether this is a matter of principle or of economy, then a different theory will be adopted. Obviously judgments as to when the same or different theories should be used will depend upon the state of the art as well as upon evaluation of the data.

One may begin from different macro theories--or first-order approximations--in order to explain the same micro events. Thus by suitable parameter adjustments, one may explain the events subsequent to 1870 either from the standpoint of a theory of international politics or from the standpoint of a theory of foreign policy. The different theories, however, will answer different sets of questions concerning these events. Thus, starting from a model or theory of international politics, we learn how public opinion produces behavior inconsistent with system stability and security optimization. If we start with a theory of foreign policy that explains how different types of politics produce different styles of external policy, we learn how regime considerations under special circumstances produce variations in that style of behavior. In this case our primary focus is upon the foreign-policy process as a process within the national political system.

Whether we can use either or both levels of explanation depends upon the benignity of the data. Theories can be used only when their relevance is not swamped by external disturbances. There are cases where international behavior may be dominated by national considerations or, alternatively, where national regime considerations may be dominated by an international occurrence that overwhelms the actors. In the usual case, however, both sets of constraints operate and each first-order approximation serves as the best starting point to answer a particular set of questions.

One cannot build a theory of international relations, however, upon a theory of foreign policy; nor can one begin with a theory of international relations and derive theories of foreign policies from it. Either attempt would assume a general theory that encompasses both viewpoints. Either theory, however, may be useful in understanding the specific parameters to which the other theory needs to be adjusted.

There is still a third point of view which for some sets of questions may be the most productive. This is the cross-sectional or biographical slicing of history in order to delineate the dynamics of social change as a particular sequence of events unrolls and places inconsistent pressures upon actors holding critical role functions in several important systems. Revolutions occur when critical numbers of key individuals respond to the demands of their other role functions at the expense of their functions in the state system. The conflicts that arise in cases of this kind are, along with the environmental changes that stimulate the problem, the cause of social change. This kind of cross-sectional problem can best be studied by examining how critical roles in different systems make critically inconsistent demands upon the critical actors who participate in the relevant set of systems.

VI. There are two kinds of research procedures being applied to international systems models at the present time. The first, resulting from the deficiency of historical materials, involves a series of comparative case studies of previous historical systems. We do not know, for instance, the characteristic behavior of the Greek city-state system or how it differs from behavior of the Italian city-state system. We do not have good ideas as to why the differences occur. We do not understand how the patterns of alignment differed or why. We do not know how or under what conditions wars were waged or peace made. It is indeed difficult even to ask questions such as these except within the framework of an international systems theory. In our comparative studies, we have found that the Italian system was stable until members of the system invited the intervention of France. The Macedonian (Greek) system was quasi-stable until members of the system got involved in the affairs of Rome. The Chinese warlord system persisted only for a very short period of time and was rolled up by a peripheral actor with a superior form of organization and ideology. In three of our historical cases--the Kuomintang roll-up of China, the Macedonian roll-up of Greece, and the later Roman roll-up of Greece--a peripheral actor rolled up what might be regarded from some perspectives as a central system. Toynbee hypothesizes that such roll-ups are examples of classical civilizations that are conquered by ruder and more warlike systems. Macedonians clearly were ruder and more warlike than the Greeks and also possessed a superior military organization. It is difficult to say whether the Romans, whose genius lay in law, were less civilized than the Greeks, whose genius lay in philosophy. Clearly the Kuomintang leaders were culturally more advanced than the more traditional Chinese warlords. Moreover, the Roman roll-up of Italy constituted a roll-up by a central rather than by a peripheral actor. We are more impressed with the fact that in each of the three cases where a "peripheral" actor conquered a "balance of power" system, it did not participate in the wars of that system until the system had run itself down. Although the actors within the "balance of power" system maintained reasonable relative positions, the series of wars ran down the absolute resources of the system while the "peripheral" actor either husbanded its resources or actually gained resources as a consequence of military gains in outside systems. Then, when for one reason or another the "peripheral" actor became involved in the affairs of the "central" system, it was able to roll it up.

The computer explorations enable us to bypass some of the imprecision of using words in the verbal models. For instance, the injunction in the essential rules of some systems to increase capabilities did not specify by how much or under what risk conditions. Similar ambiguities necessarily occur elsewhere in the general statements of the models. For instance, using words it is virtually impossible to discriminate between the behavior of a system of nine nations and a system of seven.

Only when we were able to play out realizations of the "balance of power" model on a computer were we able specifically to link outcomes to the parameters that produced them. A realization of a model or theory involves building the features of the theory into the computer program as parameters of that program. These parameters can then be varied to explore the sensitivity of the computer program to changes in the parameters. Thus it is possible to explore changes in the number of players, the battle exchange ratios, the motivations of the players, and so forth. If the outcome is unstable, we can ask how to reintroduce stability and by what changes in the parameters. We also can ask ourselves what relation each change has to the initial verbal model and why it helps shed light on that model; for the objective of this work is not merely to produce a stable computer realization.

Computer analysis is used not to prove any specific propositions but to explore the interrelationship of propositions concerning the strategic structure of a model of the international system. For instance, consider the proposition: Would a "balance of power" system operate differently if the actors were security-oriented than it would if they were hegemony-oriented? To explore this, we constructed a computer model instructing the national actors of the international system to optimize over each war cycle according to the appropriate utility schedule designed for them. We discovered, that if there was a hegemony-inclined actor in the system, the system became unstable. Initially, a "balance-oriented" actor was the victim of his stability. Pursued to the end, however, it became clear that according to the logic of the model the hegemony-inclined actor would never succeed in becoming the greatest actor in the system and would eventually be eliminated by the remaining "balance-oriented" players. Thus there was an inconsistency between short-run optimization and long-run optimization for the hegemony-inclined actor. Consequently, for the pilot model, if hegemony-inclined actors could optimize over the long run, they would behave the same way that more security-oriented players do. This, however, is not a conclusion about the real world but about the logic of the model. Therefore the next step was to explore the conditions under which the hegemony-inclined actor could succeed in obtaining hegemony. Introducing imperfect information, uncertainty, and nonsimultaneous commitment to war is believed to permit success for the hegemony-inclined actor. On the other hand there is no reason to assume that this result is the truth about the real international system; for we can explore those further counter-deviancy measures that would be sufficient to prevent the hegemony-inclined actor from exploiting his deviant tendencies.

The general literature asserts that in a system of five states wars will tend to be three against two and that wars of four against one will tend to destabilize the system.. In our initial pilot runs the wars were almost invariably four against one; yet some of our runs remained stable for hundreds of war cycles with no indication that continuation of the runs would provide instability. We then had the problem of modifying our model to produce the three-to-two wars that are more characteristic.of history. There are probably at least two ways to accomplish this: by reducing the cost of wars or by permitting side payments within coalitions. We intend to try still other modifications to see how they affect this factor of alignment size.

The two examples given should illustrate how systems theory utilizes computers. The computer is used to explore the relationships between assumptions. It is thus capable of assigning outcomes to causes, at least with respect to the structure of thought we have established to account for the real world of international politics. Thus, if we attempt to make inferences concerning the real world of international politics, we at least know how and why our hypotheses are related to our premises. We also have a ground for asserting that the real world outcome might be related to the assigned cause if in exploring the external world we find those conditions that produced the same outcome in the computer model and no other conditions (at least that we can think of) that would counteract this outcome were we to place them in the computer model.

There is still a third way in which international systems models can be used. As designs for stable or quasi-stable international systems, they constitute a set of perspectives for the future. Depending upon whether and how much we prefer one kind of system to another, we can take account of the ways in which policies designed for more specific issues may affect macro political system prospects either by favoring some lines of development or by inhibiting others. Thus, for instance, a decision to utilize present opportunities to develop international management for the resources of the seas, before the development of technologies for exploiting those resources reinforces competitive tendencies, could affect the prospects for more universally-oriented international regimes--regimes that reinforce the roles of universal organizations such as the United

Nations. Decisions at national-policy-making levels seem notably deficient in this respect and are probably so to some extent, because the currency of macro international system models is not yet sufficiently high to orient attention and affect perspectives in a systematic manner.